Green Energy and Technology

Climate change, environmental impact and the limited natural resources urge scientific research and novel technical solutions. The monograph series Green Energy and Technology serves as a publishing platform for scientific and technological approaches to "green"—i.e. environmentally friendly and sustainable—technologies. While a focus lies on energy and power supply, it also covers "green" solutions in industrial engineering and engineering design. Green Energy and Technology addresses researchers, advanced students, technical consultants as well as decision makers in industries and politics. Hence, the level of presentation spans from instructional to highly technical.

Indexed in Scopus.

Indexed in Ei Compendex.

More information about this series at https://link.springer.com/bookseries/8059

Mohamed Lahby · Ala Al-Fuqaha · Yassine Maleh
Editors

Computational Intelligence Techniques for Green Smart Cities

Springer

Editors
Mohamed Lahby
ENS
Hassan II University
Casablanca, Morocco

Yassine Maleh
Sultan Moulay Slimane University
Beni-Mellal, Morocco

Ala Al-Fuqaha
Hamad bin Khalifa University
Doha, Qatar

ISSN 1865-3529 ISSN 1865-3537 (electronic)
Green Energy and Technology
ISBN 978-3-030-96428-3 ISBN 978-3-030-96429-0 (eBook)
https://doi.org/10.1007/978-3-030-96429-0

This Springer imprint is published by the registered company Springer Nature Switzerland AG
The registered company address is: Gewerbestrasse 11, 6330 Cham, Switzerland

Preface

In recent years, the use of smart city technology has rapidly increased through the successful development and deployment of Internet of Things (IoT) architectures. The citizens' quality of life has been improved in several sensitive areas of the city, such as transportation, buildings, health care, education, environment, and security, thanks to these technological advances. Although there are important smart services deployed in cities worldwide and advanced technologies to develop these services, cities are encountering so many challenges linked directly with the environment. Indeed, most of these solutions were conducted without considering information about the environment, such as the energy consumption by Information and Communication Technologies (ICT) and the carbon footprint of ICT devices.

The green city paradigm becomes a necessity to overcome these limitations. The key objective of this paradigm is promoting a sustainable and liveable city to the citizens. This goal cannot be achieved without considering three essential elements: reducing energy consumption, improving the quality of citizens' daily life, and making citizens active and proactive actors of green smart city solutions. Computational intelligence techniques and algorithms enable a computational analysis of enormous data sets to reveal recurring patterns. This information is used to inform and improve decision-making at the municipal level to build smart computational intelligence techniques and sustainable cities for their citizens. Machine intelligence allows us to identify trends (patterns). The smart city could better integrate its transportation network, for example. By offering a better public transportation network adapted to the demand, we could reduce personal vehicles and energy consumption. A smart city could use models to predict the consequences of a change, such as pedestrianizing a street or adding a bike lane. A city can even create a 3D digital twin to test hypothetical projects.

This book comprises many state-of-the-art contributions from scientists and practitioners working in machine intelligence and green smart cities. It aspires to provide a relevant reference for students, researchers, engineers, and professionals working in this area or those interested in grasping its diverse facets and exploring the latest advances in machine intelligence for green and sustainable smart city applications.

This book contains a total of 19 chapters classified into five main parts. The first part presents a state of the art related to machine learning techniques and their applications in renewable energy forecasting, green smart home, and smart waste management. The second part explores the exploitation of machine learning techniques in the context of smart transportation. The third part focuses on machine learning techniques for green health. The fourth part provides some chapters related to green environment based on machine learning techniques. Finally, the last part presents some smart applications and valuable case studies that can be used in green smart city.

We want to take this opportunity to express our sincere thanks to the contributors to this volume and the reviewers for their outstanding efforts in reviewing and providing interesting feedback to the authors of the chapters. The editors would like to thank Anthony Doyle (Executive Editor and Series Editor-in-Chief) and Mr. Padma Subbaiyan (Springer Project Coordinator), for the editorial assistance and support to produce this important scientific work. Without this collective effort, this book would not have been possible to be completed.

Casablanca, Morocco — Prof. Mohamed Lahby
Doha, Qatar — Prof. Ala Al-Fuqaha
Khouribga, Morocco — Prof. Yassine Maleh

Contents

Machine Learning and Green Health

Machine Learning and Green Environment

Case Studies and Smart Applications

About the Editors

Mohamed Lahby is working as Assistant Professor Computer Science at the Higher Normal School (ENS) University Hassan II of Casablanca, Morocco. His did Ph.D. in Computer Science from the Faculty of Sciences and Technology of Mohammedia, University Hassan II of Casablanca, in 2013. His research interests are wireless communication and network, mobility management, QoS/QoE, Internet of things, smart cities, optimization, and machine learning. He has published more than 35 papers (chapters, international journals, and conferences), three edited books, and two authored books. He has served and continues to serve on executive and technical program committees of numerous international conferences such as IEEE PIMRC, ICC, NTMS, IWCMC, WINCOM, and ISNCC. He also serves as Referee of many prestigious Elsevier journals: *Ad Hoc Networks*, *Applied Computing and Informatics*, and *International Journal of Disaster Risk Reduction.* He organized and participated in more than 40 conferences and workshops. He is Chair of many international workshops and special sessions such as MLNGSN'19, CSPSC'19, MLNGSN'20, AI2SC '20, WCTCP'20, and CIOT'2022.

Ala Al-Fuqaha is Professor at the Computer Science Department, Hamad Bin Khalifa University, Qatar. His research interests include the use of machine learning in general and deep learning, in particular, in support of the data-driven and self-driven management of large-scale deployments of Internet of things and smart city infrastructure and services, wireless vehicular networks, cooperation and spectrum access etiquette in cognitive radio networks, and management and planning of software-defined networks. He is Senior Member of the IEEE and ABET Commissioner. He serves on editorial boards of multiple journals including *IEEE Communications Letter*, *IEEE Network Magazine*, and Springer *AJSE*. He also served as Chair, Co-chair, and Technical Program Committee Member of multiple international conferences including IEEE VTC, IEEE Globecom, IEEE ICC, and IWCMC.

Yassine Maleh is Associate Professor at the National School of Applied Sciences at Sultan Moulay Slimane University, Morocco. He received his Ph.D. degree in Computer Science from Hassan 1st University, Morocco. He is Cybersecurity and

Information Technology Researcher and Practitioner with industry and academic experience. He worked for the National Ports Agency in Morocco as IT Manager from 2012 to 2019. He is Senior Member of IEEE and Member of the International Association of Engineers IAENG and The Machine Intelligence Research Labs. He has made contributions in the fields of information security and privacy, Internet of things security, and wireless and constrained networks security. His research interests include information security and privacy, Internet of things, networks security, information system, and IT governance. He has published more than 50 papers (chapters, international journals, and conferences/workshops), seven edited books, and three authored books. He is Editor-in-Chief of the *International Journal of Smart Security Technologies* (IJSST). He serves as Associate Editor for *IEEE Access* (2019 Impact Factor 4.098), the *International Journal of Digital Crime and Forensics* (IJDCF) and the *International Journal of Information Security and Privacy* (IJISP). He was also Guest Editor of a special issue on Recent Advances on Cyber Security and Privacy for Cloud-of-Things of the *International Journal of Digital Crime and Forensics* (IJDCF), Volume 10, Issue 3, July–September 2019. He has served and continues to serve on executive and technical program committees and as Reviewer of numerous international conferences and journals such as Elsevier *Ad Hoc Networks*, *IEEE Network Magazine*, *IEEE Sensor Journal*, *ICT Express*, and Springer *Cluster Computing*. He was Publicity Chair of BCCA 2019 and General Chair of the MLBDACP 19 symposium and Data Management.

State of the Art

Machine Learning Techniques for Renewable Energy Forecasting: A Comprehensive Review

Rajae Gaamouche, Marta Chinnici, Mohamed Lahby, Youness Abakarim, and Abdennebi El Hasnaoui

Abstract Over the past decade, renewable energy resources, such as wind, solar, biomass, ocean energy and other kinds of energy, are becoming attractive technologies for building green smart cities. These new forms of energy can complete the world's energy demand, protect the environment and provide energy security. Statistics have shown that renewable energy resources offer between 15 and 30% of the world's energy. Moreover, the production and consumption of different kinds of renewable energy are constantly increasing every year. However, forecasting renewable resources in terms of production and consumption is becoming more vital for the decision-making process in the energy sector. Indeed, the accurate forecasting of renewable energy permits to ensure optimal management of energy. In this context, machine learning techniques represent a promising solution to deal with forecasting issues. Several solutions and forecasting models based on machine learning have been extensively proposed in the literature for predicting power energy that should be deployed for future smart cities. This chapter aims to conduct a systematic mapping study to analyze and synthesize studies concerning machine learning techniques for forecasting renewable resources. Therefore, a total number of 86 relevant papers published on this subject between January 1, 2007, and December 31, 2021, were carefully selected. The selected articles were classified and analyzed according to the following criteria: channel and year of publication, research type, study domain, study context, study category and machine learning techniques used for forecasting renewable resources. The results showed that wind energy and solar energy were used massively in selected papers, and the forecasting of power production based on hourly forecast model and minutely forecast was the primary interest in the majority

R. Gaamouche · A. Hasnaoui
National Superior School of Mines, Rabat, Morocco
e-mail: elhasnaoui@enim.ac.ma

M. Chinnici
ENEA—C.R. Casaccia Via Anguillarese, Rome, Italy
e-mail: marta.chinnici@enea.it

M. Lahby (✉) · Y. Abakarim
University Hassan II, Higher Normal School of Casablanca, Casablanca, Morocco
e-mail: lahby@ieee.org

M. Lahby et al. (eds.), *Computational Intelligence Techniques for Green Smart Cities*, Green Energy and Technology, https://doi.org/10.1007/978-3-030-96429-0_1

of selected papers. Furthermore, artificial neural network (ANN) and deep neural network (DNN) were the most regression algorithms used to predict renewable energy sources.

Keywords Renewable energy · Smart cities · Machine learning · Forecasting · SMS

1 Introduction

Currently, the demand for electrical energy needs is increasing along with the number of activities and services in smart cities using ICT devices. Increasing the capacity to supply electrical energy became an important issue. Several power plants (PPs) can do the additional electrical energy capacity sourced from fossil and renewable energy. However, the development of PP sourced from fossil energy is very limited and is starting to integrate with renewable energy resources (RERs) through smart grid networks in the cities. The adoption of RERs opens a new scenario for energy production and, in the meantime, represents an essential and urgent resource to consider for future world scenarios correlated to climate change. Indeed, the RERs refer to the energy that can be recycled in nature, such as solar energy, wind power, tidal energy and geothermal energy, produce clean, green and low emission, and thus benefit the protection of the environment. According to the EC climate and energy framework [1] and the report of the international renewable energy agency, the contribution of RERs to electricity generation is projected to reach 85% by 2050, which is mainly dyed to the growth of solar produced and wind produced power [2] to build green smart cities. Although RERs are considered to be the most promising alternative to fossil fuels because they are highly efficient, clean and pollutant-free, naturally replenished in a wide geographical area and inexpensive to produce and distribute, they lack consistency [3]. Unlike the capability of generating conventional resources (e.g., fossil fuels) according to consumption and at specific and accurate schedules, the production of renewable energies is variable and rely on seasonal and weather conditions such as temperature, pressure, wind speed and visibility [4]. Hence, they bring un-schedulable uncertainty, threatening the reliability and stability of energy systems, especially with the large-scale integration of renewable energy.

On the one hand, renewable energy exhibits intense volatility, intermittent and randomness, which will undoubtedly increase the reserve capacity of the electric energy systems, thereby increasing the cost of power generation. On the other hand, the use of renewable energy involves a large number of power electronics, which reduces the rotational inertia of the power system and thus reduces the stability margin of the system [5]. Therefore, renewable energy forecasting as a practical measure is essential for mitigating related uncertainties, conducive to the planning, management and operation of electrical power and energy systems [6]. However, accurate renewable energy forecasting remains a challenging task due to renewable energy data's intermittent, chaotic and random nature. Indeed, these chaotic conditions can change dramatically during the time, which enforces difficulties in the schedule and management of optimal electricity generation and imposes concerns about the electricity quality and stability. Indeed, if the integration of RERs into the electricity sector is not handled and controlled adequately, it could cause an imbalance and excess power production, which may increase government expenses instead of reducing the costs [7].

Moreover, this unpredictable stochastic nature of RERs resulted in serious unit commitment issues [8]. Therefore, an accurate prediction of RERs has become an enduringly worldwide interest in embryonated literature studies. Thus far, various research studies have been employed to tackle the unreliability and inaccurateness of renewable power forecasting models, and different algorithms have been used to provide accurate renewable energy predictions for the next few minutes to the next few days.

The studies and hence the associated algorithms can be usually divided into four categories: persistence models, physical models, statistical models and artificial intelligence (AI) models. At present, among these forecasting methodologies, the AI-based models, particularly machine learning (ML) models, have gained the interest of researchers. Indeed, unlike statistical models, ML techniques can capture the nonlinearity in power data. They can be applied for several purposes with only minor modifications. Therefore, ML forecasters could outperform and alternate the conventional forecasters [8, 9].

Nevertheless, despite the nascent studies about the ML techniques for renewable energy forecasting models, a systematic review that summarizes forecasting models from energy and computer science perspectives and evaluates their performance from a systematic mapping for green smart cities has not been investigated yet. This chapter provides a systematic mapping study (SMS) of the recently published and proposed forecasting ML-based models. We notice that a SMS is a method that consists of searching the literature for all articles published in a given field, in order to carry out a statistical study based on research questions [10].

This chapter is organized as follows. In Sect. 2, we present an overview of the existing survey and reviews in literature related to renewable energy forecasting based on machine learning techniques. In Sect. 3, we describe the research methodology used in our study to conduct this review. In Sect. 4, we report the results and we discuss the findings of this SMS. In Sect. 5, we give the implications for researchers. Finally, we conclude our work and look into our future research in Sect. 6.

2 Background of Forecasting Methods

In this section, recent studies in a few works of literature for renewable energy resources forecasting are reviewed. Various models, methodologies, technologies and AI algorithms and tools are discussed and described.

With the rapid development of global industrialization, it has been recognized that excessive consumption of fossil fuels will accelerate the reduction in fossil fuel reserves and have an adverse impact on the environment. These influences will result in increased health risks and threats of global climate change. In addition to fossil fuels and nuclear energy, renewable energy is currently the fastest-growing energy source. Renewable energy refers to reusable energy that can be recovered in nature, such as solar energy, wind power, hydropower, biomass energy, waves, tides, and geothermal energy. With characteristics of sustainability and low environmental pollution, the issue of renewable energy has attracted attention, and plenty of related studies have been performed recently. One of the most critical challenges of renewable energy shortly is the energy supply. Renewable supply is the integration of renewable energy sources into the existing or future energy supply structures [10]. The development of renewable energy systems will cope with essential issues of current energy problems, such as improving the reliability of energy supply and solving regional energy shortages. However, due to the enormous volatility and the intermittent and random nature of renewable energy, various energy sources are intermittent and chaotic. Therefore, accurately dealing with the randomness of renewable energy data is still a work to be conquered. High-precision energy monitoring can improve the efficiency of the energy system. Energy forecasting technology plays a vital role in the development, management and policy-making of energy systems. As ways of utilization and production of electrical energy from renewable energy sources are continuously increasing, it became necessary to develop proper technologies to store renewable energy [11].

Many studies have revealed that various machine learning models have been employed in renewable energy predictions. The data-driven models do provide practical ways of renewable energy predictions. In addition, hybrid machine learning models were designed to increase the prediction accuracy of renewable energy. Various time intervals, such as minutes, hours, days, and weeks, were employed to predict renewable energy according to different purposes of predictions. Forecasting accuracy and efficiency were typically utilized to evaluate the performance of machine learning models in renewable energy predictions [12].

An overview of the different approaches and related works for forecasting renewables (mainly wind power and solar power) is shown within the work of Santhosh and Venkaiah [9]. As aforementioned, the models associate to forecasting issue can be divided into four categories: persistence models, physical models, statistical models and artificial intelligence models. These approaches are reviewed and discussed in the following paragraphs.

2.1 Persistence Models

These models assume that the values of power data in the next step are similar to those in the current step. Although these methodologies are not very practical for long-term forecasting, they perform well in very short-term and short-term forecasting (from a few seconds to 6 h-ahead) [13].

2.2 Physical Models

Besides geographical locations and physical characteristics and layouts of wind turbines or solar panels, these models depend on numerical weather predictions (NWPs), e.g., temperature, pressure, wind speed, wind density, roughness, turbulence intensity, etc. [14]. Although these models are reliable for medium-term and long-term forecasting, they cannot perform accurately for short-term forecasting [15]. In addition, they fail to adopt interferences and are computationally expensive, and require advanced computing machines [16]. In work [17], an exhaustive review of the available studies on tackling short-term forecasting using NWP models has been done. This study summarizes the applications of NWP-based models even if the implication of these models was no longer attractive for researchers, and many other recent methodologies started to flourish and outperform physical models [14].

2.3 Statistical Models

Statistical-based forecasting models are the mathematical models that attempt to map and recognize the relationship between time series historical data and target outputs [18]. In detail, they describe the linear relationship of data based on elementary mathematical equations [9]. Furthermore, since they can be formulated easily, they can deliver timely predictions. Thus, in a few literature studies, these forecasters are mainly used for short-term forecasting [19].

A comprehensive literature review on statistical approaches for time series and renewable energy resources forecasting was presented within [20]. Autoregressive (AR) and moving average (MA) models are well-known examples of statistical forecasting systems [21]. The hybrid integration of these two techniques is known as the autoregressive moving average (ARMA). ARMA is widely used for forecasting and provides models with high accuracy for different applications. Erdem and Shi [22] compared four other ARMA-based models for the forecasting of wind speed and direction. Gomes and Castro [23] presented a comparative study between ARMA and artificial neural networks (ANNs) for the prediction of wind speed and power. They concluded that both approaches provide the similar results; however, the ARMA performance is slightly better. In Fentis et al. [24], a nonlinear autoregressive (NAR) model was suggested for the forecasting of short-term photovoltaic (PV) power utilizing only historical data of the PV power (without using the NWP data). When comparing the performance of a NAR model with the autoregressive with an exogenous input (ARX) model, it was determined that NAR gives better results than ARX. This conclusion contrasts with the result obtained by Bacher et al. [25] where ARX performed better.

Another robust approach known as autoregressive integrated moving average (ARIMA) is widely employed for different purposes in a few literature studies to date. For example, Atique et al. [26] used the ARIMA approach to predict the daily solar energy production. It is noted that the application of ARIMA models requires the utilized data to be stationary; therefore, in their work, the nonstatic seasonal data are transformed into stationary ones. For longer-term forecasting, Pasari and Shah [27] used the ARIMA model for one-year ahead forecasting of wind speed and temperature. According to their conclusion, this generated model is generic, and having some minor modifications such as increasing the size of the input data, this model can be applied for two-year ahead forecasting. A specific type of ARIMA, namely fractional-ARIMA, was studied by Kavasseri and Seetharaman [28] for wind forecasting. It is computationally simple and can capture time series relations for both long-term and short-term forecasting horizons.

In general, statistical models are considered attractive to researchers because they are inexpensive and straightforward to apply. They presented acceptable accurate results for short-term horizons up to 2 days; however, they fail in forecasting and result in very unstable predictions for longer-term horizons [29]. Indeed, the statistical models see their usability in renewable energy power generation forecasting in the energy management process on microgrid [30].

2.4 *Artificial Intelligence (AI) Models*

AI-based forecasting models accelerate decision-making, data mining, and clustering problems because they can robustly handle big data fitting and develop good representations. In addition, they can employ complex tasks with moderately short time and without being explicitly programmed. Under the AI umbrella, many techniques can be utilizing such as: machine learning (ML), data assimilation (DA), deep learning, etc.

Machine learning techniques have been widely applied to many fields associated with data-driven problems and include many interdisciplinary areas, such as statistics, mathematics, artificial neural networks, data mining, optimization, and artificial intelligence. Machine learning techniques try to seek the relations between input data and output data with or without mathematical forms of problems [31]. After the machine learning models are well-trained by the training dataset, decision makers can obtain satisfying forecasting output values by feeding the forecasting input data into the well-trained models. The data preprocessing procedure plays an essential role in machine learning and can improve the performance of machine learning efficiently [32]. Basically, machine learning technology mainly uses three learning methods: namely, supervised learning, unsupervised learning, and reinforcement learning. Supervised learning takes advantage of labeled data in the training phase. Unsupervised learning is to automatically categorize input data into clusters by certain criteria for training data that has not been labeled in advanced. Thus, the number of clusters generally depends on the clustering criteria used. Reinforcement learning is learning through interaction with the external environment to obtain feedback in order to maximize the expected benefits. By ways of three basic learning principles, many theoretical mechanisms and applications have been proposed [33]. Deep learning can realize characteristic nonlinear attributes and high-level invariant data configurations and, therefore, has been applied in many fields to obtain satisfying performances [34]. Additionally, some studies have focused on forecasting renewable energy using a single machine learning model [35]. However, due to the diversified datasets, time steps, prediction ranges, settings, and performance indicators, it is not easy to improve the forecasting performance using a single machine learning model. Therefore, some studies developed hybrid machine learning models or overall prediction methods in renewable energy predictions to improve the prediction performance.

Recently, support vector machines (SVM), artificial neural network (ANN) and deep-learning processes have been prevalent in machine learning [36]. Unlike statistical based models, ML techniques can generally capture the nonlinearity and adapt instability in data, resulting in more reliable predictors [21]. Therefore, ML models and algorithms are introducing to solve various forecasting problem such as RERs forecasting. ML models are widely used in the energy sector for energy load forecasting. In the reference [37], the authors surveyed different computational optimization methods applied to renewable and sustainable energy. Maimouna et al. [38] highlighted the existing techniques applicable to forecast solar irradiance in order to select the appropriate forecast method according to needs. In [39], Cyril et al. provided an overview of forecasting methods of solar irradiation using machine learning approaches. Utpal et al. [40] performed a systematic and comprehensive literature review on direct photovoltaic power forecasting models and techniques. These techniques include statistical models, machine learning models and hybrid models. In [41], Huaizhi et al. provided a comprehensive and extensive review of renewable energy forecasting methods based on deep learning to explore its effectiveness, efficiency and application potential. The authors Dylan et al. [42] conducted a systematic literature review of deep learning-based solar and wind energy forecasting research published during the last five years.

Although there are several review articles and research papers based on machine learning techniques that are recently published for dealing with forecasting issues, there is no existing literature review that focuses on renewable energy prediction based on machine learning techniques. Thus, in this chapter, we carry out a systematic mapping study (SMS) with 86 relevant articles published between January 1, 2007, and December 31, 2021. The main purpose of this SMS is having a clear vision about the recent advances in renewable energy forecasting based on machine learning techniques. Particularly, we aim to give readers the opportunity to understand the followings keys: (1) the most frequently used machine learning techniques to predict different types of renewable energy, (2) different forms of renewable energy used in the selected papers, (3) the domain fields that have been targeted the literature, (4) where the literature has been published, (5) which kind of forecasting tasks have been covered in the literature and (6) which the forecast period used in the selected papers.

Fig. 1 The systematic mapping study process [44]

3 Research Methodology

In order to have an overview of the state of the art related to the recent research published in various fields and disciplines [43], the structuring of all articles published in a specific topic is now mandatory. For that, many guidelines have attempted to structure a literature study. Systematic mapping study (SMS) and systematic literature review (SLR) represent the most approaches that can be used to conduct any survey at any research field [43]. In this context, SMS and SLR have recently started appearing in various fields and disciplines.

According to [43], the SMS process provides a superficial overview of a particular research field by applying statistical analysis in order to classify different contributions published in this field. While, the SLR process offers a deep analysis and synthesis. We note that both SMS and SLR have based on rigorous research questions [43] for performing such studies and are often performed visually using graphics. In this way, we can consider that an SLR is an SMS that contains some additional steps; for example, it reviews the methodology adopted in each work and evaluates the obtained results [43].

Figure 1 presents the SMS process as it was described in Ref. [44]. The guidelines used in this process is composed of five steps: (1) definition the research questions (RQ) of the study; (2) formulation the search string; (3) selection of relevant studies; (4) extraction the data from selected papers and (5) analysis and classification of data.

In the following, we give the principle of each step and how it can be used in the context of renewable energies forecasting.

3.1 Mapping Questions

This step permits the preparation of research questions in order to structure the body of knowledge related to a specific topic. In this study, our goal is obtaining a comprehensive overview of published works about the use of machine learning techniques for renewable energies forecasting. For that, we define eight RQ that are presented in in Table 1. Each research question is associated with an explanation that provides the rationale behind its adoption for this study.

Table 1 Definition and rationale of different research questions

ID	Research question	Rationale
RQ1	In which years, sources and publication channels papers were published?	To indicate where articles concerning this research topic can be found and whether there are specific publication channels. It also determines when efforts in this field of research have been made
RQ2	Which research types are adopted in selected papers?	To highlight the different types of research published in the literature regarding the use of machine learning to forecast different types of renewable energy
RQ3	Which contexts are targeted in selected papers?	To identify in which context different studies were carried out and published in literature regarding the application of ML for forecasting different types of renewable energy
RQ4	What kinds of renewable energy are targeted in selected papers?	To determine the different forms of renewable energy used in the selected papers, such as wind, solar, tidal currents, geothermal and biomass
RQ5	Which domain fields are targeted in selected papers?	Various domains are used forecasting models such as smart home, smart grid, building, industry, transportation, residential and commercial sector
RQ6	Which forecasting tasks are used in selected papers?	To determine different forecasting tasks used in the selected papers such as production, consumption and management
RQ7	Which the forecast period is used in selected papers?	To determine different the forecast period used in the selected papers such as yearly, monthly, weekly, daily, hourly and minutely
RQ8	Which ML models, tasks and techniques are used to forecast renewable energy?	To provide an overview of different ML techniques used in the selected papers to forecast different types of renewable energy

3.2 Search Strings

This step allows the identification of different keywords from the search questions obtained in the first step, in order to formulate the search strings. The population, interventions, comparison and outcomes (PICO) model developed by Kitchenham and Charters [1] is widely adopted to achieve this step. According to this model, each question be separated in terms of four key components: population (P), intervention (I), comparison (C) and outcome (O). The rationale and the value of each

PICO component varies according to each domain. In this chapter, we investigate the suitability of the PICO model as a knowledge representation for renewable energies forecasting questions.

As shown in Table 2, for each component, we have used some sentences that describe the subject of this SMS. Table 3 defines the search terms extracted from each PICO element by using the roles proposed in [15].

After obtaining the set of keywords, we classify them into seven groups as shown in Table 4. Each group contains a set of keywords that are characterized by same synonyms or same semantic meaning. After that, we apply the following two rules in order to formulate the final search string:

- Rule 1 : the Boolean OR is applied to concatenate terms in the same group and
- Rule 2 : the Boolean AND is applied to join groups of terms.

The definitive search string obtained is (renewable energy OR clean energy OR green energy OR solar energy OR wind energy OR geothermal energy OR biomass energy OR marine energy) AND (collect OR data) AND (predict OR forecast OR forecasting OR predicting) AND (approach OR technique OR method OR algorithm) and (benchmark OR application OR tool OR framework OR solution) AND (optimization OR accuracy OR performance OR management OR production OR consumption).

Table 2 Pico definition for our SMS study

Element	Description
Population	This element refers to renewable energy forecasting studies
Intervention	The aim of this SMS is to provide an overview related to the benefits when machine learning algorithms be applied for predicting power energy that should be deployed for future smart cities. For that, we envisage to collect data used by each solution proposed in published works
Comparison	We compare different machine learning algorithms used for predicting renewable energy
Outcome	In the context of this study, the outcomes represent the factors that will be used to compare the interventions for forecasting renewable energy resources

Table 3 The extracted keywords from the PICO definition

Element	Description
Population	Renewable energy, clean energy, green energy, solar energy, wind energy, geothermal energy, biomass energy, marine energy
Intervention	Collect, data, analyze, learn, predict, forecast
Comparison	Approach, technique, method, algorithm, benchmark, application, tool, framework, solution
Outcome	Forecasting, prediction, optimization, accuracy, performance, production, consumption, management

Table 4 Classification of different keywords according to their similarity

Terms	G1	G2	G3	G4	G5	G6
Renewable energy	*					
Clean energy	*					
Green energy	*					
Solar energy	*					
Wind energy	*					
Geothermal energy	*					
Biomass energy	*					
Marine energy	*					
Analyze		*				
Learn		*				
forecast		*				
Predict		*				
Collect			*			
Data			*			
Approach				*		
Technique				*		
Method				*		
Algorithm				*		
Benchmark					*	
Application					*	
Tool					*	
Framework					*	
Solution					*	
Forecasting		*				
Prediction		*				
Optimization						*
Performance						*
Accuracy						*
Production						*
Consumption						*
Management						*

3.3 Selection of Papers

This step allows to extract the relevant papers related to renewable energies forecasting from three digital libraries most used in the literature: IEEE Xplore, ScienceDirect, and Springer Link. We have adapted the search string according to the advanced search option corresponding each digital library [43], before the lunching

Table 5 Inclusion and exclusion criteria

Category	Criteria
Inclusion	Studies presenting methods and techniques to develop intelligent solution for predicting renewable energy resources that should be deployed for future smart cities
	Studies presenting an overview or a comparison between different machine learning techniques used in the literature to forecast renewable energy resources
	Studies published between 2007 and 2021
Exclusion	Studies not accessible in full-text
	Studies not presented in English
	Books and gray literature
	Studies that are duplicates of other studies

of the query. As result, there are several possible ways to carry out our search process in different digital libraries. For instance, we can search our query in title, in author, in abstract, in full-text, etc. For this SMS, we apply our query within full-text option for all digital libraries. All searches were restricted to the studies published between January 01, 2007, and December 31, 2021. We note that the three digital libraries do not provide the same research results after the execution of our query.

In order to retain only the relevant papers that match the subject of this SMS, we apply inclusion and exclusion criteria to the candidate papers retrieved by executing our query. During this process, we included some papers and we excluded other based on titles and abstracts, as well as full-text reading and quality assessment. In this chapter, the quality assessment is mainly associated to the citation number of each candidate paper. In Table 5, we present different inclusion and exclusion criteria used for each selected paper.

3.4 *Data Extraction*

This step permits to collect the relevant data from the selected papers retrieved in the previous step. For that, we use the template presented in Table 6. Each data extraction field is characterized by three keys which are: data item, a value and a research question to which they refer. In the following, we describe each data item and we provide some values related to each data item as example.

Publication channel indicates any source of communication used to publish each selected paper such as scientific journal, conference, symposium or workshop .

Publication source refers to the effective name of the journal or the academic events (conference, symposium or workshop) that have used to publish each selected paper.

Research type refers to a classification of the selected papers based on the nature of the contribution made in each paper. Petersen et al. [43] and Wieringa et al. [45]

Table 6 Data extraction template

Data	Item value	RQ
Authors	Set of names of the authors	RQ1
Title	Title of the paper	RQ1
Publication channel	Kind of publication channel	RQ1
Publication source	Name of publication source	RQ1
Year of publication	Calendar year	RQ1
Research type	Which research strategy was followed	RQ2
Study context	In which context the research was conducted	RQ3
Study category	Which forms of renewable energy studied in the selected papers	RQ4
Study domain	In which domain fields the research was applied	RQ5
Study task	Which forecasting task research was used in the selected papers	RQ6
Study period	Which forecasting period was considered in the selected papers	RQ7
ML model	Which machine learning approach was adopted	RQ8
ML task	Which data mining tasks were used by selected paper	RQ8
ML technique	Which data mining techniques were used by selected papers	RQ8

have classified research into five categories: (i) Validation research (VR), (ii) solution proposal (SP), (iii) evaluation research (ER), (iv) philosophical papers (PP) and (v) opinion papers (OP). Validation research (VR) presents a novel technique, approach or strategy that has not been implemented in practice, but whose effectiveness has been evaluated in-depth through experiments, simulations, prototyping, etc. Solution proposal (SP) contains studies that propose a novel solution or an improvement of an existing solution and argues for its relevance without a validation. Evaluation research (ER) contains studies that empirically evaluate a technique, approach, or strategy in practice. Philosophical papers (PP) contain studies which provide a new vision or concept of looking at things or a new conceptual framework. Finally, opinion papers (OP) contain studies that give an opinion about what is good or wrong related to something, how we should do something.

Study context refers to the context in which the study was performed. According to the reference [46], the authors have introduced four contexts: academic, organization, government and industrial.

Study domain refers to the domain of the application of the study such as smart home, smart building, smart grid, microgrid, transportation, residential and industry. A study that does not report the application domain is considered as generic.

Study category refers to renewable energy kind used in each selected paper such as wind, solar, tidal currents and biomass.

Study task refers to forecasting task used in the selected papers such as production, consumption and management.

Study period refers to forecasting period considered in the selected papers such as yearly, monthly, weekly, daily, hourly and minutely .

ML model refers to the model of learning that is used to extract knowledge from data in order to predict the availability of renewable energy sources. In this chapter, we use four categories of ML models: supervised learning (SL) [47], unsupervised learning (UL) [48], ssemi-supervised learning (SSL) [49] and reinforcement learning (RL) [50]. Supervised learning is a machine learning type that learns from input data also known as training examples, comes with a label and generalize results as a model. This model can be used to predict the label for new, unforeseen examples. Unsupervised learning is a machine learning type that learns from data that has not been labeled. The goal of unsupervised learning is to detect patterns in the data. Semi-supervised learning is a machine learning technique that falls between supervised and unsupervised learning. The algorithms of this category include some labeled data with a large amount of unlabeled data in order to build a new model. Finally, reinforcement learning is a machine learning type that does not need prior knowledge. It can autonomously get optional policy with the knowledge obtained using interaction with dynamic environment. The algorithms of this category attempt to discover an association between the goal and the sequence of events that leads to a successful outcome. The main characteristic of reinforcement learning is the use of trial and error mechanism in order to achieve a goal or to maximize the total reward.

ML techniques refer to the algorithms that are used to perform the learning task in order to solve and optimize any problem. According to each machine learning model, we can classify different machine learning techniques into several machine learning tasks. For instance, for supervised learning models, we find regression task classification task and. For Unsupervised Learning models we have both tasks: asso-

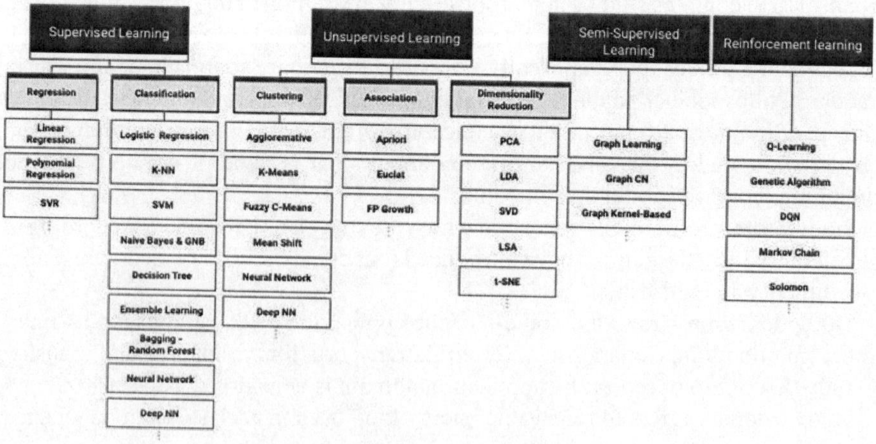

Fig. 2 The taxonomy of machine learning algorithms [44]

ciation and clustering. Finally, for semi-supervised learning models, we use the same ML tasks of supervised learning models, because the SSL models are considered as variants of the SL models. The most popular ML techniques for RERs forecasting are reported in Fig. 2.

3.5 Analysis and Classification

The main goal of this step consists to analyze and classify the results of this SMS that deals with renewable energies forecasting based on machine learning techniques. For that, the items obtained in the previous step are presented with many tables and illustrated with numerous graphical representations. After that, based on these tables and graphical representations, we discussed the results of our SMS related to the questions presented in Table 1. The results achieved in this step are provided in the next section "Results and Discussion."

4 Results and Discussion

In this section, the main findings of this mapping process are highlighted. Firstly, we discuss in depth the selection process from the search in the bases, to the selection of relevant papers for extraction and analysis of the results. Secondly, we discuss the answers related to different mapping questions shown in Table 1.

4.1 Overview of the Selected Studies

After the execution of our query on the three digital libraries (IEEE Xplore, ScienceDirect and Springer Link), we have retrieved 12779 candidate papers between January 01, 2007, and December 31, 2021. Afterward, we have filtered these retrieved papers by using the inclusion and exclusion criteria process. The outcome of this process leading to the identification of 86 relevant papers that have focused on forecasting renewable energy by using machine learning techniques.

Table 7 presents the number of selected papers according to each step of the selection process. In fact, we have updated the selected papers by withdrawing the duplicate papers as well as the papers reporting the same study. Besides, we have considered only all papers written in English and accessible in full-text. Then, we have excluded also papers that have added to each data base after December 31, 2021. Finally, we have applied a full-text review for many papers in order to decide about their inclusion or exclusion in our study. It is clear that the number of selected papers proposed for conducting this SMS is very reasonable and reflects the importance and the relevance of this research study. Moreover, the number of 86 selected papers

Table 7 The number of retained papers after each step of selection process

Database	Returned studies	Year filter (2007–2021)	Title/Abstract review	Full-text review	Add manually	Retained studies
IEEE	4384	4148	109	32	6	38
Springer	4134	3708	84	10	1	11
SCDirect	5329	4923	124	33	4	37

will permit us to ensure a credible and a relevant overview about renewable energy forecasting using machine learning techniques. The list of 86 selected papers used in this SMS is referenced between [54–136].

4.2 *RQ1: In Which Years, Sources, and Publication Channels Papers Were Published?*

Figure 3 shows the variation of relevant studies published in the period January 1, 2007, and December 31, 2021. Based on this figure, we remark that the annual number of publications related to renewable energy forecasting increases significantly every year. Moreover, in the last five years between 2017 and 2021, we have seen a high rate of publication with more than 50% of the total selected papers. In addition, 14 papers were published in 2021, which is a significant number if it is compared to other years. As result, it is very clear that researchers are becoming more and more interested in this research field. The growth in terms of amount of publications can be explained due to two major factors. The first one is that the renewable energy sources have attracted much attention, in recent years. The second factor is related to the trend of machine learning usage in different domains over the last decade.

From Fig. 4, we can see that 67.4% of selected papers were published in journals, while 27.9% were published in conferences, 3.5% were published in symposium and 1.2% were published in workshop. It is clear that the number of articles published in journals is very large compared to other publication channels. This high percentage of journal papers is justified by the fact that the the most of the selected studies have high quality and the majority of papers published in conferences have a lower quality than journal papers. Moreover, as we have already mentioned above, our goal is ensuring a credible and a relevant SMS study related to renewable energy forecasting. For that, the most of the selected studies have been published in publication sources with a high ranking. From 86 selected papers, 94.2% have published in prestigious conference or prestigious journal.

Table 8 presents the most frequent publication sources concerning our selected studies. Thus, the most frequent publication sources are journals Q1 with a percentage of 65.12 and 2.33% represent journals Q2. While 8.14% of selected papers were published in conferences with ranking A, 5.81% in conferences with ranking B and

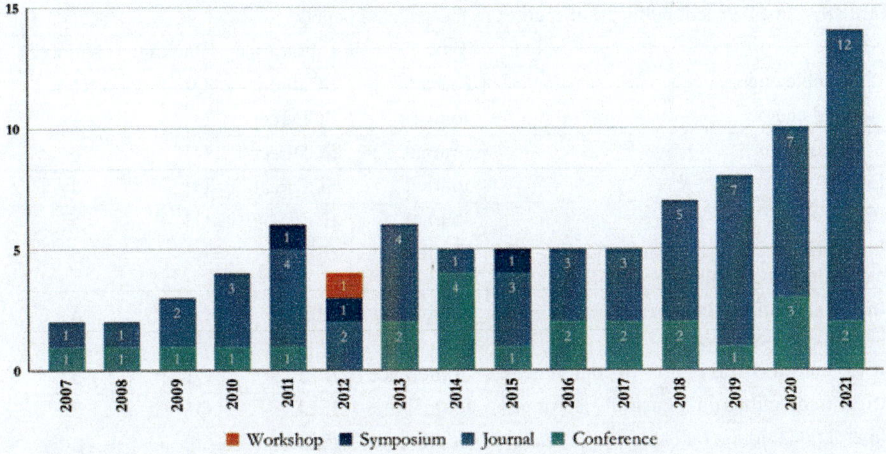

Fig. 3 The distribution of selected papers during the years 2007–2021

Fig. 4 The percentage of each publication channel in selected papers

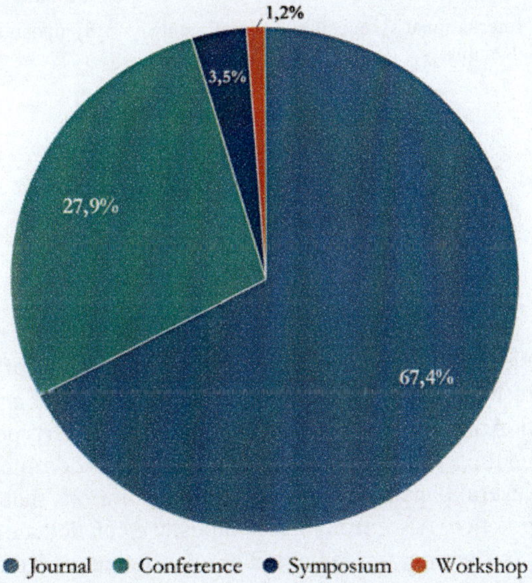

11.63% in conferences with ranking C. It is clear that all these selected studies. have a high ranking and also published in journals specializing in the energy fields such as renewable energy, applied energy, solar energy, etc.

Table 8 Most frequent publication sources in the selected papers

Name	Type	Libraries	Ranking	Number
Renewable energy	Journal	SCDirect	Q1	15
Applied enegy	Journal	SCDirect	Q1	8
Solar energy	Journal	SCDirect	Q1	6
Energy	Journal	SCDirect	Q1	4
IEEE access	Journal	IEEE	Q1	5
International conference on industrial electronics and applications (ICIEA)	Conference	IEEE	A	5
IEEE international conference on machine learning and applications	conference	IEEE	Q1	4
IEEE tencon (IEEE region 10 conference)	Conference	IEEE	Q1	3
IEEE transactions on industrial electronics	Journal	IEEE	Q1	2
International joint conference on neural networks	Conference	IEEE	C	2
Neural computer and applications	Journal	Springer	Q1	1
Soft computing	Journal	Springer	Q1	1
International symposium on intelligent data analysis	Symposium	Springer	B	1

4.3 RQ2: Which Research Types Are Adopted in Selected Papers?

Figure 5 presents the percentage of each research type in selected papers during the years 2007–2021. From this figure, we observe that the evaluation research (ER) is the most adopted research type with a value of 55.8%. The second most adopted research type is validation research (VR) with a percentage of 29.1%. Finally, solution proposal (SP) is the less adopted research type with a percentage of 15.1%. Based on these values, it is clear that ER type is dominant over other search types, because several papers have proposed final solutions that are evaluated and experimented in practice. Also, from these values, we can deduce that the majority of proposed solutions are not implemented or experimented in real context. Finally, there is no such philosophical paper (PP) published in the literature that presents a new conceptual framework solution without any implementation.

Figure 6 shows the percentage obtained for each type of research adopted in journal and conference papers published during the years 2007–2021. We observe that the three types of research ER, VR and SP are often published in journals during this period. This result is justified by the fact that 67.4% of the selected papers were published in journals.

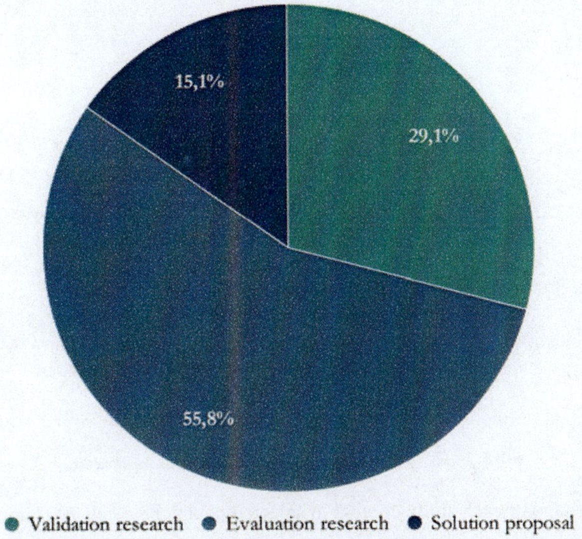

Fig. 5 The percentage of each research type in selected papers between 2007–2021

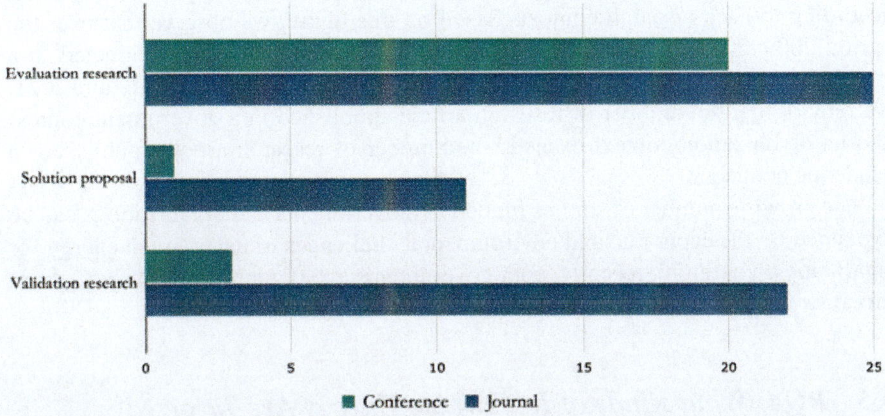

Fig. 6 The number of research types in journals and conferences

4.4 *RQ3: Which Contexts Are Targeted in Selected Papers?*

Figure 7 shows the percentage of each study context in selected papers over the period between 2007 and 2021. From this figure, we can see that 57.7% of selected papers were conducted in an academic context, 23.9% were adopted by government institutes, 15.5% were funded by organizations and only 2.8% of selected papers were conducted in an industrial context. Based on these values, it is clear that the majority of selected studies are done in academic context.

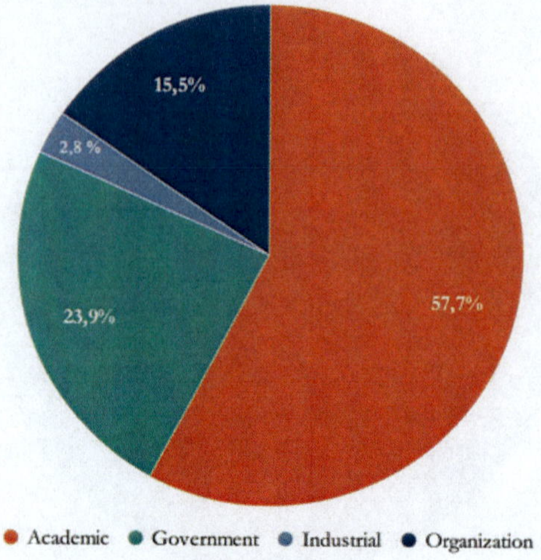

Fig. 7 The percentage of each study context in selected papers between 2007 and 2021

Figure 8 presents the distribution of selected studies during the years 2007–2021, according to each research context. Based on this figure, we observe that over the period 2009–2021, there is at least one study per year that were conducted in a government context or in an organization context. Moreover, between 2019 and 2021, we remark that the number of research articles published in a government context and an organization context exceeds the number of research articles published in academic context.

The growing number of studies funded by governments and organizations can be explained by the economic and environmental challenges of using green energy for smart cities. Therefore, recently, both governments and organizations have begun to invest strongly in research related to renewable energy projects.

4.5 RQ4: What Kinds of Renewable Energy Are Targeted in Selected Papers?

Figure 9 presents the percentage of each kind of renewable energy considered in selected papers. Based on this figure, we observe that the most selected papers have considered only two kind of energy, wind and solar. Both wind energy and solar energy are used with equitable manner in selected papers with a value of 48.3%. Moreover, tidal current energy is less used in the selected papers with a value of 3.4%. Therefore, we confirm that researchers are focusing more on wind energy and solar energy because these sources of energy have the potential to assist with energy production at any time.

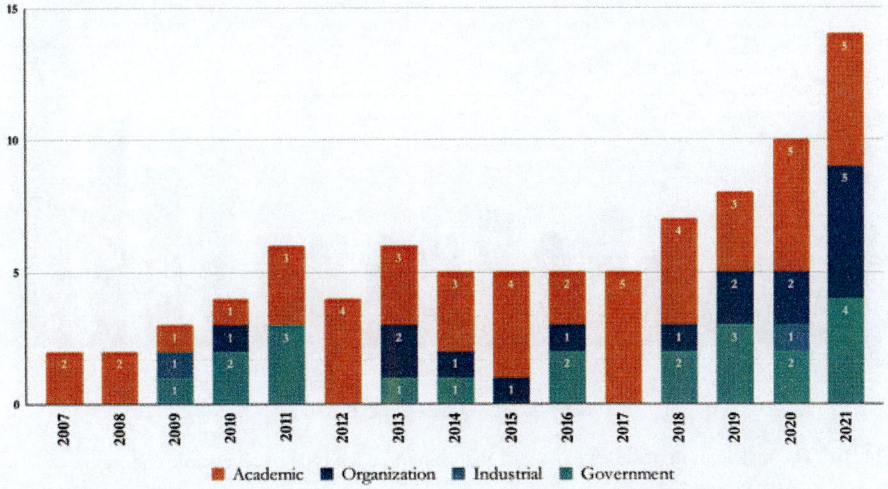

Fig. 8 Distribution of selected papers between 2007 and 2021 for each research context

Fig. 9 The percentage of each kind of renewable energy in selected papers

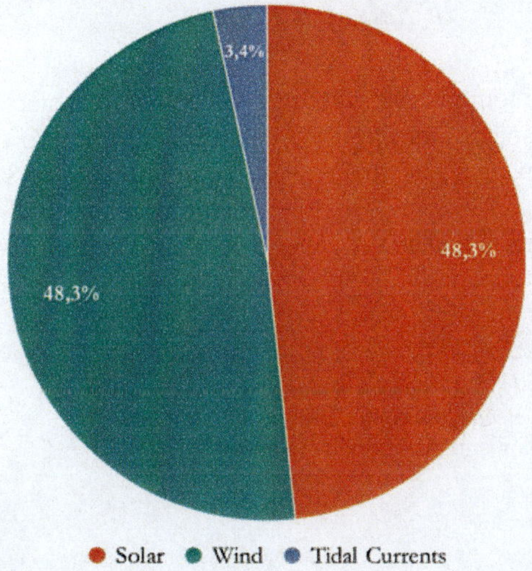

Figure 10 presents the distribution of selected papers according to each type of each type of renewable energy during the years 2007–2021. Based on this figure, we observe that wind energy is dominant during the period 2007–2015, while solar energy is dominant during the period 2016–2021.

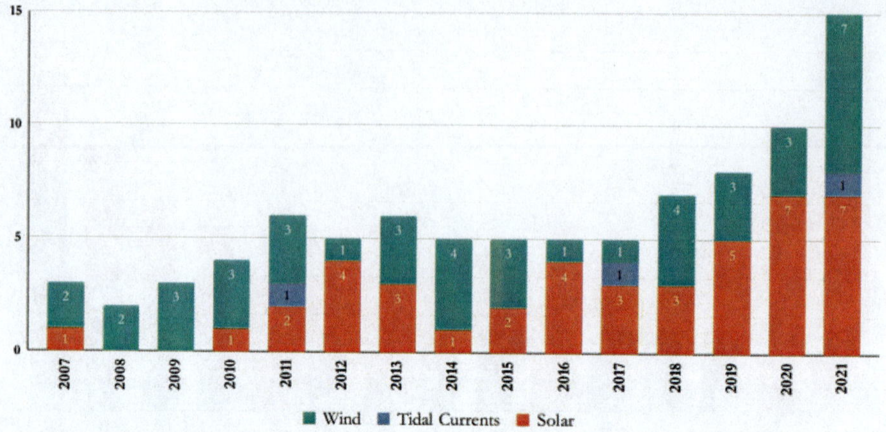

Fig. 10 The distribution of selected papers according to each type of renewable energy

4.6 *RQ5: Which Domain Fields Are Targeted in Selected Papers?*

Figure 11 presents the percentage of selected papers for both domains generic or specific during the years 2007–2021. We can see that the percentage of selected papers that focus on a generic domain is 87.3%, while the percentage of selected papers that focus on specific domain is 12.7%. It is clear that the higher value is provided by the generic domain. Therefore, we confirm that researchers are focusing more and more on generic domain.

Fig. 11 The percentage of selected papers for both domains generic and specific

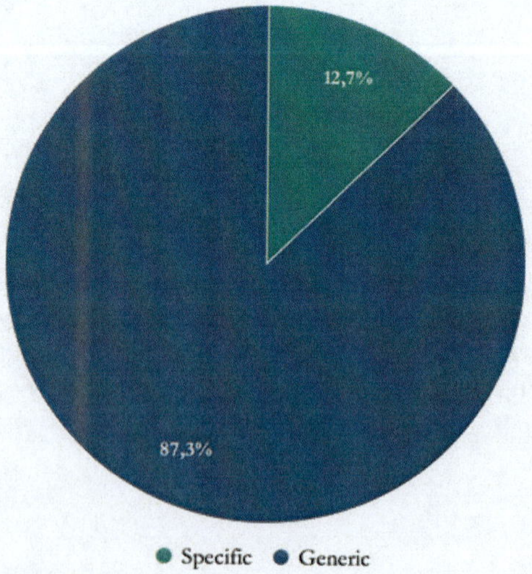

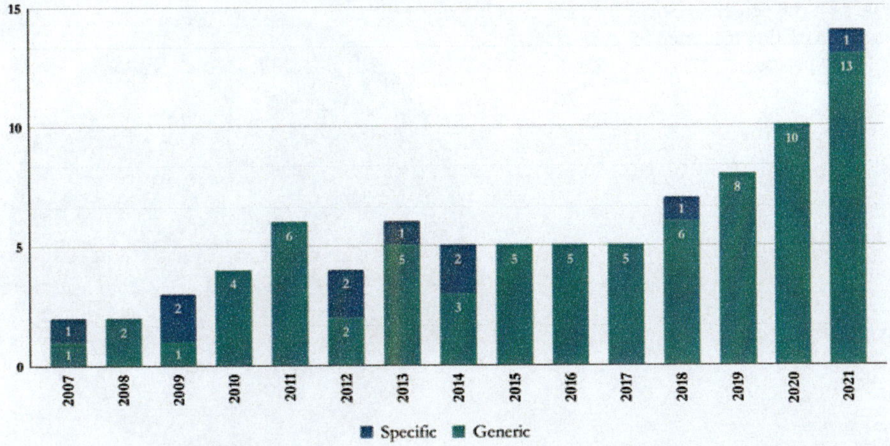

Fig. 12 Distribution of selected papers for generic and specific domains

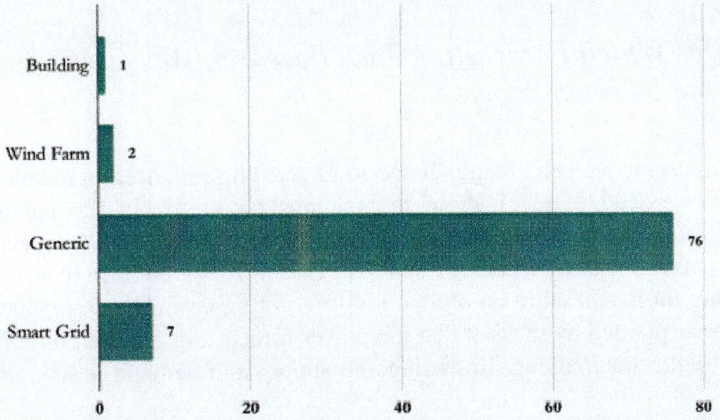

Fig. 13 The number of selected papers for both domains generic or specific

Figure 12 presents the distribution of selected papers for generic and specific domains in the period between 2007 and 2021. We observe for all year, the number of published papers that focus on a generic field is greater than those that are specific.

Figure 13 shows the variation in specific domains adopted in selected papers. It shows the most targeted domains affected by renewable energy forecasting are smart grid domain with seven papers (8.13%), wind farm domain with two papers (2.32%) and building domain with one paper (1.16%). The dominance of smart grid domain can be explained by the fact that many researchers have been interested in scheduling and energy management in a smart grid environment.

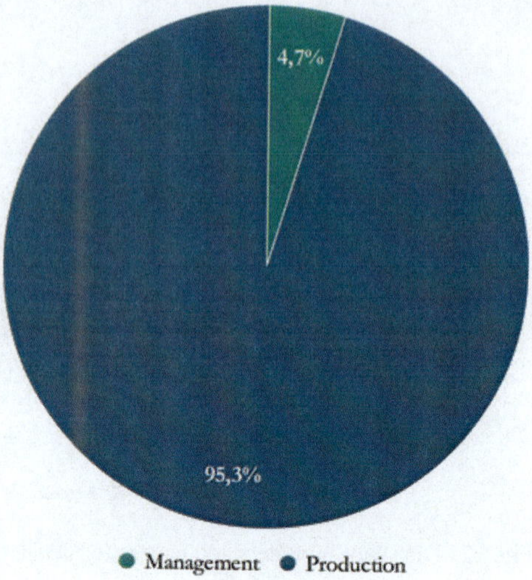

Fig. 14 The percentage of each forecasting task used in selected papers

4.7 RQ6: Which Forecasting Task Research Was Used in the Selected Papers?

Figure 14 presents the percentage related to forecasting task used in selected papers during the years 2007–2021. It shows that the most targeted tasks affected by renewable energy forecasting are production task with a percentage of 95.3%, succeeded by management task with a percentage of 4.7%. Therefore, we confirm that researchers are focusing more and more on production forecasting task. The dominance of this task can be explained by the fact that the power forecasting plays an important role in dealing with the challenges of balancing supply and demand in any smart grid system.

Figure 15 presents the evolution usage of both tasks production and management during the years 2007–2021. It can be observed from this figure that the number of papers published based on production forecasting increases every year. Moreover, we remark the use of management task only on four years 2007, 2012, 2013 and 2020. Finally, we notice the absence of consumption task during the years 2007–2021.

4.8 RQ7: Which Forecasting Period Was Considered in the Selected Papers?

Figure 16 presents the percentage of each forecasting period considered in selected papers. Based on this figure, we observe that the hourly forecast model is the most used with a percentage of 58.1%, succeeded by minutely forecast model with 22.1%.

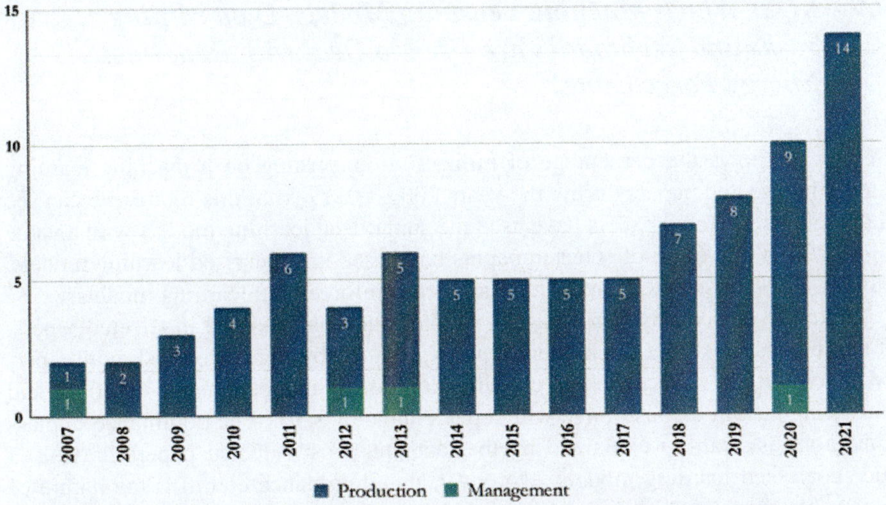

Fig. 15 The evolution usage of production and management during 2007 and 2021

Fig. 16 The percentage of each forecasting period used in selected papers

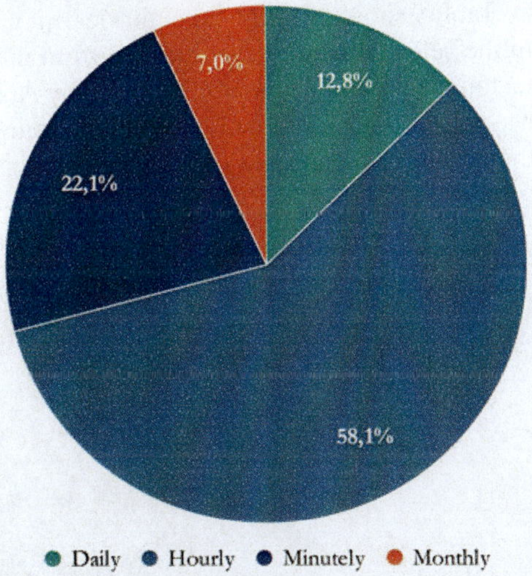

Moreover, daily forecast model is used with a value of 12.8%. Finally, monthly forecast model is less used in the selected papers with a value of 3.4%. Therefore, we confirm that hourly forecast model and minutely forecast model are the most used because they provide providing accurate prediction in short time interval.

4.9 RQ8: Which Machine Learning Models, Data Mining Tasks and Techniques are Used to Deal with Renewable Energy Forecasting?

Figure 17 shows the percentage of utilization concerning each machine learning model in selected papers during the years 2007–2021. From this figure, we can see that the most selected papers have used the supervised learning models with a value of 89.7%. While 6.9% of selected papers have used unsupervised learning models. Finally, 3.6% of selected papers have applied reinforcement learning models.

Figure 18 presents the percentage of adopted data analysis tasks in selected papers. It can be observed that the higher value is provided by regression task with a percentage of 70.2%, succeeded by classification task with a percentage of 21.4% and clustering task in the last range with a percentage of 8.3%. The dominance of classification task can be explained by the fact that most selected papers have used the supervised learning models. Moreover, this dominance confirms the nature of renewable energy forecasting which is regression task. Although classification task is supervised learning task, this task is less used in the context of renewable energy prediction.

Table 9 shows different machine learning algorithms that are used at least two time in the selected studies. We can see from the table that DNN is the most frequently technique used in 22 papers, followed by ANN technique that was used in 21 studies. Moreover, both SVM and ensemble learning were used in eight papers for each of them, NB, K-means and ELM were used in four papers for each of them, and GPR

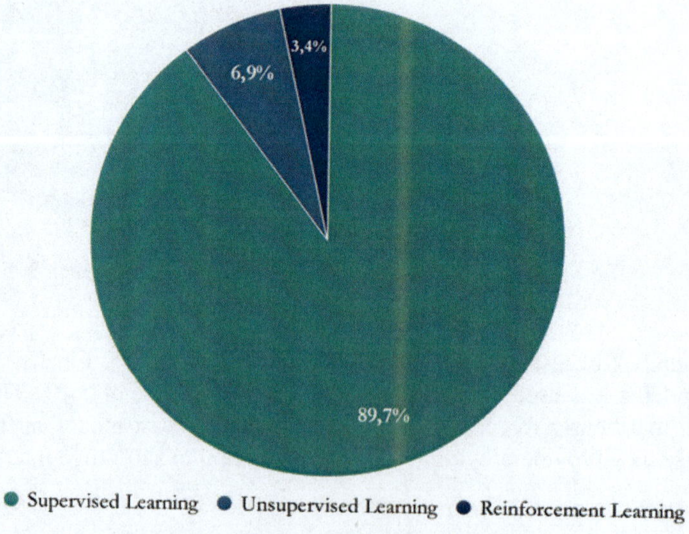

Fig. 17 The percentage utilization of each machine learning model

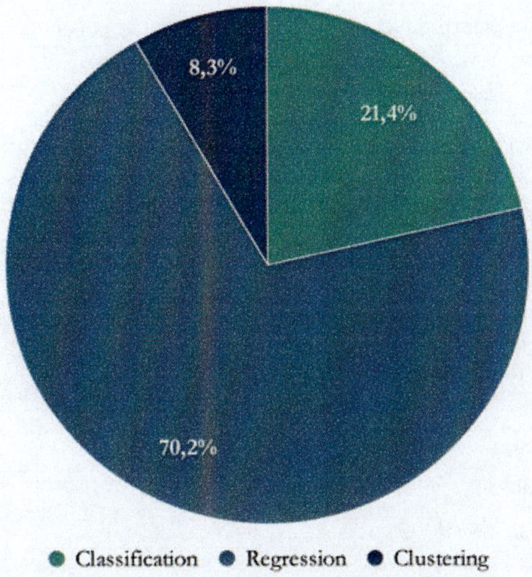

Fig. 18 The percentage of adopted data analysis tasks in selected papers

was used in three studies. Finally, we point out that RF, KNN, BDT and MDP were used only in two studies for each of them.

Figure 19 presents the evolution of frequent machine learning techniques usage in selected papers between 2010 and 2021. It can be seen that ANN technique is the only technique that was used in all years from 2007 until 2021. Moreover, this technique is the most dominant algorithm in selected studies between 2007 and 2015. Figure 10 shows also the majority of studies that were based on SVM and Ensemble learning have been published during 2007–2014 and 2015–2018, respectively. Finally, we remark that deep neural network (DNN) is the most dominant technique in selected studies during the period 2016–2021. Therefore, we confirm that researchers are focusing more and more on DNN for predicting different renewable energy sources.

We conclude that the use of supervised learning models and precisely the regression task is intuitive due to the nature of renewable energy forecasting issue. Indeed, our goal is predict different renewable energy sources that can be integrated into smart grid, for that we need to forecast each attribute in dataset. In addition, the regression techniques are known by their simplicity because they attempt to learn from the training dataset in order to find the suitable model. Finally, it is clear that unsupervised learning and reinforcement learning are less used for forecasting renewable energy because of their complexity by comparing them with supervised learning.

Table 9 Frequent machine learning techniques used in selected papers

Techniques	Number of papers
DNN (deep neural network)	22
ANN (artificial neural network)	21
SVM (support vector machine)	8
Ensemble learning model	8
NB (Nave Bayesian)	4
K-means	4
ELM (extreme learning machine)	4
GPR (Gaussian process regression)	3
RF (random forest)	2
KNN (k-nearest neighbors)	2
BDT (boosted decision tree)	2
MDP (Markov decision process)	2

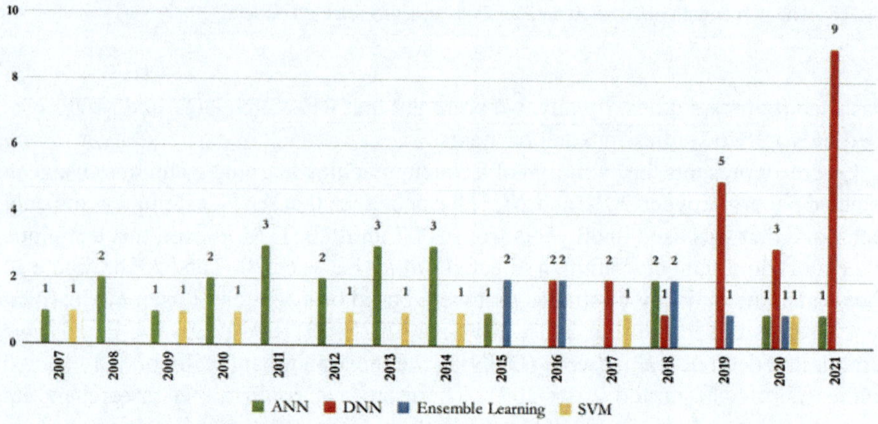

Fig. 19 The evolution of frequent machine learning techniques usage in selected papers

5 Implications for Researchers

In this section, we highlight some implications and recommendations for further research. These implications and recommendations are structured according to each research question as follows.

5.1 *RQ1*

In this study, the selected studies were retrieved from three digital libraries: IEEEx-plore, SpringerLink and Science direct. In addition, these studies have been published in publication sources with a high ranking. However, we did not consider the studies published in ACM data base. As result, researchers must extend their researches into this relevant data base.

5.2 *RQ2*

The majority of solutions proposed in selected papers are not implemented or experimented in real context. As result, we recommend that researchers may work for implementing these solutions. In addition, there is no such philosophical paper (PP) published in the literature that present a new conceptual framework solution. We encourage researchers to focus on philosophical paper related to renewable energy forecasting.

5.3 *RQ3*

The majority of selected studies are done in academic context, and recently, both governments and organizations have begun to invest strongly in research related to renewable energy projects. However, there is a lack in terms of collaboration with industry partners. In this context, we encourage researchers to collaborate with industry for the future works.

5.4 *RQ4*

The majority of selected studies have considered only two kind of energy: wind and solar and tidal currents energy is less used. So, it is recommended for researchers to focus on tidal currents energy and other kind of energy such as geothermal energy and biomass energy.

5.5 *RQ5*

The majority of selected papers were focused on a generic domain. It is therefore recommended that researchers focus more on specific domains. Moreover, the specific domains that most used in selected papers are smart grid. We encourage researchers to focus on other specific domain such as smart building, wind farm and transportation.

5.6 *RQ6*

For the majority of selected studies, we found that the authors focus on forecasting of power production. So, it is strongly recommended for researchers to make more effort in order to forecast the power management and the power consumption.

5.7 *RQ7*

The majority of selected studies have based short-time forecasting (hourly and minutely). As result, we encourage researchers to focus on other forecasting models that are able to predict daily, weekly as well as monthly.

5.8 *RQ8*

The majority of selected studies have based on supervised learning models, and there is a lack in terms of algorithms based on unsupervised learning models and reinforcement learning models. It is therefore recommended for researchers to conduct more research on renewable energy forecasting by using unsupervised learning models or reinforcement learning models.

6 Conclusion

Machine learning and renewable energy are considered among the most promising technologies that can be adopted in the next generation of smart cities. The utilization of machine learning techniques for renewable energy forecasting has shown great promise in terms of profitability over the last decade. In this chapter, we have performed a systematic mapping study regarding the use of machine learning techniques for renewable energy forecasting. A total of 86 relevant articles published between January 1, 2007, and December 31, 2021, were retrieved from three relevant digital databases in order to provide different responses related to different mapping questions shown in Table 8. To achieve this task, the selected papers were analyzed by year and publication source, research by year, sources and publication channel, research type, study context, study category, study domain, forecast task, forecast period and finally machine learning models adopted, and techniques used.

The obtained results show that the number of published papers in the field of renewable energy forecasting has increased significantly over the last decade, and particularly in the last three years. These papers were published in high quality journals (Q1 and Q2) and prestigious conferences (A and B). The majority of the selected studies were adopted evaluation search methodology and were done in academic context. Moreover, the majority of selected papers focused on generic domain and there are only some studies that focused on smart grid domain. Both wind energy and solar energy were used massively in selected papers than other kinds of energy. The forecasting of power production from different forms of energy was the primary interest in the majority of selected papers. Both hourly forecast model and minutely forecast model were used massively in selected papers than other kinds of energy. In addition, the majority of selected articles have used supervised learning models and more specifically regression task. Both ANN and DNN are the most used machine learning techniques to deal with renewable energy forecasting issue.

Finally, we propose two future directions for our research in the area of renewable energy forecasting. The first one, we intend to perform a systematic mapping study in order to analyze all recent studies of renewable energy forecasting based on deep learning techniques. The second future work consists to perform a systematic literature review (SLR) that will make an in depth analysis all selected papers published recently on renewable energy forecasting topic.

References

1. European Commission: 2030 Climate and Energy Framework. https://ec.europa.eu
2. Global energy transformation: a roadmap to 2050 (2019). Available online at https://www.irena.org/publications/2019/Apr/Global-energy-transformation-A-roadmap-to-2050-2019. Accessed August 17, 2020
3. Alkabbani, H., Ahmadian, A., Zhu, Q., Elkamel, A.: Machine learning and metaheuristic methods for renewable power forecasting: a recent review. Front. Chem. Eng. **26** (2021)
4. Zerrahn, A., Schill, W.P., Kemfert, C.: On the economics of electrical storage for variable renewable energy sources. Eur. Econ. Rev. 108, 259–279 (2018)
5. Wang, H., Lei, Z., Zhang, X., Zhou, B., Peng, J.: A review of deep learning for renewable energy forecasting. Energy Convers. Manage. 198, 111799 (2019)
6. Frías-Paredes, L., Mallor, F., Gastón-Romeo, M., León, T.: Assessing energy forecasting inaccuracy by simultaneously considering temporal and absolute errors. Energy Convers. Manage. 142, 533–46 (2017)
7. Lara-Fanego, V., Ruiz-Arias, J.A., Pozo-Vazquez, D., Santos-Alamillos, F.J.: Evaluation of the WRF model solar irradiance forecasts in Andalusia. Solar Energy 86, 2200–2217 (2012)
8. Chakraborty, S., et al.: A fuzzy binary clustered particle swarm optimization strategy for thermal unit commitment problem with wind power integration. IEEJ Trans. Electr. Electron. Eng. 7(5), 478–486 (2012)
9. Santhosh, M., Venkaiah, C.: Sustainable energy, grids and networks short-term wind speed forecasting approach using ensemble empirical mode decomposition and deep Boltzmann machine. Sustain. Energy Grids Netw. 19, 100242 (2019)
10. Kitchenham, B.: Procedures for Performing Systematic Reviews, vol. 33, pp. 1–26. Keele, UK, Keele University (2004)
11. Olabi, A.G.: Renewable and energy storage system. Energy 136, 1–6 (2017)

12. Zendehboudi, A., Baseer, M.A., Saidur, R.: Application of support vector machine models for forecasting solar and wind energy resources: a review. J. Clean. Prod. 199, 272–285 (2018)
13. Nielsen, T.S., Joensen, A., Madsen, H.: A new reference for wind power forecasting. Wind Energy 34, 29–34 (1998)
14. Lei, M., Shiyan, L., Chuanwen, J., Hongling, L., Yan, Z.: A review on the forecasting of wind speed and generated power. Renew. Sustain. Energy Rev. 13, 915–920 (2009)
15. Giebel, G., Brownsword, R., Kariniotakis, G., Denhard, M., Draxl, C.: The State-of-the-Art in Short-Term Prediction of Wind Power (2011)
16. Murata, A., Ohtake, H., Oozeki, T.: Modeling of uncertainty of solar irradiance forecasts on numerical weather predictions with the estimation of multiple confidence intervals. Renew. Energy 117, 193–201 (2018)
17. Giebel, G., Kariniotakis, G., and Brownsword, R., The state-of-the- art in short term prediction of wind power from a danish perspective. In: 4th International Workshop on Large Scale Integration of Wind Power and Transmission Networks for Offshore Wind Farms (Billund) (2018)
18. Ahmed, A., Khalid, M.: A review on the selected applications of forecasting models in renewable power systems. Renew. Sustain. Energy Rev. 100, 9–21 (2019)
19. Ezzat, A.A., Jun, M., Ding, Y., Member, S.: Spatio-temporal asymmetry of local wind fields and its impact on short-term wind forecasting. Trans. Sustain. Energy X 9, 1437–1447 (2018)
20. Ghofrani, M., and Alolayan, M.: Time series and renewable energy forecasting. In: Time Series Analysis and Applications, pp. 78–92 (2018)
21. Jiang, Y., Huang, G., Peng, X., Li, Y., Yang, Q.: Journal of wind engineering and industrial aerodynamics a novel wind speed prediction method: hybrid of correlation-aided DWT, LSSVM and GARCH. J. Wind Eng. Industrial Aerodynamics 174, 28–38 (2018)
22. Erdem, E., Shi, J.: ARMA based approaches for forecasting the tuple of wind speed and direction. Appl. Energy 88, 1405–1414 (2011)
23. Gomes, P., Castro, R.: Wind speed and wind power forecasting using statistical models: AutoRegressive moving average (ARMA) and artificial neural networks (ANN). Int. J. Sustain. Energy Dev. 1, 41–50 (2012)
24. Fentis, A., Bahatti, L., Tabaa, M., Mestari, M.: Short-term nonlinear autoregressive photovoltaic power forecasting using statistical learning approaches and in-situ observations. Int. J. Energy Environ. Eng. 10, 189–206 (2019)
25. Bacher, P., Madsen, H., Nielsen, H.A.: Online short-term solar power forecasting. Solar Energy 83, 1772–1783 (2009)
26. Atique, S., Noureen, S., Roy, V., Subburaj, V., Bayne, S., MacFie, J., Forecasting of total daily solar energy generation using ARIMA: a case study. In: IEEE 9th Annual Computing and Communication Workshop and Conference. CCWC (Las Vegas, NV), pp. 114–119 (2019)
27. Pasari, S., Shah, A.: Time Series Auto-Regressive Integrated Moving Average Model for Renewable Energy Forecasting. Springer International Publishing, Pilani (2020)
28. Kavasseri, R.G., Seetharaman, K.: Day-ahead wind speed forecasting using f -ARIMA models. Renew. Energy 34, 1388–1393 (2009)
29. Widodo D.A., Iksan N., Udayanti E.D.: Renewable energy power generation forecasting using deep learning method. IOP Conf. Ser. Earth Environ. Sci. 700, 012026 (2021)
30. https://www.discoverdatascience.org/industries/clean-energy/
31. Chang, J.-P., Lai, Y.-M., Chen, C.-H., Pai, P.-F.: A survey of machine learning models in renewable energy predictions. Appl. Sci. 10(5975), 2020 (2020)
32. Kotsiantis, S.B.: Supervised machine learning: a review of classification techniques. Informatica 31, 249–268 (2007)
33. Qiu, J., Wu, Q., Ding, G., Xu, Y., Feng, S.: A survey of machine learning for big data processing. EURASIP J. Adv. Signal Process. 2016, 67 (2016)
34. Gu, J., Wang, Z., Kuen, J., Ma, L., Shahroudy, A., Shuai, B., Liu, T., Wang, X., Wang, L., Wang, G., et al.: Recent advances in convolutional neural networks. Pattern Recognit. 1, 1–24 (2017)

35. Amasyali, K., El-Gohary, N.M.: A review of data-driven building energy consumption prediction studies. Renew. Sustain. Energy Rev. 81, 1192–1205 (2018)
36. Wang, H.Z., Lei, Z.X., Zhang, X.: A review of deep learning for renewable energy forecasting. Energy Convers. Manage. 198, 111799 (2019)
37. Banos, R., et al.: Optimization methods applied to renewable and sustainable energy: a review. Renew. Sustain. Energy Rev. 15(4), 1753–1766 (2011)
38. Diagne, M., et al.: Review of solar irradiance forecasting methods and a proposition for small-scale insular grids. Renew. Sustain. Energy Rev. 27, 65–76 (2013)
39. Voyant, C., et al.: Machine learning methods for solar radiation forecasting: a review. Renew. Energy 105, 569–582 (2017)
40. Das, U.K., et al.: Forecasting of photovoltaic power generation and model optimization: a review. Renew. Sustain. Energy Rev. 81, 912–928 (2018)
41. Wang, H., et al.: A review of deep learning for renewable energy forecasting. Energy Convers. Manage. 198, 111799 (2019)
42. Alkhayat, G., Mehmood, R.: A review and taxonomy of wind and solar energy forecasting methods based on deep learning. Energy AI, 100060 (2021)
43. Petersen, K., Vakkalanka, S., Kuzniarz, L.: Guidelines for conducting systematic mapping studies in software engineering: an update. Inf. Softw. Technol. 64, 1–18 (2015)
44. Lahby, M., Aqil, S., Yafooz, W., Abakarim, Y.: Online Fake News Detection Using Machine Learning Techniques: A Systematic Mapping Study. Combating Fake News with Computational Intelligence Techniques, pp. 3–37 (2022)
45. Wieringa, R., Maiden, N., Mead, N., Rolland, C.: Requirements engineering paper classification and evaluation criteria: a proposal and a discussion. Requirements Eng. 11(1), 102–107 (2006)
46. Kitchenham, B.A.: Systematic review in software engineering: where we are and where we should be going. In: Proceedings of the 2nd International Workshop on Evidential Assessment of Software Technologies, pp. 1–2 (2012)
47. Marsland, S. (2011). Machine Learning: An Algorithmic Perspective. Chapman and Hall/CRC
48. Han, J., Pei, J., Kamber, M.: Data Mining: Concepts and Techniques. Elsevier (2011)
49. Van Engelen, J.E., Hoos, H.H.: A survey on semi-supervised learning. Machine Learning 109(2), 373–440 (2020)
50. Sutton, R.S., Barto, A.G.: Reinforcement Learning: An Introduction. MIT Press
51. Chakraborty, S., Weiss, M.D., Simoes, M.G.: Distributed intelligent energy management system for a single-phase high-frequency AC microgrid. IEEE Trans. Ind. Electron. 54(1), 97–109 (2007)
52. Zhou, B., Du, S., Li, L., Wang, H., He, Y., Zhou, D.: An explainable recurrent neural network for solar irradiance forecasting. In: 2021 IEEE 16th Conference on Industrial Electronics and Applications (ICIEA), pp. 1299–1304 (2021)
53. Al-Dahidi, S., Louzazni, M., Omran, N.: A local training strategy-based artificial neural network for predicting the power production of solar photovoltaic systems. IEEE Access 8, 150262–150281 (2020)
54. Ji, G.R., Han, P., Zhai, Y.J.: Wind speed forecasting based on support vector machine with forecasting error estimation. Int. Conf. Mach. Learn. Cybern. 5, 2735–2739 (2007)
55. Cellura, M.A.U.R.I.Z.I.O., Cirrincione, G., Marvuglia, A., Miraoui, A.: Wind speed spatial estimation for energy planning in Sicily: introduction and statistical analysis. Renew. Energy **33**(6), 1237–1250 (2008)
56. Sanz, S.S., Perez-Bellido, A., Ortiz-Garcia, E., Portilla-Figueras, A., Prieto, L., Paredes, D., Correoso, F.: Short-term wind speed prediction by hybridizing global and mesoscale forecasting models with artificial neural networks. In: 2008 Eighth International Conference on Hybrid Intelligent Systems, pp. 608–612. IEEE (2008)
57. Ramirez-Rosado, I.J., Fernandez-Jimenez, L.A., Monteiro, C., Sousa, J., Bessa, R.: Comparison of two new short-term wind-power forecasting systems. Renew. Energy 34(7), 1848–1854 (2009)
58. Fan, S., Liao, J.R., Yokoyama, R., Chen, L., Lee, W.J.: Forecasting the wind generation using a two-stage network based on meteorological information. IEEE Trans. Energy Convers. 24(2), 474–482 (2009)

59. Colak, I., Demirtas, M., Bal, G., Kahraman, H.T.: A parameter determination system for wind turbines based on nave bayes classification algorithm. In: 2009 International Conference on Machine Learning and Applications, pp. 611–616. IEEE (2009)
60. Zhao, P., Xia, J., Dai, Y., He, J.: Wind speed prediction using support vector regression. In: 2010 5th IEEE Conference on Industrial Electronics and Applications, pp. 882–886. IEEE (2010)
61. Li, G., Shi, J.: On comparing three artificial neural networks for wind speed forecasting. Appl. Energy 87(7), 2313–2320 (2010)
62. Paoli, C., Voyant, C., Muselli, M., Nivet, M.L.: Forecasting of preprocessed daily solar radiation time series using neural networks. Solar Energy 84(12), 2146–2160 (2010)
63. Kusiak, A., Li, W.: Short-term prediction of wind power with a clustering approach. Renew. Energy 35(10), 2362–2369 (2010)
64. Mora-Lpez, L., Martnez-Marchena, I., Piliougine, M., Sidrach-de-Cardona, M.: Binding statistical and machine learning models for short-term forecasting of global solar radiation. In: International Symposium on Intelligent Data Analysis, pp. 294–305. Springer, Berlin, Heidelberg (2011)
65. Jahromi, M.J., Maswood, A.I., Tseng, K.J.: Long term prediction of tidal currents. IEEE Syst. J. 5(2), 146–155 (2010)
66. Catalo, J.P.D.S., Pousinho, H.M.I., Mendes, V.M.F.: Short-term wind power forecasting in Portugal by neural networks and wavelet transform. Renew. Energy 36(4), 1245–1251 (2011)
67. Erdem, E., Shi, J.: ARMA based approaches for forecasting the tuple of wind speed and direction. Appl. Energy 88(4), 1405–1414 (2011)
68. Chen, C., Duan, S., Cai, T., Liu, B.: Online 24-h solar power forecasting based on weather type classification using artificial neural network. Solar Energy 85(11), 2856–2870 (2011)
69. Lorenzo, J., Mndez, J., Castrilln, M., Hernndez, D.: Short-term wind power forecast based on cluster analysis and artificial neural networks. In: International Work-Conference on Artificial Neural Networks, pp. 191–198. Springer, Berlin, Heidelberg (2011)
70. Ferrari, S., Lazzaroni, M., Piuri, V., Salman, A., Cristaldi, L., Rossi, M., Poli, T.: Illuminance prediction through extreme learning machines. In: 2012 IEEE Workshop on Environmental Energy and Structural Monitoring Systems (EESMS), pp. 97–103. IEEE (2012)
71. Santos, N.I., Said, A.M., James, D.E., Venkatesh, N.H.: Modeling solar still production using local weather data and artificial neural networks. Renew. Energy 40(1), 71–79 (2012)
72. Shi, J., Lee, W.J., Liu, Y., Yang, Y., Wang, P.: Forecasting power output of photovoltaic systems based on weather classification and support vector machines. IEEE Trans. Ind. Appl. 48(3), 1064–1069 (2012)
73. Bonanno, F., Capizzi, G., Gagliano, A., Napoli, C.: Optimal management of various renewable energy sources by a new forecasting method. In: International Symposium on Power Electronics Power Electronics, Electrical Drives, Automation and Motion, pp. 934–940. IEEE (2012)
74. Quan, D.M., Ogliari, E., Grimaccia, F., Leva, S., Mussetta, M.: Hybrid model for hourly forecast of photovoltaic and wind power. In: 2013 IEEE International Conference on Fuzzy Systems (FUZZ-IEEE), pp. 1–6. IEEE (2013)
75. Wytock, M., Kolter, J.Z.: Large-scale probabilistic forecasting in energy systems using sparse gaussian conditional random fields. In: 52nd IEEE Conference on Decision and Control, pp. 1019–1024. IEEE (2013)
76. Kuznetsova, E., Li, Y.F., Ruiz, C., Zio, E., Ault, G., Bell, K.: Reinforcement learning for microgrid energy management. Energy 59, 133–146 (2013)
77. Marquez, R., Pedro, H.T., Coimbra, C.F.: Hybrid solar forecasting method uses satellite imaging and ground telemetry as inputs to ANNs. Solar Energy 92, 176–188 (2013)
78. Chen, S.X., Gooi, H.B., Wang, M.Q.: Solar radiation forecast based on fuzzy logic and neural networks. Renew. Energy 60, 195–201 (2013)
79. Hu, J., Wang, J., Zeng, G.: A hybrid forecasting approach applied to wind speed time series. Renew. Energy 60, 185–194 (2013)
80. Heinermann, J., Kramer, O.: Precise wind power prediction with SVM ensemble regression. In: International Conference on Artificial Neural Networks, pp. 797–804. Springer, Cham (2014)

81. Mellit, A., Pavan, A.M., Lughi, V.: Short-term forecasting of power production in a large-scale photovoltaic plant. Solar Energy 105, 401–413 (2014)
82. Khan, G.M., Ali, J., Mahmud, S.A.: Wind power forecasting an application of machine learning in renewable energy. In: 2014 International Joint Conference on Neural Networks (IJCNN), pp. 1130–1137. IEEE (2014)
83. Li, J., Mao, J.: Ultra-short-term wind power prediction using BP neural network. In: 2014 9th IEEE Conference on Industrial Electronics and Applications, pp. 2001–2006. IEEE (2014)
84. Pravilovic, S., Appice, A., Lanza, A., Malerba, D.: Wind power forecasting using time series cluster analysis. In: International Conference on Discovery Science, pp. 276–287. Springer, Cham (2014)
85. Pedro, H.T., Coimbra, C.F.: Nearest-neighbor methodology for prediction of intra-hour global horizontal and direct normal irradiances. Renew. Energy 80, 770–782 (2015)
86. Wang, J., Hu, J.: A robust combination approach for short-term wind speed forecasting and analysis-combination of the ARIMA (autoregressive integrated moving average), ELM (extreme learning machine), SVM (support vector machine) and LSSVM (least square SVM) forecasts using a GPR (Gaussian process regression) model. Energy 93, 41–56 (2015)
87. Duran, M.A., Filik, Ü.B.: Short-term wind speed prediction using several artificial neural network approaches in Eskisehir. In: 2015 International Symposium on Innovations in Intelligent SysTems and Applications (INISTA), pp. 1–4. IEEE (2015)
88. Ak, R., Fink, O., Zio, E.: Two machine learning approaches for short-term wind speed time-series prediction. IEEE Trans. Neural Netw. Learn. Syst. 27(8), 1734–1747 (2015)
89. Silva, C.V., Lim, L., Stevens, D., Nakafuji, D.: Probabilistic models for one-day ahead solar irradiance forecasting in renewable energy applications. In: 2015 IEEE 14th International Conference on Machine Learning and Applications (ICMLA), pp. 1163–1168. IEEE (2015)
90. Gensler, A., Henze, J., Sick, B., Raabe, N.: Deep Learning for solar power forecasting an approach using AutoEncoder and LSTM neural networks. In: 2016 IEEE International Conference on Systems, Man, and Cybernetics (SMC), pp. 002858–002865. IEEE (2016)
91. Melzi, F.N., Touati, T., Same, A., Oukhellou, L.: Hourly solar irradiance forecasting based on machine learning models. In: 2016 15th IEEE International Conference on Machine Learning and Applications (ICMLA), pp. 441–446. IEEE (2016)
92. Golestaneh, F., Pinson, P., Gooi, H.B.: Very short-term nonparametric probabilistic forecasting of renewable energy generation With application to solar energy. IEEE Trans. Power Syst. 31(5), 3850–3863 (2016)
93. Li, J., Ward, J.K., Tong, J., Collins, L., Platt, G.: Machine learning for solar irradiance forecasting of photovoltaic system. Renew. Energy 90, 542–553 (2016)
94. Wang, H.Z., Wang, G.B., Li, G.Q., Peng, J.C., Liu, Y.T.: Deep belief network based deterministic and probabilistic wind speed forecasting approach. Appl. Energy 182, 80–93 (2016)
95. Bayindir, R., Yesilbudak, M., Colak, M., Genc, N.: A novel application of naive bayes classifier in photovoltaic energy prediction. In: 2017 16th IEEE International Conference on Machine Learning and Applications (ICMLA), pp. 523–527. IEEE
96. Neo, Y.Q., Teo, T.T., Woo, W.L., Logenthiran, T., Sharma, A.: Forecasting of photovoltaic power using deep belief network. In: Tencon 2017-2017 IEEE Region 10 Conference, pp. 1189-1194. IEEE (2017)
97. Bouzgou, H., Gueymard, C.A.: Minimum redundancy-maximum relevance with extreme learning machines for global solar radiation forecasting: toward an optimized dimensionality reduction for solar time series. Solar Energy 158, 595–609 (2017)
98. Kavousi-Fard, A., Su, W.: A combined prognostic model based on machine learning for tidal current prediction. IEEE Trans. Geosci. Remote Sensing 55(6), 3108–3114 (2017)
99. Shi, Z., Liang, H., Dinavahi, V.: Direct interval forecast of uncertain wind power based on recurrent neural networks. IEEE Trans. Sustain. Energy 9(3), 1177–1187 (2017)
100. Li, C., Xiao, Z., Xia, X., Zou, W., Zhang, C.: A hybrid model based on synchronous optimisation for multi-step short-term wind speed forecasting. Appl. Energy 215, 131–144 (2018)
101. Sun, S., Wang, S., Zhang, G., Zheng, J.: A decomposition-clustering-ensemble learning approach for solar radiation forecasting. Solar Energy 163, 189–199 (2018)

102. Shi, Z., Liang, H., Dinavahi, V.: Wavelet neural network based multiobjective interval prediction for short-term wind speed. IEEE Access 6, 63352–63365 (2018)
103. Nespoli, A., Ogliari, E., Dolara, A., Grimaccia, F., Leva, S., Mussetta, M.: Validation of ANN training approaches for day-ahead photovoltaic forecasts. In: 2018 International Joint Conference on Neural Networks (IJCNN), pp. 1–6. IEEE (2018)
104. Rodrguez, F., Fleetwood, A., Galarza, A., Fontn, L.: Predicting solar energy generation through artificial neural networks using weather forecasts for microgrid control. Renew. Energy 126, 855–864 (2018)
105. Yang, Z., Wang, J.: A combination forecasting approach applied in multistep wind speed forecasting based on a data processing strategy and an optimized artificial intelligence algorithm. Appl. Energy 230, 1108–1125 (2018)
106. Domingo, A.J., Garcia, F.C., Salvaña, M.L., Libatique, N.J., Tangonan, G.L.: Short term wind speed forecasting: a machine learning based predictive analytics. In: TENCON 2018—2018 IEEE Region 10 Conference, pp. 1948–1953. IEEE (2018)
107. Lin, K.P., Pai, P.F., Ting, Y.J.: Deep belief networks with genetic algorithms in forecasting wind speed. IEEE Access 7, 99244–99253 (2019)
108. Huang, C.J., Kuo, P.H.: Multiple-input deep convolutional neural network model for short-term photovoltaic power forecasting. IEEE Access 7, 74822–74834 (2019)
109. Zhao, J., Wang, J., Guo, Z., Guo, Y., Lin, W., Lin, Y.: Multi-step wind speed forecasting based on numerical simulations and an optimized stochastic ensemble method. Appl. Energy 255, 113833 (2019)
110. Liu, D., Sun, K.: Random forest solar power forecast based on classification optimization. Energy 187, 115940 (2019)
111. Prasad, R., Ali, M., Kwan, P., Khan, H.: Designing a multi-stage multivariate empirical mode decomposition coupled with ant colony optimization and random forest model to forecast monthly solar radiation. Appl. Energy 236, 778–792 (2019)
112. Abdel-Nasser, M., Mahmoud, K.: Accurate photovoltaic power forecasting models using deep LSTM-RNN. Neural Comput. Appl. 31(7), 2727–2740 (2019)
113. Deng, Y., Jia, H., Li, P., Tong, X., Qiu, X., Li, F.: A deep learning methodology based on bidirectional gated recurrent unit for wind power prediction. In: 2019 14th IEEE Conference on Industrial Electronics and Applications (ICIEA), pp. 591–595. IEEE (2019)
114. Wen, L., Zhou, K., Yang, S., Lu, X.: Optimal load dispatch of community microgrid with deep learning based solar power and load forecasting. Energy 171, 1053–1065 (2019)
115. Devi, A.S., Maragatham, G., Boopathi, K., Rangaraj, A.G.: Hourly day-ahead wind power forecasting with the EEMD-CSO-LSTM-EFG deep learning technique. Soft Comput. 24(16), 12391–12411 (2020)
116. Faraji, J., Ketabi, A., Hashemi-Dezaki, H., Shafie-Khah, M., Catalao, J.P.: Optimal day-ahead scheduling and operation of the prosumer by considering corrective actions based on very short-term load forecasting. IEEE Access 8, 83561–83582 (2020)
117. Pan, M., Li, C., Gao, R., Huang, Y., You, H., Gu, T., Qin, F.: Photovoltaic power forecasting based on a support vector machine with improved ant colony optimization. J. Clean. Prod. 277, 123948 (2020)
118. Hai, T., Sharafati, A., Mohammed, A., Salih, S.Q., Deo, R.C., Al-Ansari, N., Yaseen, Z.M.: Global solar radiation estimation and climatic variability analysis using extreme learning machine based predictive model. IEEE Access 8, 12026–12042 (2020)
119. Shawon, M.M.H., Akter, S., Islam, M.K., Ahmed, S., Rahman, M.M.: Forecasting PV panel output using prophet time series machine learning model. In: 2020 IEEE Region 10 Conference (Tencon), pp. 1141–1144. IEEE
120. Theocharides, S., Makrides, G., Livera, A., Theristis, M., Kaimakis, P., Georghiou, G.E.: Day-ahead photovoltaic power production forecasting methodology based on machine learning and statistical post-processing. Appl. Energy 268, 115023 (2020)
121. Wang, L., Li, K., Ji, Z., Zhang, C.: An ultra-short-term prediction method for wind speed series based on Gaussian process median regression. In: 2020 15th IEEE Conference on Industrial Electronics and Applications (ICIEA), pp. 495–499. IEEE (2020)

122. Fraccanabbia, N., da Silva, R.G., Ribeiro, M.H.D.M., Moreno, S.R., dos Santos Coelho, L., Mariani, V.C.: Solar power forecasting based on ensemble learning methods. In: 2020 International Joint Conference on Neural Networks (IJCNN), pp. 1–7. IEEE (2020)
123. Pang, Z., Niu, F., O'Neill, Z.: Solar radiation prediction using recurrent neural network and artificial neural network: a case study with comparisons. Renew. Energy 156, 279–289 (2020)
124. Jahangir, H., Tayarani, H., Gougheri, S.S., Golkar, M.A., Ahmadian, A., Elkamel, A.: Deep learning-based forecasting approach in smart grids with microclustering and bidirectional LSTM network. IEEE Trans. Ind. Electron. 68(9), 8298–8309 (2020)
125. Huang, H., Jia, R., Shi, X., Liang, J., Dang, J.: Feature selection and hyper parameters optimization for short-term wind power forecast. Appl. Intell. 1–19 (2021)
126. Jalali, S.M.J., Ahmadian, S., Khodayar, M., Khosravi, A., Ghasemi, V., Shafie-khah, M., Catalo, J.P.: Towards Novel Deep Neuroevolution Models: Chaotic Levy Grasshopper Optimization for Short-Term Wind Speed Forecasting. Engineering with Computers, pp. 1–25 (2021)
127. Jumin, E., Basaruddin, F.B., Yusoff, Y.B.M., Latif, S.D., Ahmed, A.N.: Solar radiation prediction using boosted decision tree regression model: a case study in Malaysia. Environ. Sci. Pollut. Res. 28(21), 26571–26583 (2021)
128. Vidya, S., Janani, E.S.V.: Wind speed multistep forecasting model using a hybrid decomposition technique and a selfish herd optimizer-based deep neural network. Soft Comput. 25(8), 6237–6270 (2021)
129. Bento, P.M.R., Pombo, J.A.N., Mendes, R.P.G., Calado, M.R.A., Mariano, S.J.P.S.: Ocean wave energy forecasting using optimised deep learning neural networks. Ocean Eng. **219**, 108372 (2021)
130. Kılıç, F., Yılmaz, İ.H., Kaya, Ö.: Adaptive co-optimization of artificial neural networks using evolutionary algorithm for global radiation forecasting. Renewable Energy 171, 176–190 (2021)
131. Hassan, M.A., Bailek, N., Bouchouicha, K., Nwokolo, S.C.: Ultra-short-term exogenous forecasting of photovoltaic power production using genetically optimized non-linear auto-regressive recurrent neural networks. Renew. Energy 171, 191–209 (2021)
132. Jeong, J., Kim, H.: DeepComp: Deep reinforcement learning based renewable energy error compensable forecasting. Appl. Energy 294, 116970 (2021)
133. Wang, J., Wang, S., Li, Z.: Wind speed deterministic forecasting and probabilistic interval forecasting approach based on deep learning, modified tunicate swarm algorithm, and quantile regression. Renew. Energy 179, 1246–1261 (2021)
134. Kolodziejczyk, W., Zoltowska, I., Cichosz, P.: Real-time energy purchase optimization for a storage-integrated photovoltaic system by deep reinforcement learning. Control Eng. Practice 106, 104598 (2021)
135. Wang, J., Yang, Z.: Ultra-short-term wind speed forecasting using an optimized artificial intelligence algorithm. Renew. Energy 171, 1418–1435 (2021)
136. Knol, D., de Leeuw, F., Meirink, J.F., Krzhizhanovskaya, V.V.: Deep learning for solar irradiance nowcasting: a comparison of a recurrent neural network and two traditional methods. In: International Conference on Computational Science, pp. 309–322. Springer, Cham (2021)

Machine Learning for Green Smart Homes

Brian O'Regan, Fábio Silva, Paula Carroll, Xavier Dubuisson, and Pádraig Lyons

Abstract Smarter approaches to data processing are essential to realise the potential benefits of the exponential growth in energy data in homes from a variety of sources, such as smart metres, sensors and other devices. Machine learning encompasses several techniques to process and visualise data. Each technique is specifically suited to certain data types and problems, whether it be supervised, unsupervised or reinforcement learning. These techniques can be applied to increase the efficient use of energy within a home, enable better and more accurate home owner decision-making and help contribute to greener building stock. This chapter presents the state of the art in this area and looks forward to potential new uses for machine learning in renewable energy data.

Keywords Machine learning · Smart home · Energy management · Energy modelling · Green buildings · Big data

B. O'Regan (✉) · F. Silva · P. Lyons
International Energy Research Centre (IERC), Tyndall National Institute (UCC), Lee Maltings, Dyke Parade, Cork, Ireland T12 R5CP
e-mail: brian.oregan@ierc.ie

F. Silva
e-mail: fabio.silva@ierc.ie

P. Lyons
e-mail: padraig.lyons@ierc.ie

P. Carroll
University College Dublin (UCD), UCD Energy Institute, Q250, College of Business, Belfield, Dublin, Ireland D04 V1W8
e-mail: paula.carroll@ucd.ie

X. Dubuisson
XD Sustainable Energy Consulting Ltd, Rocksavage, Clonakilty, Cork, Ireland
e-mail: xavier@xdconsulting.eu

M. Lahby et al. (eds.), *Computational Intelligence Techniques for Green Smart Cities*, Green Energy and Technology, https://doi.org/10.1007/978-3-030-96429-0_2

1 Introduction

In the past, humankind had no other choice than to build sustainable habitations. Whatever they made was out of natural and local materials. People were forced to plan accordingly to survive from the resources available. With time, human groups started to gather together in small tribes, villages and farms, and then bigger and bigger cities . By the 1970s, it was obvious that the ever-growing demand for power and materials, inefficient power systems, poorly planned buildings and the demand for more comfortable homes with heating and cooling systems, would not be sustainable and climate change became a real concern.

Over the years, the idea of smart green (sustainable) cities has developed. An United Nations Economic Commission for Europe (UNECE) and International Telecommunication Union (ITU) [70] joint initiative (with a consortium of over 300 international experts) coined the concept of smart sustainable cities as *a city that explores Information and Communication Technology (ICT) and other available resources to improve efficiency of urban operations and/or services, enhance competitiveness to improve its citizens quality of life*. Such a city targets the social, economic, cultural and environmental needs of the present and for generations to come. This concept is well structured and documented with its own key performance indicators (KPIs) [21].

Additionally, with energy prices rising (e.g. fossil fuels and natural gas), the pressure for energy consumption reduction is a reality. The concept of the green (or *eco*) home addresses both concerns: climate change and energy consumption reduction. Green smart homes are an integral part of the smart green cities concept and a very important tool to address several of its core objectives.

The energy sector alone is responsible for roughly 75% of all greenhouse gas (GHG) emissions. Residential energy consumption represents 20% of the overall energy sector, as depicted in Fig. 1.

Buildings alone are one of the biggest emitters of GHGs [67], with domestic buildings responsible for approximately 21% of energy consumption and 21% of GHG emissions [5]. It is vital that we address this to help meet the challenges of the sustainable development goals (SDG), EU Climate Targets and the various climate actions plans throughout the world [69].

Despite the efforts to tackle climate change, CO_2 emissions (from industry and energy sectors) have increased nearly 60%—since the United Nations Framework Convention on Climate Change (UNFCCC) [71] in 1992, parent treaty of the 2015 Paris Agreement [68].

The IEA report "Net Zero by 2050" [32] establishes a roadmap of around 400 milestones with necessary global actions and commitments detailing when and what to do to decarbonise the economy in the next three decades to limit the average global temperature increase to 1.5 °C.

Such audacious goals demand a complete revolution on how we produce, transport and consume energy. This chapter explores how machine learning can help improve the performance of green smart homes and contribute to the creation of a cleaner and

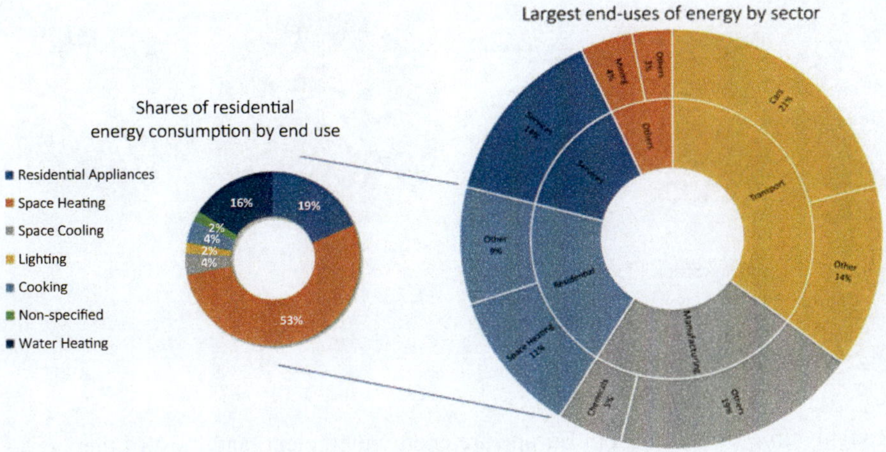

Fig. 1 Energy efficiency indicators—IEA 2018/2020 [31]

more resilient energy system. To contextualise the smart home concept evolution, let us dig a little back in time and explore the technology history from the early twentieth century up to today.

1.1 A Little History

It all started in the first two decades of the twentieth century with appliances that would not be considered smart as we understand them today, but were nevertheless revolutionary. The first models of vacuum cleaners (engine-powered in 1901 and an electricity-powered model six years later, in 1907), refrigerators, dishwashers, washing machines and toasters were just the beginning [11].

From the 1960s, electronics became more popular and accessible with the touch screens [36], computer-aided home operators [49, 63] and the ARPAnet project [13] with Tim Bernes-Lee [72] paving the way for the Internet. The following decades witnessed many more innovations with environmentally friendly washers and dryers in the 1990s and later robotic vacuum cleaner prototypes.

One of the first examples, of what would become "smart tech", comes from "Gerontechnology" (early 90s) [28], with special devices (trip/fall buttons) to improve senior citizens' lives. In parallel, mobile data exchange on 2G networks in Finland was followed by the standardised IEEE 802.11 Wi-Fi model [33].

The convergence of groups of technologies in hardware (more reliable and smaller devices), the Internet and wireless communication made it possible to create the smart homes we know today. Well-known examples of these technologies are Amazon®'s Echo and Alexa, Google® Home and Apple®'s Siri voice activated devices and also Google® Nest (Learning) thermostat.

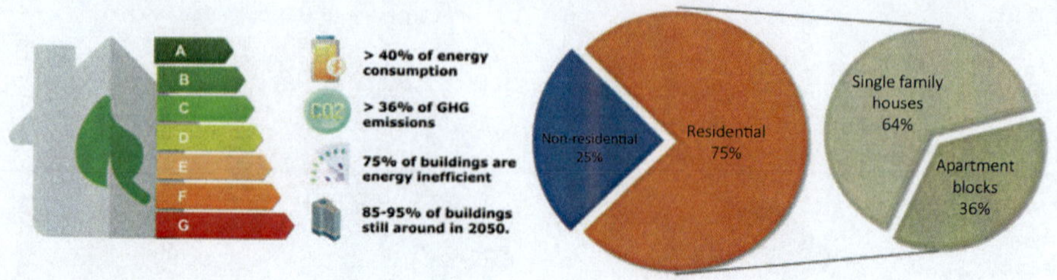

Fig. 2 EU Building Stock [23, 24]

1.2 Where Are We Today?

Roughly 75% of buildings in Europe are energy inefficient, and most of them (85–95%) will still be around by 2050 [23, 24], see Fig. 2. Building floor areas globally are estimated to double by 2060. As urbanisation in emerging markets increases, the demand for housing in urban areas increases and concentrates growth in residential construction [59]—with an average of 25% of non-residential against 75% of residential units.

These scenarios pose very challenging goals as the building stock energy consumption is responsible for roughly 36% of greenhouse gas (GHG) emissions. Green smart homes can offer significant improvements for the building stock energy performance to address these challenges. By implementing intelligent ways to manage and control the homes' energy consumption, green smart homes can provide ways to minimise energy consumption/carbon emission and, by doing so, reduce the overall home energy costs and raise the property value. This enhanced management is dependent on the capacity of the green smart home to collect data from its environment (e.g. temperature, luminosity and occupancy) and consumption behaviours (e.g. smart metres and appliances usage), and to interact with smart appliances and other devices.

Processing the data collected can be a challenge in itself. As smart home systems evolve and integrate, big data issues (explored in Sect. 4) become more obvious. The four V's of big data are usually present:

- Volume: the more data a system can collect, the more accurate its measurements and model predictions will be;
- Variety: data come from diverse types of hardware and protocols, making data integration harder;
- Velocity: the velocity of data collection depends on the time resolution and the number of sensors and devices monitored. It can present a real challenge to integrate and synchronise the monitored units into a single platform;
- Veracity: is concerned with the data quality, reliability and availability.

Finally, artificial intelligence and machine learning can help leverage value from the data by properly exploring the home energy data and adapting the energy consumption dynamically to new conditions and demands [59] using data-oriented decision-making for better planning, management and policy-making.

2 Smart Green Homes

The term "smart" has become a generic term over the years. Frequently used frivolously as an adjective to qualify some new or innovative product or service (not necessarily attached to the use of machine learning or artificial intelligence, but at least having some level of computer-aided control) [29], the term has become a very popular marketing tool. However, in our scope, the term "smart" is a technology capable of collecting information about its surroundings and offering some level of intelligent reaction to the scenario described by the data—in combination (or not) with other data sources [44].

What once was limited to control "environmental" systems in a home (e.g. heating) has evolved to encompass almost every electrical device in use in a house [57], from rooftops (photovoltaic panels and solar water heating), bedrooms and commons areas (heating and cooling systems, doors and windows sensors, smart lighting and thermostat), kitchen (energy-efficient appliances operation), smart metres (fro electricity and gas consumption) and smart metres to manage all resources consumption (e.g. electricity, gas, water) [59]. As the devices become "*smarter and smarter*", they become capable of communicating with each other and operating based on predefined behaviour.

As these devices (e.g. sensors, controls and appliances) are getting interconnected in networks, smart homes grow in complexity, interoperability and data collection become easily available. The "green" smart homes come from the potential that smart homes have to address the reduction of energy consumption and emission of associated GHG—and help in the green energy transition.

Also, the volume of data grows exponentially and big data (refer to Sect. 4) connected to smart homes becomes a necessity. One example that justifies the urge in dealing with such data volumes of data is illustrated in Fig. 3. The amount of time between data collection and the proper action has a direct impact on the energy savings of a household.

Moreover, machine learning applied to residential data becomes viable (refer to Sect. 5) and, in conjunction with energy modelling techniques and home energy management (refer to Sects. 3 and 6), green smart homes now can grow into a valuable tool to significantly impact the carbon footprint reduction.

Therefore, most of the potential for energy saving comes from monitoring (usage and behaviour) and developing proper automation to make the most of the energy available. Occupancy-based lighting and smart thermostats, optimisation and/or recommendations on better energy consumption profiles are only a few examples of what can be achieved. Data-driven decisions, in both local (operational) level and in

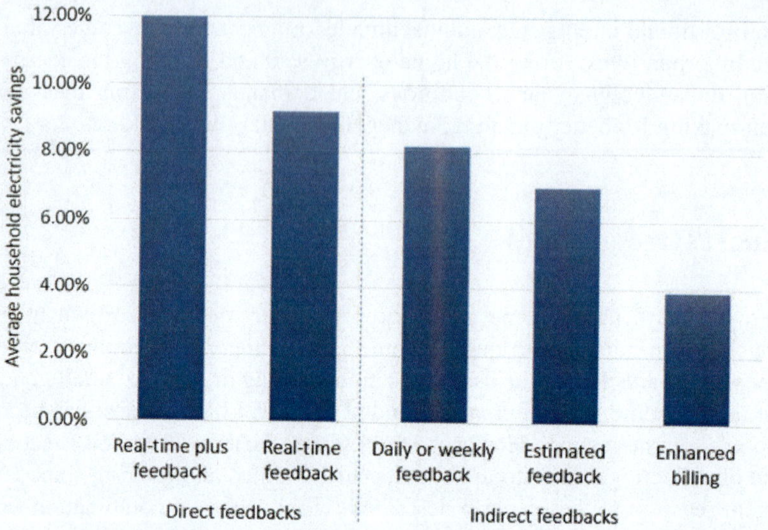

Fig. 3 Average household electricity savings by feedback type [42]

policy-making (planning and investments) level, will lead to an improved building stock energy consumption.

In summary, green smart homes are a group of network-connected devices and applications capable of integrating sensors, appliances and an ever-growing list of other devices to allow remote monitoring, user comfort convenience and energy consumption performance improvement of households in the green energy transition.

3 Home Energy Management

A Home Energy Management System (HEMS) provides us with the means to integrate, monitor, automate and control the household's smart appliances and also renewable energy-related devices (e.g. batteries, photovoltaic (PV) panels and so on). Additionally, the HEMS can also encompass a broader integration including smart sensors, a whole heterogeneous universe of Internet of Things (IoT) devices, security, external partners and the power grid [39] (Fig. 4).

By doing so, the HEMS addresses the energy performance issue on households and also make it possible to explore eventual surplus energy generation that can be available (e.g. PV panels, wind power and/or battery energy storage) from and to interconnected partners in a distributed energy generation scheme. Typical examples of household appliances include, but are not limited to washing machines, dishwashers, electric vehicles (charging stations), dryer machines, air conditioning, water heaters, lamps, televisions and even battery/energy storage systems where local generation is present.

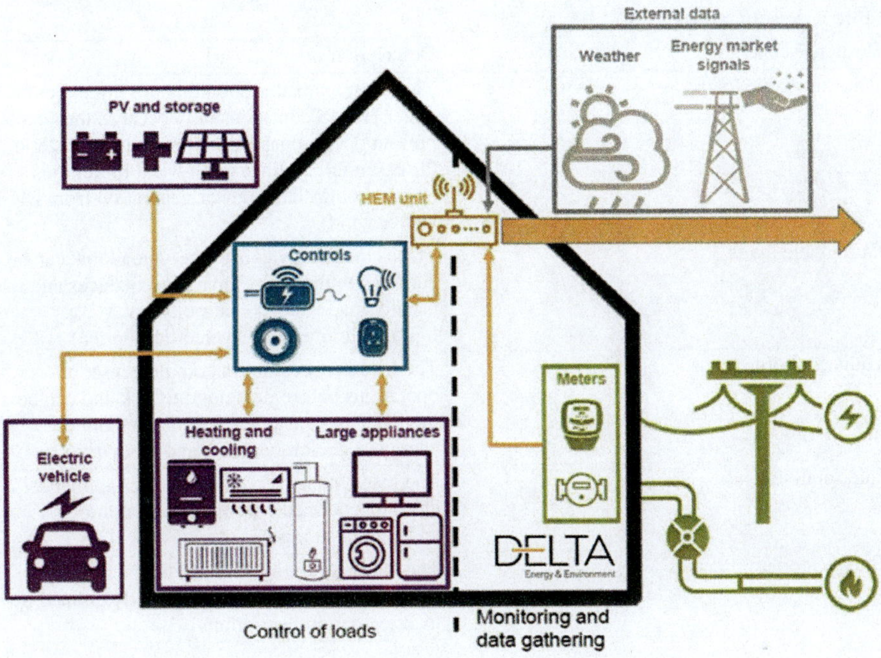

Fig. 4 Delta home energy management (HEM) [37]

However, not all appliances and devices in a household allow the same level of interaction and control (if any at all), which can limit scheduling objectives due to physical and/or operational constraints. Table 1 shows a list of possible response classes of devices, as follows.

Additionally, HEMSs must manage and coordinate loads in a scheduled manner. To achieve that—beyond the obvious communication and interconnection challenges—the HEMS has to deal with difficult integration issues. The lack of standardisation amongst diverse manufacturers in what concerns the devices and appliances communication/integration in a household (given that HEMS is a relatively new concept) poses significant barriers for the development and deployment of HEMSs.

For example, if a homeowner adds a new device to the house, the HEMS most likely will have difficulties to dynamically identifying and incorporating the new device. If it is a new electric vehicles (EV) in the garage, a new battery for renewable energy storage or an additional heater, the modelling and control of a myriad of distinct devices will demand reasonable flexibility from HEMSs. However, as home automation technologies advance and become more accessible/affordable, the research institutes and the market focus their efforts on the development of standardised ways for integration [34, 61] and the strengthening of HEMSs as a whole.

The goal is to promote a coordinated operation of these components and to create some level of integration between HEMSs. Even though its operation is usually on

Table 1 Response classes list [3]

Response class	Description
Uncontrollable loads	Those are loads that should not or could not be under HEMS control usually because they can present operational limitations (e.g. a freezer or the essential lighting) or physical limitations (e.g. uncontrollable power generation from PV and/or wind)
Curtailable loads	This is the energy consumption load that can be curtailed without any later consequences and is usually made due to electricity price (e.g lighting dimmed due to electricity price)
Uninterruptible loads	The load is necessary to complete a set of operations before finishing the task and can be modelled to consume a specific amount of energy (e.g. cloth washers and cloth dryers)
Interruptible loads	The loads must keep the devices or appliances close to a defined state (e.g. air conditioning and energy storage)
Energy storage	The loads are used to control how energy is stored or dispensed, and it is usually defined by load regulation restrictions

a demand-side (e.g. local energy savings), HEMS must interact somehow with the energy grid for proper supply in a dynamic energy load scheme. For that, a HEMS is usually composed of these components:

Smart appliances: household devices (e.g. heaters and air conditioning) and/or energy generators (e.g. PV and wind turbines), both improved with communication and (at least basic) computing functionalities, allow data exchange and interaction for operation coordination.

Sensing and measuring: sensing (e.g. temperature and motion detection) and measuring (e.g. smart metres for collecting data about energy consumption in general or about specific appliances and devices) [39].

User interaction: as HEMS is a user-centric technology from design, it demands a way to allow users/residents to interact with the HEMS functionalities. For that reason, it is expected user access, via some sort of interface (e.g. web interfaces or touch-screen devices), to check energy consumption, define operation preferences, comfort parameters and so on.

Centralised services: help the user to manage and use HEMS functionalities (especially in what concerns remote access), integrate the households (sharing data) and coordinate renewable energy exchange (or trading) [60].

The integrated operation of HEMSs seeks to provide the necessary data to the grid (so it can guarantee the energy supply), avoid (as much as possible) unnecessary demand stress and eventual outages. This coordinates operation can be developed in a centralised or in a distributed approach.

Each approach has its positive and negative points. The centralised approach concentrates all operations in a single processing point (usually giving access to all sorts of data, including private information) and demands heavy computational power to deal with the necessary analysis. However, as all necessary data is present, the results are expected to be more effective. Differently, the distributed approach expects the collaboration of distinct and independent components that will work together to develop demand-side plans and grid operation. Usually, the distributed approach does not share private data but, at the same time, it can have its performance negatively affected because it works with "*incomplete*" information.

Nevertheless, the integration of the HEMSs into to grid infrastructure is essential to achieve higher levels of grid stability and reliability—and it is especially important in distributed renewable energy (DRE) generation schemes where each HEMS can be a potential renewable energy generator with "*erratic*" outputs.

In summary, HEMSs allows householders to manage and monitor appliances and devices energy consumption (in real-time or near real-time) in a planned operation manner. It can be understood as an evolution of the traditional smart features into a more effective role in what concerns power grid interconnected operations.

4 Big Data

Globally, data is increasing at an exponential rate and energy data is no different. As a result of the significant growth in ICT in buildings such as sensors and metres, data in buildings has the potential to change the way we monitor and manage building stock. Unfortunately, the data gathered is often not stored in a manner to maximise the opportunity for potential benefits from the data and this is a huge cause for concern given the rise in dark data [52].

The importance of finding good data is paramount to the success of research, and to meet the ambitious European Union 80% reduction goal in primary energy consumption by 2050 [22]. In support of the transition to an environmentally sustainable society, a significant amount of good or useful relevant data, is required that is capable of informing decisions on channelling future energy research, investment opportunities, in-depth policy analysis and delivering better national policies. Providing this data will be a considerable undertaking, requiring careful analysis of the V-based characterisations to achieve this.

V-based Characteristics: Big data uses a variety of "V" values when describing the V-based characteristics, for example IBM coined the 4Vs of big data for qualifying and quantifying the important factors but up to 14Vs have been identified by others [54]. In the following list, we have captured the 14Vs in no particular order, in order to ensure that all are considered:

- **Volume**: As we are dealing with big data, volume refers to the quantity of data.
- **Variety**: Due to the growing amount of data that is generated that is either structured or unstructured, variety refers to this.

- **Velocity**: The speed at which big data is generated, created, produced or refreshed
- **Veracity**: Unfortunately, due to the volume of the data that is generated, created, produced or refreshed, confidence in such data is reduced.
- **Validity**: It is obviously important that the data we are using is valid, i.e. real data from a reliable source, using invalid data can skew results and lead to poor models.
- **Value**: The primary reason for data collection and processing is to extract value from it, without value then it would be a pointless exercise.
- **Variability**: If data was always the same, then there would be no need to repeatedly gather it, data that is gathered typically changes over time.
- **Venue**: Where the actual data science work takes place, also where the data came from and where it was stored.
- **Vocabulary**: Refers to data terminology, including models and structures.
- **Vagueness**: Confusion over the data that was gathered and difficulty extracting meaning from it.
- **Visualisation**: With all data, it is a struggle to visualise it in a meaningful way.
- **Virality**: Refers to how fast data is shared from source to another.
- **Volatility**: How long the data stored is of actual use to a user.
- **Viscosity**: The time lag between an event and when information was shared.

At present, the level of effort required in producing good data is placing huge demands on the producers and users of data, with a talent gap and skills shortage adding to the increasing list of the challenges that we are facing. In the short term, it is of critical importance that we address the rapidly increasing amount of data being generated from buildings and to devise effective and reliable tools and methods that can be promptly put in place for its evaluation and use to secure a low carbon economy.

This problem can be addressed by applying a variety of methods, such as for example data classification, data optimisation, data mapping and standardisation. To meet the skills gap, we need to develop a knowledge sharing structure, a network of interconnected research centres, organisations and individuals helping one another to clean their data and maximise its usage for the global benefit. The availability of a database of building-related data providing relevant information quickly will have substantial benefits in both the building sector and a variety of sectors that contribute to the increasing greenhouse gas emissions (Fig. 5).

It is extremely difficult to obtain high-quality data quickly or easily and obtaining buildings data is no exception. Additionally, data, by itself, is not useful until it is organised and processed for the extraction of meaningful information and, furthermore, refined to become a knowledge base. It must be noted that while having good knowledge of our buildings does not necessarily mean good decisions, it does provide the means to make more informed decisions. At present, with all data, we struggle to get it right and have unfortunately put the *cart before the horse*, so to speak. Too often, we have placed more importance on devices that create data, than on the data quality created by such devices itself, which has led us to our current predicament, whereby we have an estimated 90% of all data from Internet of things devices never

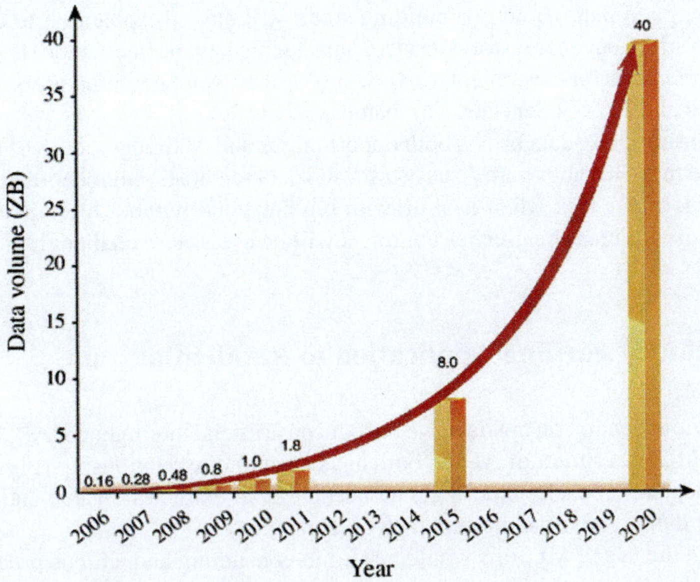

Fig. 5 Big data growth 2006–2020 [40]

used [30], meaning decisions are based on approximately 10% of available data, which is astonishing to say the least.

A white paper published by the International Data Corporation (IDC) in 2018 informed us that we recently reached 18 zettabytes of data, and by 2025, this is expected to reach a phenomenal 175 zettabytes [56]. While building-related data is not anywhere near this "global data sphere" figure, it is still in the petabytes range and will undoubtedly require artificial intelligence (AI) to release its full potential.

Advances in artificial intelligence (AI) will, without doubt, be one of the keys to meeting these challenges and while data analytics itself is not new and has been around since databases technologies first came on the scene, digital transformation, having been highlighted as the core of the ongoing industrial revolution [17] requires the need for software applications to be rolled-out quickly to address the challenges of achieving a 40% reduction of energy use in buildings and 36% of CO_2 emissions in the EU [18]. These software applications have the potential to provide greater insights into our building stock, enable quicker more qualified decisions to be made, thus reducing energy and other costs, reduce CO_2 emissions, meeting EU targets and ensure greater alignment with the sustainable development goals.

Crucially, greater awareness of our buildings data will lead to more accurate policy decisions, and with the availability of real-time data, there is the potential to perform analysis and monitoring of policies enacted and currently underway around the EU such as the renewable electricity support scheme (RESS). The aim of this section is to provide governments with the tools to monitor and measure their initiatives and meet their individual targets.

Finally, good data about our building stock will have the potential to drive not only innovation, but energy-as-a-service, introducing new business models and business opportunities for several enterprises, such as hardware manufacturers, software companies and Energy Services Companies (ESCOs).

In summary, Big data brings both opportunities and challenges, each of the 14Vs mentioned in the list above are no easy task, and data scientists, engineers and analysts have an enormous task when dealing with big data, and without high-quality data, our potential for creating greener buildings will be even more challenging.

5 Machine Learning Application to Residential Data

In the previous section on big data, we introduced artificial intelligence (AI), machine learning (ML) is a subset of AI and both have a symbiotic relationship (Fig. 6) [20]; i.e. they depend on each other, big data needs ML to extract the value and ML the volume of data to increase the accuracy of its results.

Without the use of ML, this would be a time-consuming and tedious processes to complete this task, and in the majority of cases involving buildings energy-related data, it would lead to the data being of little or no use in optimising systems to reduce consumption; i.e. once analysis was carried out, the situation (such as weather, temperature or occupancy) would have changed. It is important to extract meaningful value from the data quickly in order to create actions while they are still relevant. ML is divided into several different types: (1) supervised, (2) unsupervised and (3) reinforcement learning; each of these types has several methods (or algorithms); in

Fig. 6 Artificial intelligence subsets [8]

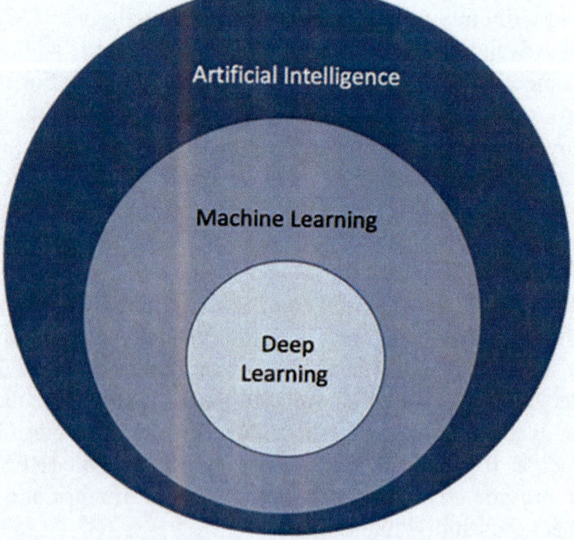

the rest of this section. we will discuss these types and methods and describe how they can be of use in extracting value from building energy data.

5.1 *Machine Learning Algorithms*

- **Linear Regression**: linear regression aims to find the best-fitting line, also referred to as the regression line [55], it is represented by the linear equation $y = mx + b$. Linear regression is one of the most common ways of estimating real values by modelling the dependent (y) and independent variables (x). For example, heating degree days (HDD) and energy use by creating a slope (m) and intercept (b) which could form the baseline for a measurement and verification (M & V) project.
- **Logistic Regression (LR)**: while the name implies that it is a type of regression algorithm, it is in fact a classification algorithm. In short, LR looks at data such as 1/0, true/false, on/off, etc. to predict the probability of occurrence [64]. LR can be broken down into three types: (1) Binary, (2) multinomial and (3) ordinal and in relation to building energy data is of particular use, for example, to determine whether a window is open or closed, a piece of equipment is on or off, or if a space is occupied or unoccupied. This can be of benefit in making adjustments to heating/cooling of a space.
- **Decision Tree (DT)**: to create actions in buildings (such as HVAC) based on other factors (such as weather or occupancy), it is important to have a method for making decisions, this is where DT comes in. As a supervised learning algorithm, DT enables, as the name suggests, decisions to be made based on if the answer to an if/else type statement is true or false, for example, is it raining? If it is raining is the window open or closed? If the window is closed, then the space temperature increased above a specific temperature? If it has, then reduce heat to the space or increase cooling.
- **Support Vector Machines (SVM)**: SVM is a supervised learning algorithm for two-group classification problems; after plotting the data points, a best hyper-plane (decision line) is calculated, with these data points falling on either one side or the other of the data plane. Predicting energy consumption is just one of the benefits of SVM, the ability to take variables such as equipment data, sensor data, weather data, occupancy and the time of year to predict building energy consumption and performance.
- **Naive Bayes (NB)**: based on the Bayes' theorem [4], NB predicts membership probabilities for each class. This means that it determines if a data point is in one class or another. In conjunction with other algorithms mentioned previously, NB can be used in buildings to predict HVAC energy [41].
- **k-Nearest Neighbors (KNN)**: KNN is a supervised classifier, which can be used for both regression and classification and is based on the principle that data points in proximity to one another fall under the same class. It begins by classifying a value for k, for example, 5 and looks for the 5 nearest neighbour to k [16]. In

building energy, KNN is used, amongst other things, for predicting daily energy demand for example.

- **K-Means**: K-means applies random centroids in plotted data; it clusters data around the nearest centroid, creating an average of all points within that centroid. There are many variables that can affect energy consumption, such as occupants. Occupants are generally unpredictable and harder to create accurate models for. In building energy, the ability to cluster different occupants can help predict energy consumption [1] or applying behaviour recommendations to occupants with similar behaviour patterns (such as when they use electricity, arrive at or leave a building or what their preferred environmental conditions are), NB can be of particular use in such cases.
- **Random Forest (RF)**: RF gets its name from a collection of decision trees and is a particularly easy to use algorithm, although there are some differences between RF and DT. Similar to KNN, it can be used for both regression and classification tasks. RF starts by deciding the n_estimators hyperparameter, simply put this is the number of trees. RF can provide similar services to those mentioned previously.
- **Dimensionality Reduction Algorithms (DRA)**: DRA refers to several techniques, such as principal component analysis, or PCA, and linear discriminant analysis, or LDA, that are used in training data to reduce the number of input variables. As occupants spend 90% of their time indoors, it is important to ensure thermal comfort is assured, and DRA is one of the methods used to help analyse data to achieve this, by identifying significant values from the huge quantity of data gathered and works in combination with some of the previously mentioned algorithms to achieve this.
- **Gradient Boosting algorithms (GBA)**: GRA employs the prediction of several base estimators to improve robustness over a single estimator and is used in building-related energy data to take values from several weak sources to help make a better decision, for example, in cases where there are several sensors gathering information on temperature, for example, GBA can help increase the confidence of the predicted temperature based on the values of all of these combined.

In summary, as described in the list above, there are several options available to data scientists and engineers, some are costly from an energy use point of view, or time-consuming depending on the quantity of data being processed. It is important when choosing one method over another, that we monitor the actual results of the actions taken based on such results. In the following section on eenergy modelling, we will see how the algorithms above can be used. Without ML, the extracting value from the huge volume of data being generated will be enormously challenging, even impossible for time sensitive tasks.

6 Energy Modelling

Historically, residential buildings and their occupants have played a passive role as electricity and energy consumers. Residential electricity demand could be reliably forecast at an aggregate level based on the time of year, time of day and the day of the week. The time of the year is related to weather, while occupant behaviour and usage activities are related to time of day and day of the week.

Representative load profiles are derived for similar groups of customers exhibiting characteristic diurnal pattern with morning and evening peaks [25, 47]. It is important to understand the similarities and differences between groups of customers and to understand their coincident demand because electricity networks are designed to meet the maximum coincident demand connected at the same substation. Paatero et al. note the importance of understanding and modelling residential load for planning medium and low voltage networks in residential areas [53]. They also note the need for extensive data about consumers, their appliances and the households in general to create useful models. They describe a bottom-up statistical modelling and simulation approach based on two Finnish sets of hourly data from appliances and lighting for blocks of 1082 households, but excluding heating and cooling appliances. They fit probability density functions to the data and simulate additional load for domestic appliances to create sample load profiles for a portfolio of residential consumers.

The focus in [73] is on load patterns for consumer groups using frequency domain analysis. They analyse the average load for groups of customers in the frequency domain, noting that the individual load patterns are smoothed by taking the average over the group. They also note that they made no attempt to re-cluster the customers in the sample. They rely on the customer class group defined by the utility company. Customer classification is important in developing demand-side products suitable for the customers' needs and lifestyle.

The adoption of low carbon heating and transport technologies will drive the demand for electricity up, while distributed renewable energy sources such as solar thermal or solar photovoltaic panels will drive demand from the grid down by allowing electricity to be generated locally. There is a need to adapt electricity supply and demand forecast models taking the adoption of low carbon technologies and the potential for self-consumption into account at the level of the individual home for the HEMS described in Sect. 3.

Short-, medium- and long-term electricity forecasts are needed so that the grid can be operated and managed efficiently. Short-term high resolution forecasts are needed for operational purposes such as the unit commitment problems. Medium-term forecasts facilitate maintenance scheduling, while longer-term forecasts and estimates of demand are needed for strategic investment and planning decisions. These forecasts are used by system operators to dimension the network to meet the maximum average coincident demand.

The objective of all forecasting is to create as accurate a forecast as possible. The accuracy of predictions decreases as the forecast horizon increases. Multiple linear regression is often used for medium- and long-term forecasting to estimate a trend based on historic demand and influenced by socioeconomic factors. For higher resolution short-term forecasts, additional approaches include traditional time series statistical approaches ((A)ARIMA), machine learning (ML) such as SAX, or neural network (NN) or deep learning methods.

A comparison of time series approaches to electricity forecasting is given in [46]. A machine learning artificial NN approach is used in [58], while [26] give an overview of the main methods described in the academic literature between 2005 and 2015. In all cases, the time series data need to be reduced to a smaller set of explanatory variables or features without loosing important information.

The residential sector accounts for about a quarter of the energy used in Ireland. With the availability of smart metre data, individual homes or appliance can be modelled. Characterising electricity demand at an aggregate level poses different challenges to individual dwelling level [46]. The signal from an individual device or home can be noisy. Individual appliance level can be aggregated to a household, then to building, district and smart city level. Alternatively, aggregate forecasts can be created at geographic regional areas for (smart) homes connected at the same substation, or for local or renewable energy communities.

Households are heterogeneous customers that consume different amounts of electricity at different times of the day for different purposes such as heating, cooking and electrical appliances. Understanding the timing and amplitude of the demand and the relationship with household characteristics is important in planning production and grid capacities and in designing policies [2].

The increased demand from homes heated by air source heat pumps (ASHPs) is explored in [7] while the real-world operations of ASHPs in retrofitted homes are analysed in [6]. Multiple linear regression is used by [45] on smart metre data from consumer behaviour trials in Ireland, [9]. They explore two bottom-up approaches: a model based on dwelling and occupant characteristics that explains usage patterns for different types of households; and a model based on electrical appliances that explains how electricity is consumed. They find that time of use for maximum electricity demand is strongly influenced by occupant characteristics, such as head of household age and household composition.

An understanding at the individual household level offers opportunities to support high demand or inefficient high peak users with recommended solutions. Householders have different motivation to engage in energy efficiency measures. Many may simply wish to use greener energy, improve the efficiency and comfort of their home or just reduce their energy costs. After "efficiency first" measures such as insulation and low wattage lighting, smart approaches can help householders to understand their usage patterns and highlight opportunities to contribute to a green energy transition. Changes to market structures, policies to achieve climate action plans, and the availability of low carbon technologies mean the role of households and buildings is changing. The availability of data, statistical and machine learning software tools, and smart grid infrastructure enables smart green homes and creates opportunities for householders to transition from passive end-user consumers of energy, to active prosumer roles—both producing and consuming energy.

Developing computationally tractable control-oriented models, which adequately represent the complex and nonlinear thermal-dynamics of individual buildings, is a significant challenge [38]. Data-driven predictive control models may replace traditional mechanistic physics-based models. The smart metre and smart device data may give a better understanding of the performance of devices in the real world in contrast to simulations or laboratory tests.

In summary, energy modelling is important to understand energy consumption from the customers' perspective and to support their transition from a passive role to a user-centric scheme. Individual behaviours when aggregated in groups of customers represent coincident demands and impact performance and/or forecasting directly.

7 Use Cases

7.1 CENTS

The Cooperative ENergy Trading System (CENTS) framework [34] is a collaborative project coordinated by the International Energy Research Centre (IERC), Tyndall National Institute, in partnership with industry experts (e.g. Smart MPower [62] and mSemicon [48]), research organisations (e.g. University College Cork (UCC) [66], National University of Ireland Galway (NUIG) [50] and Technological University Dublin (TUD) [65]) and community energy groups associated with the sustainable energy sector (e.g. Community Power [10].

CENTS's main purpose is to deliver a blockchain-enabled peer-to-peer (P2P) energy trading platform, hardware requirements prototypes, and market and regulatory strategies. Additionally, through its integration platform, CENTS is capable of addressing important aspects related to energy poverty [43].

Figure 7 describes CENTS high-level layered framework to address important topics for the establishment of the smart grid (SG) integration with green smart homes and the development of a user-centric platform of services.

At layer 1 in Fig. 7, CENTS describes briefly a typical green smart home that can produce its own renewable energy and, if there is a surplus generation (remaining available energy after the household consumption took place and it is properly stored in batteries), make it available to the grid. This green smart home would be classified as a prosumer or, in other words, someone that can act as a consumer and/or a producer of energy. All sorts of IoT components (e.g. smart metres, sensors and actuators) are at the core of the CENTS platform to provide monitoring and control over the local renewable power generation and usage.

This way, CENTS can provide the data to make SG integrated operation (at local, community and wider grid level). Nevertheless, such IoT apparatus lacks standardisation and the CENTS platform offers all the services necessary to integrate such a diverse range of equipment/devices, protocols and the operation itself—including the communication and protocols issues.

In this implementation, the integration components work as a hub (concentrator) interconnecting all IoT devices and services present in the green smart home and its local distributed energy generation infrastructure to the remaining layers of the framework.

The multi-layered framework used for the CENTS project provides a decoupled architecture to guarantee faster adaptation to new scenarios, regulations and new technical demands. Most importantly, CENTS integrates prosumers, their energy production capabilities and usage profile. With the use of machine learning algorithms to define the best policies of generation, consumption and energy trading (including energy poverty policies [43], the platform can offer data-centric decision-making decision and provide interaction between its functionalities and the green smart home local infrastructure.

7.2 *BIM4EEB/BIMcpd*

The project BIM-based fast toolkit for Efficient rEnovation of residential Buildings (BIM4EEB) [12], Fig. 8, is funded by Horizon 2020 and targets the building stock renovation industry by developing a powerful Building Information Models (BIM) management system toolset to support designers in the construction planning phase and services development for building retrofitting. It facilitates decision-making and asset management (for both public and private owners) through the use of augmented reality (AR) and digital logbooks. BIM4EEB provides a BIM management

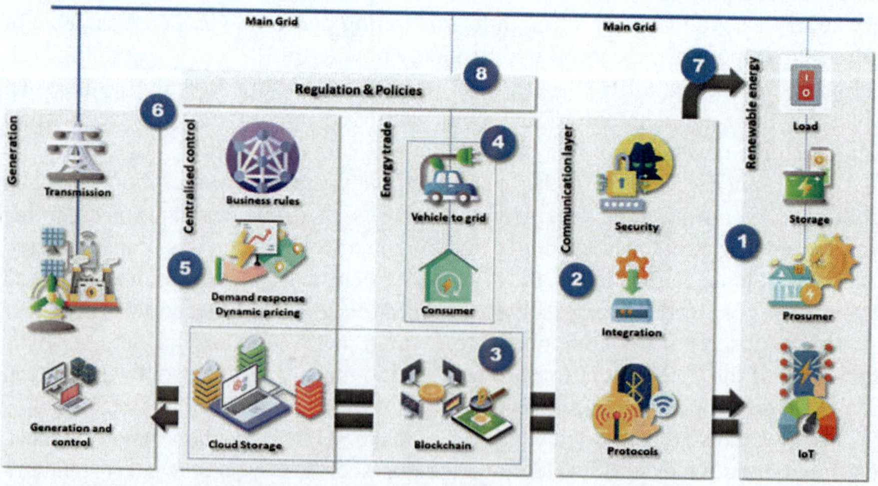

Fig. 7 CENTS high level framework [60]

Fig. 8 BIM4EEB

system consisting of six tools: Fast Mapping of Buildings Toolkit, BIMeaser tool, BIM4Occupants tool, Auteras tool, BIM4EEB BIMPlanner tool, and the BIM Constraint Checking, performance analysis and data management (BIMcpd) tool [51].

BIM4EEB is a comprehensive BIM toolset and integrates data from diverse sources and partners. As discussed previously in this chapter, the data integration issue is a critical point in green smart homes integration. One of BIM4EEB's tools, the BIMcpd, delivers (amongst several other functionalities) the important task of data integration automation.

By providing the necessary toolset for data integration (e.g. file mapping and translation, and application programming interface (API) integration), BIMcpd integrates data originated from residential building apartments in different sites/countries (e.g. Italy and Poland). It collects data from several sensors types and metres (e.g. energy, multisensor metres, air quality, motion detection, humidity, illuminance and temperature) and saves the data in a structured format into BIM4EEB database.

BIMcpd is capable of providing important data management functionalities to support constraint checking and performance evaluation services (e.g. BIM designer and energy auditor). By doing so, BIMcpd brings user interaction and behaviour aspects into BIM4EEB and expands its functionalities to properly develop user-centric services.

7.3 H2020: InterConnect

The use case "InterConnect" (Interoperable Solutions Connecting Smart Homes, Buildings and Grids) [19] describes one of the several projects funded by the Horizon 2020 (H2020) Framework Programme—the European Union (EU) flagship initiative for the development and maintenance of Europe's competitiveness in the global scenario (innovation, sustainable economic growth and job creation) and was supported by Europe's leaders and the Members of the European Parliament. H2020 was the biggest EU Research and Innovation programme with roughly €80 billion of funding invested in the period from 2014 to 2020.

The InterConnect project was funded under the call "Digitising and transforming European industry and services: digital innovation hubs and platforms" and targeted the integration of renewable energy sources (RES) and how smart homes (including buildings) could promote energy efficiency, as it was considered by the call an

Fig. 9 Project reach

essential crucial element for the flexible consumption, optimisation and integration of DERs, and storage.

The project started on 1 October 2019 is planned to last until 30 September 2023 and had an overall budget of €35.8 million (with an EU contribution of €30 million). It is a pan-European project coordinated by INESC TEC (a research organisation from Portugal) [35] and with a consortium composed by other pan-European research organisations and private for-profit entities from several EU member states: Austria, Belgium, Germany, Greece, Italy, Netherlands, Poland, Serbia and Slovenia—as illustrated in Fig. 9.

The partners in the consortium are important stakeholders in the new energy paradigm represented by smart homes and smart grids revolution and comprises competencies in all key sectors (e.g. ICT, IoT, energy and data science) and relevant associations with ICT and the energy sector.

Interconnect seeks to develop an interoperable ecosystem bringing demand-side flexibility with effective advantages for the end-user. Although control over appliances is not a new concept, neither is the problem with interoperability. End-users should not be penalised (change their appliances and other interoperability issues) every time they choose to move to another technology or service provider to benefit from new technologies to improve their sustainable behaviour.

As the energy sector moves towards digitalisation and becomes more user-centric (and market-driven), the number of energy service providers increases and so does

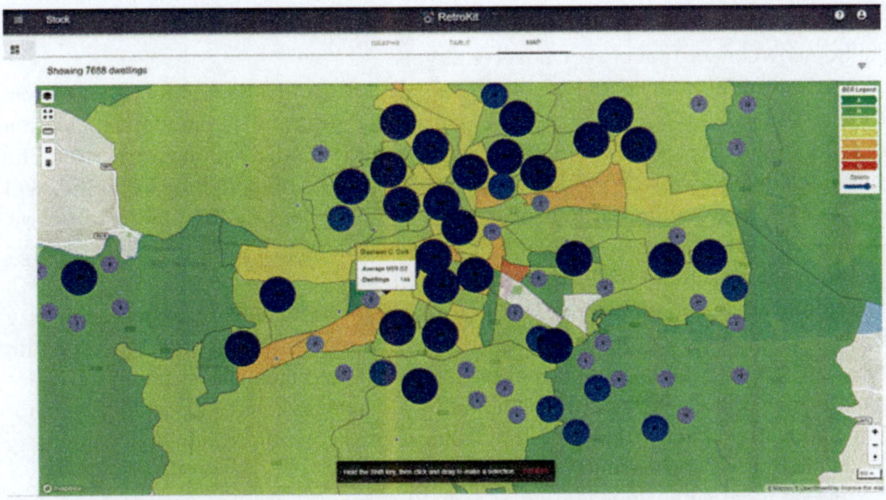

Fig. 10 Dwellings mapping

the number of improvements for monitoring and controlling—and it makes the interoperability problem even more challenging.

Interconnect is still a work in progress, but it aims to prove that it is possible to develop a digital market for the energy sector with considerable contributions of demand-side flexibility (DSF), more viable investments and accessible operational costs, and help the EU attain its energy efficiency goals. For that, seven large-scale pilots will be distributed in diverse countries to hit several types of end-users and green smart homes setups (appliances, services and interoperability).

7.4 Retrokit

RetroKit® [15] is a company that illustrates how green smart homes can be integrated to develop data-driven decisions and support policy-making and guide investment in the housing stock retrofit. To do that, RetroKit® create a software-based decision-support tool to aggregate all data collected from potentially hundreds of associated smart homes—an example of the dwelling mapped is shown in Fig. 10.

It then creates a customised database with the baseline energy performance of a housing stock, including energy use and expenditure, CO_2 emissions and Building Energy Rating (BER), Ireland's Energy Performance Certificate (PEC) rating, at the whole-stock level and per relevant dwelling cohorts. RetroKit® can pull a dwelling's EPC data on behalf of its owner from the Sustainabel Energy Authority of Ireland (SEAI) BER National Administration System, allowing easy access and analysis of this data by the user. The data is then analysed to model and compare a wide range of energy renovation scenarios, helping decide the best route to meeting the user's objectives, whether these are based on budget, CO_2 emissions, the health of the homeowners or fuel/energy poverty targets.

Next, a multi-criteria analysis approach is used to identify optimal renovation scenarios considering a range of KPIs and ML algorithms automate the selection of energy renovation measures and improve scenario modelling outcomes, from the individual dwelling level to whole-stock level. This type of modelling capability will also help policy-making, at the local and national level (e.g. to support the cost-optimal energy renovation calculations according to Article 5 of the recast Energy Performance of Buildings Directive (EBPD).

With this data and knowledge, the housing stock owner can develop their energy renovation roadmap, define bespoke packages of energy conservation measures with budget estimates, funding opportunities and an action plan. This structure provides accurate evidence-based decision-making by key stakeholders supported by a mapping application that facilitates spatial planning of energy renovation in housing (e.g. for an area-based approach to project development).

Finally, local authorities and other social housing landlords can access the data to identify options to facilitate funding applications and reporting and can integrate other datasets with its API (e.g. stock condition surveys) for improved evidence-based decision-making and planning.

In summary, a common characteristic in all use cases is the importance of the user-centric approach. Although the integration of green smart homes (both at local, community or grid level) offers significant integration challenges, the hardware (e.g. IoT and appliances) and data collection and integration is fundamental for the development of any solution targeting green smart homes in the broader carbon footprint reduction.

8 Conclusion

Green smart homes offer enormous potential to help address the carbon footprint issue and the 2030 Climate Target Plan goals. As smart home technologies evolve to encompass more household appliances, advanced sensor devices, disseminated IoT adoption, flexible integration and pervasive communication, added to their potential for DRE generation, green smart homes offer unprecedented support to user-centric and data-centric decision- and policy-making schemes.

Big data is growing at a exponential rate and this will continue to rise as we have become dependent on data for every facet of our lives, this puts huge demands on our electricity systems (data centres), and we must ensure that data is gathered and used to ensure that we reduce our energy consumption and carbon emissions. Data offers major opportunities to optimise building stock, but requires skilled data scientists, engineers and analysts, in order to extract value from big data. There are several AI techniques available, and these must be used properly to ensure that they do not add to the resource demands on the electrical system and most importantly that AI is used for good, i.e. from an ethical point of view.

The United Nations, the European Union and several global governments are committed to climate change, with several goals and targets in place. It was encouraging to see world leaders at COP26 take on the climate change challenge and it is imperative that the entire planet take on this challenge together in the same or even greater manner that it is currently doing to get a handle on the global pandemic.

Green smart homes are vital in helping reduce the energy consumption and carbon emissions as mentioned previously in this chapter, and without machine learning, we would face an uphill challenge in decarbonising global building stock.

Acknowledgements The authors want to acknowledge all the support of the Department of Business, Enterprise and Innovation, via its Disruptive Technologies Innovation Fund (DTIF) [14] which provided funding for the CENTS project under the Government of Ireland's Project 2040 Plan [27].

References

1. Amri, Y., et al.: Analysis clustering of electricity usage profile using K-means algorithm. IOP Conf. Ser.: Mater. Sci. Eng. **105**, 012020 (2016). ISSN: 1757-8981, 1757-899X. https://doi.org/10.1088/1757-899X/105/1/012020.
2. Andersen, F., et al.: Residential electricity consumption and household characteristics: an econometric analysis of Danish smart-meter data. Energy Econ. 105341 (2021)
3. Beaudin, M., Zareipour, H.: Home energy management systems: a review of modelling and complexity. In: Renew. Sustain. Energy Rev. **45**, 318–335 (2015). ISSN: 13640321. https://doi.org/10.1016/j.rser.2015.01.046.
4. Brownlee, J.: A Gentle Introduction to Bayes Theorem for Machine Learning, Oct 2019
5. C2ES. Home Energy Use. https://www.c2es.org/content/home-energy-use/. Institutional Website (2021)
6. Chesser, M. et al.: Air source heat pump in-situ performance. Energy Build. **251**, 111365 (2021). ISSN: 0378 7788. https://doi.org/10.1016/j.enbuild.2021.111365., https://www.sciencedirect.com/science/article/pii/S0378778821006496
7. Chesser, M. et al.: Probability density distributions for household air source heat pump electricity demand. In: Proceedings of Computer Science, vol. 175, The 17th International Conference on Mobile Systems and Pervasive Computing (MobiSPC), The 15th International Conference on Future Networks and Communications (FNC), The 10th International Conference on Sustainable Energy Information Technology, pp. 468–475. ISSN: 1877-0509 (2020). https://doi.org/10.1016/j.procs.2020.07.067., https://bit.ly/3HcKXNk
8. Ciptadi, A.: What Is Deep Learning and How Is It Different from Machine Leearning. https://bit.ly/3xsjjr4. Institutional Website, May 2019
9. Commission for Energy Regulation (CER): CER Smart Metering Project—Electricity Customer Behaviour Trial, 2009–2010 [dataset]. Irish Social Science Data Archive. SN: 0012-00, 1st edn. Accessed Jan 2018 (2012). www.ucd.ie/issda/CER-electricity
10. CommunityPower: Community Power
11. Constable, G., et al.: A Century of Innovation: Twenty Engineering Achievements that Transformed Our Lives. Joseph Henry Press, Washington. DC, USA (2021). ISBN: 0-309-08908-5
12. Daniotti, B., et al.: Workshop: BIM4EEB: A BIM-based toolkit for efficient innovation in buildings. Proceedings **65**(1), 17 (2021). ISSN: 2504-3900. https://doi.org/10.3390/proceedings2020065017
13. DARPA: ARPANET. https://www.darpa.mil/about-us/timeline/arpanet. Institucional (2021)

14. DTIF: Disruptive Technologies Innovation Fund. https://bit.ly/3sjuz6v. Government
15. Dubuisson, X.: Retrokit Software Platform (2021). https://retrokit.eu/. Institutional Website
16. Dwivedi, R.: How Does K-Nearest Neighbor Works In Machine Learning Classification Problem? https://www.analyticssteps.com/blogs/how-does-k-nearest-neighbor-works-machine-learning-classification-problem. Institutional Website, July 2021
17. EC: Communication from the Commission to the European Parliament, The European Council, The Council, The European Economic and Social Committee, The Committee Of the Regions and the European Investment Bank, Sept 2017
18. EC: Energy Performance of Buildings Directive (EPBD) Compliance Study, Dec 2015
19. EC: Interoperable Solutions Connecting Smart Homes, Buildings and Grids—Digitising and Transforming European Industry and Services: Digital Innovation Hubs and Platforms. Funding & Tender Opportunities, July 2018. https://bit.ly/3F87kS0
20. ECE. Improving Efficiency of Buildings through Digitalization—Policy Recommendations from the Task Force on Digitalization in Energy. Policy Recommendation ECE/ENERGY/GE.6/2021/5, p. 11. Economic Commission for Europe, Geneva, Switzerland, June 2021
21. ECE: Report of the Committee on Housing and Land Management on Its Seventy-Seventh Session. Committee on Housing and Land Management ECE/HBP/188, p. 9. Economic and Social Council, Geneva, Switzerland, Sept 2016
22. ECF: European Climate Foundation—Annual Report 2011. Executive Summary, p. 40. European Climate Foundation, Netherlands, Nov 2021
23. Economidou, M.: Europe's Building Under the Microscope. Technical Report Brussel, p. 132. Buildings Performance Institute Europe (BPIE), Belgium (2011)
24. Commission, European: Directorate General for Energy. Publications Office, Clean Energy for All Europeans. LU (2019)
25. Fitzpatrick, J., Carroll, P., Ajwani, D.: Creating and characterising electricity load profiles of residential buildings. In: Lemaire, V. et al. (eds.) Advanced Analytics and Learning on Temporal Data, pp. 182–203. Springer International Publishing, Cham (2020). ISBN: 978-3-030-65742-0. https://doi.org/10.1007/978-3-030-65742-0_13
26. Ghalehkhondabi, I., et al.: An overview of energy demand forecasting methods published in 2005–2015. Energy Syst. (Berlin Period.) **8**(2), 411–447 (2017). https://doi.org/10.1007/s12667-016-0203-y
27. GOI: Project Ireland 2040. https://bit.ly/3yLEVi2. Institutional, Apr 2021
28. Graafmans, J., et al.: Gerontechnology: matching the technological environment to the needs and capacities of the elderly. Technische Universiteit Eindhoven **93**(161), 13 (1993)
29. Gram-Hanssen, K., Darby, S.J.: "Home is where the smart is"? Evaluating smart home research and approaches against the concept of home. Energy Res. Soc. Sci. **37**, 94–101 (2018). ISSN: 22146296. https://doi.org/10.1016/j.erss.2017.09.037
30. IBM: Internet of Things. Institutional Website (2021). https://www.ibm.com/analytics/au/en/internet-of-things/
31. IEA: Energy Efficiency Indicators: Overview. Statistics Report Statistics report—December 2020. International Energy Agency, Paris (Dec 2020)
32. IEA: Net Zero by 2050. Flagship Report, p. 224. International Energy Agency (IEA), Paris, France (Oct 2021)
33. IEEE: IEEE Standard for Wireless LAN Medium Access Control (MAC) and Physical Layer (PHY) Specifications (IEEE Std 802.11-1997). IEEE Standard, p. 466. IEEE, New York, NY, USA, June 1997. https://doi.org/10.1109/IEEESTD.1997.85951
34. IERC: CENTS Project. http://www.centsproject.ie/. Research Project (2019)
35. INESC: INESC TEC. https://www.inesctec.pt/en#projects. Institutional Website, Nov 2011
36. Johnson, E.: Touch display—a novel input/output device for computers. Electron. Lett. **1**(8), 219 (1965). ISSN: 00135194. https://doi.org/10.1049/el:19650200
37. Jouannic, A.: Could Home Energy Management Be the next Big Connected Home Opportunity? Public Blog, Aug 2017

38. Kathirgamanathan, A., et al.: Data-driven predictive control for unlocking building energy flexibility: a review. Renew. Sustain. Energy Rev. **135**, 110120 (2021). ISSN: 1364-0321. https://doi.org/10.1016/j.rser.2020.110120., https://www.sciencedirect.com/science/article/pii/S1364032120304111
39. Leitao, J. et al.: A survey on home energy management. IEEE Access **8**, 5699–5722 (2020). ISSN: 2169-3536. https://doi.org/10.1109/ACCESS.2019.2963502
40. Liang, D., et al.: Scientific big data and digital earth. Chin. Sci. Bull. **59**(12), 1047–1054 (2014). ISSN: 0023-074X. https://doi.org/10.1360/972013-1054
41. Lin, C.-M. et al.: Applying the Naïve Bayes Classifier to HVAC Energy Prediction Using Hourly Data. Microsystem Technology (June 2019). ISSN:0946-7076, 1432-1858. https://doi.org/10.1007/s00542-019-04479-z
42. Lobaccaro, G., Carlucci, S., Löfström, E.: A review of systems and technologies for smart homes and smart grids. Energies **9**(5), 348 (2016). ISSN: 1996-1073. https://doi.org/10.3390/en9050348
43. Manhique, M., Kouta, R.: Energy inclusion in Mozambique: an approach to community energy. In: 2021 IEEE International Humanitarian. IEEE, Dec 2021
44. Marikyan, D., Papagiannidis, S., Alamanos, E.: A systematic review of the smart home literature: a user perspective. Technol. Forecasting Soc. Change **138**, pp. 139–154 (Jan 2019). ISSN: 0040-1625. https://doi.org/10.1016/j.techfore.2018.08.015
45. McLoughlin, F., Duffy, A., Conlon, M.: Characterising domestic electricity consumption patterns by dwelling and occupant socio-economic variables: an Irish case study. Energy Build. **48**, 240–248 (2012)
46. McLoughlin, F., Duffy, A., Conlon, M.: Evaluation of time series techniques to characterise domestic electricity demand. Energy **50**, 120–130 (2013). ISSN: 0360-5442
47. Motlagh, O., Berry, A., O'Neil, L.: Clustering of residential electricity customers using load time series. Appl. Energy **237**, 11–24 (2019). ISSN: 0306-2619. https://doi.org/10.1016/j.apenergy.2018.12.063., https://bit.ly/31CJlvN
48. mSemicon: mSemicon. https://www.msemicon.com/en-GB/
49. MySmartHome: Smart Home Technology in 1966. https://bit.ly/3qqL8yA. Institutional, Oct 2021
50. NUIG: NUI Galway. http://www.nuigalway.ie/
51. O'Regan, B., et al.: BIMcpd: a combined toolkit for constraint checking, performance evaluation and data management in building renovation projects. Proceedings **65**(1), 32 (2021). ISSN: 2504-3900. https://doi.org/10.3390/proceedings2020065032
52. Oliveira, M.: Dark Data—Why You Need to Know About It (SaaShollic). https://bit.ly/3HnElM0. Blog, Nov 2021
53. Paatero, J.V., Lund, P.D.: A model for generating household electricity load profiles. Int. J. Energy Res. **30**(5), 273–290 (2006). https://doi.org/10.1002/er.1136
54. Panimalar, A., Shree, V., Kathrine, V.: The 17 V's of big data. IRJET, 5. E-ISSN: 2395-0056 04.09 (Sept 2017), ISSN: p-ISSN: 2395-0072
55. Ray, S.: Commonly Used Machine Learning Algorithms (with Python and R Codes). https://bit.ly/315PZeg. Blog, Sept 2017
56. Reinsel, D., Gantz, J., Rydning, J.: The Digitization of the World—From Edge to Core. White Paper US44413318, p. 28. IDC, Framingham, USA, Nov 2018
57. Ricquebourg, V., et al.: The smart home concept: our immediate future. In: 2006 1ST IEEE International Conference on E-Learning in Industrial Electronics, pp. 23–28. IEEE, Hammamet, Tunisia, Dec 2006. ISBN: 1-4244-0323-5. https://doi.org/10.1109/ICELIE.2006.347206
58. Ringwood, J.V., Bofelli, D., Murray, F.T.: Forecasting electricity demand on short, medium and long time scales using neural networks. J. Intell. Robot. Syst. **31**(1), 129–147 (2001)
59. Saberi, O., Menes, R.: Artificial Intelligence and the Future for Smart Homes. Executive Summary Note 78, p. 8. Internation Finance Corporation, Washington, D.C., USA, Feb 2020
60. Silva, F., O'Regan, B.: An Innovative Smart Grid Framework for Integration and Trading. ICSREE2021. ICSREE2021, Strasbourg, France, May 2021

61. Silva, F., et al.: System Integration and Data Models to Support Smart Grids Energy Trading. ECRES 2021. Istanbul, Turkey, Apr 2021. ISBN: 978-605-86911-9-3
62. SmartTech: Smart Tech—Alternative Energy Solutions
63. Spicer, D.: The Echo IV Home Computer. https://bit.ly/3Ep5mNg. Institutional, May 2016
64. Swaminathan, S.: Logistic Regression. https://bit.ly/3xvZDm0, Mar 2018
65. TUD: TU Dublin. Technological University Dublin. https://www.dit.ie/
66. UCC: UCC. https://www.ucc.ie/en/
67. UN: Building Sector Emissions Hit Record High, but Low-Carbon Pandemic Recovery Can Help Transform Sector. https://bit.ly/3xYK4Ub. Institutional Website, Dec 2020
68. UN: Paris Agreement. Agreement, p. 27. United Nations, Paris, France (2015)
69. UN: Sustainable Development Goals (SDG). https://sdgs.un.org/goals. Institutional (2021)
70. UNECE: Sustainable Smart Cities—UNECE. https://bit.ly/3GhRAfR. Institutional Website (2021)
71. UNFCCC: United Nations Framework Convention on Climate Change. Framework Convention. United Nations Framework Convention on Climate Change, p. 33, Geneva, Switzerland (1992)
72. W3.org: Tim Berners-Lee-Biography. https://bit.ly/3ou3oFG (2001)
73. Zhong, S., Tam, K.-S.: A frequency domain approach to characterize and analyze load profiles. IEEE Trans. Power Syst. **27**(2), 857–865 (2012)

Artificial Intelligence Based Smart Waste Management—A Systematic Review

Nusrat Jahan Sinthiya, Tanvir Ahmed Chowdhury, and A. K. M. Bahalul Haque

Abstract Smart waste management is an approach that utilizes modern technology to manage waste materials in effective, efficient, and economical way. Artificial intelligence offers various approaches which can help to construct smart waste management systems. AI based systems are used to tackle complicated problems, handle uncertainty, and exhibit the efficiency of smart systems. This article aims to conduct systematic literature review on artificial intelligent-based smart waste management systems. In this study, we have identified and analyzed 40 research papers published between the years 2001 to 2021. These papers have proposed various frameworks and smart models for different types of waste management. The main goal of this study is to summarize the findings of selected research papers, provide comprehensive analysis and identify the future research avenues of waste management. This chapter has addressed various waste management domains like municipal solid waste management, smart bin management, domestic waste management, medical waste management, construction and industrial waste management, and so on. Furthermore, categorical representation of most extensively used machine learning and deep learning algorithms along with their contribution have been elaborately discussed as well.

Keywords Smart waste management · Artificial intelligence · Systematic literature review (SLR) · Machine learning · Deep learning

N. J. Sinthiya · T. A. Chowdhury
North South University, Dhaka, Bangladesh
e-mail: nusrat.sinthiya@northsouth.edu

T. A. Chowdhury
e-mail: tanvir.chowdhury02@northsouth.edu

A. K. M. Bahalul Haque (✉)
LUT University, Lappeenranta, Finland
e-mail: bahalul.haque@lut.fi

M. Lahby et al. (eds.), *Computational Intelligence Techniques for Green Smart Cities*, Green Energy and Technology, https://doi.org/10.1007/978-3-030-96429-0_3

1 Introduction

Green cities are required if the human species is to have a long-term future in anything like the environmental abundance that humanity now enjoys. It is important to keep in mind that inequality is not caused by a lack of wealth; rather, it is caused by an inadequacy or desire to share it equitably amongst residents. Our premise during this whole chapter will be that green cities are already emerging. Their future transformation requires significant assistance, as this positive advent must therefore transition from experiment to mainstream. Avoiding cataclysmic climate change is not a debate [1, 2]. Therefore intelligent waste management is a critical component of a smart green city. Waste generation is severe in emerging cities like- Laogang in Shanghai, China; Sudokwon in Seoul; the now-full Jardim Gramacho in Rio de Janeiro, Brazil; and Bordo Poniente in Mexico, etc. Each of them produces 10,000 tons of waste everyday [3]. According to a study, 10 billion tons of waste is generated worldwide. Domestic, commercial, and industrial construction contribute to the production of over two billion tons of garbage [4]. According to a projection of the World Bank, by 2050 the municipal waste generation can rise up to 3.40 billion. It was also said that 33% of those wastes was not effectively managed [5]. Which means that a large amount of waste is deposited at random, posing a risk to people and the environment. Because of the mismanagement of these wastes, a number of issues like groundwater pollution, land deterioration and health risks including cancer incidence, childhood mortality and birth anomalies can occur [6].

Waste management is a term that encompasses different procedures like proper exertion, waste disposal, removal, recycling, etc. These procedures have complicated operations, since there are various interrelated processes and socioeconomic aspects involved. Therefore, strategic planning is required to compete with the alarming rate of waste production. Smart waste management employs a variety of techniques. For instance, improved recycling tactics, disposal technology, optimized routing, IoT-based system, efficient sorting, reliable estimation of waste generation, etc. In light of these, Kellow Pardini et al. has stated an IoT-based solid waste management solution [7], Shwetashree Vijay et al. implemented a smart waste management system using ARDUINO [8], Similarly, Md. Shafiqul Islam et al. described a smart solid waste bin monitoring and collecting system [9]. Md. Abdulla Al Mamun et al. supervised an automated solid waste bin management [10]. Despite these, new technologies such as artificial intelligence, deep learning, and robotics are becoming more prevalent in this industry.

AI technologies involves advanced computer systems and programs which can effectively mimic human characteristics for instance—self learning, reasoning, problem-solving, and et cetera. Different AI models such as artificial neural network (ANN), expert system, genetic algorithm (GA), fuzzy logic can solve critical problems, predict reliable results, and solve complex mapping [11]. Although, the idea of expert systems for waste management was introduced long before [12]. Based on that decision support system (DSS) in order to promote the concept of clean cities by intelligent management was introduced too [13]. Artificial intelligence can be applied from garbage collection and transportation to central management, control, and adequate surveillance. Therefore, manual monitoring and traditional waste management methods can be replaced with AI-based smart solutions.

There have been few comprehensive reviews of AI based waste management. Xia W et al. presented a mini review regarding the application of machine learning algorithms in Municipal solid waste management. Different machine learning algorithms and their advantages and disadvantages along with their uses in different areas in MSW Management have been highlighted in this study [14]. Similarly, Abdallah M et al. presented an extensive discussion about the artificial intelligence application in solid waste management. This review basically focused on the assessment of various AI models used in solid waste management. The advantages and limitations of different AI applications were pointed out as well [16]. On the contrary, V. Agarwal et al. focused on the involvement of artificial intelligence in waste electronic and electrical equipment treatment. Data extraction and synergy methodology was applied to collect all the data regarding the topic and they did the analysis through graphs to illustrate the scope of artificial intelligence in E waste management [15].

However, there are currently no systematic reviews that can examine the existing artificial intelligence methods utilized in different domains of waste management. In this chapter, we have done a review of those papers which includes artificial intelligence as a major tool for different fields of smart waste management. The goal of this chapter is to collect and analyze the existing research trends, conduct a systematic literature review on the topic and answer the following research questions:

1. RQ 1) What are the waste management fields that involve AI as a major tool?
2. RQ 2) What are the AI techniques that are used as a solution for those?
3. RQ 3) What are the research gaps in the artificial intelligence and smart waste management domain?

We conduct a systematic literature review by following the methodology set by Kitchenham and Charters [17]. After all the analysis and applying the inclusion and exclusion criteria, we selected 40 research articles for this SLR. We observed each research article thoroughly and prepared a categorized representation. The rest of the chapter is structured as follows. After this introductory section, we introduce artificial intelligence technology and smart waste management briefly in Sect. 2. Next in Sect. 3, we present the SLR methodology. In Sect. 4, we present the findings from SLR, answer the research questions and provide overall discussion. In Sect. 5, we describe the contribution of this chapter and finally Sect. 6 concludes the chapter.

2 Background

Artificial intelligence (AI) is an enormous branch of computer science AI applications attempt to mimic human traits and solve different problems more efficiently and precisely. The range of AI applications includes automated reasoning, machine learning, natural language understanding, intelligent robots, automated programming and so on [18]. AI-based knowledge management software, automated systems, virtual agents, identity analytics, cognitive robotics and autonomous systems recommendation systems, speech analytics et cetera are few examples of AI applications that are used extensively [19].

Smart Waster Management

Every day, six tons of rubbish are created in Europe [20]. People generate 2.12 billion tons of garbage each year [21]. Wikipedia has listed almost 50 types of waste. One of them is E-waste (Electronic Waste). Roughly 20 to 50 million tons of E-waste are produced in a year throughout the world [22]. Another type of waste is municipal solid waste, almost 2.01 billion tons of municipal solid waste is generated in a year globally [23]. By 2050 it will increase by 70 percent reaching 3.4 billion tons a year. About 400 million tons of hazardous waste are produced every year that means 13 tons in every second [24], those wastes may do substantial harm to our health and environment because they are toxic, infectious, and radioactive. In 2017 China produced about 3.3 billion tons of industrial waste, with a total stockpiled over 60 tons over the year [25]. Another waste in construction waste and it is estimated that this industry is responsible for overall 35% of the generated waste [26]. Researchers are trying to find a way to solve this problem but still the percentage of construction waste in the US, Canada, Australia, UK are 33%, 35%, 30%, and 50%, respectively [26, 27].

Traditional waste management procedures are still used in many countries. Overcoming the flaws of traditional wastes is too crucial in order to ensure healthy living. Smart waste management is an effective solution in this regard. Unlike typical waste management systems, this method is automated by which it offers the possible shortest route, intelligent monitoring, saves unnecessary fuel cost of vehicles, and ensures time efficiency. To assure advancement and efficiency in this industry, intelligent systems are developed. Smart waste management using Internet of things (IoT) is a quite popular approach. Different frameworks and applications have been developed using this [28, 29]. Big belly smart waste and recycling system is considered as one of the finest solutions of collecting waste and recycling for public spaces [30, 31]. B. Chowdhury et al. implemented RFID-based smart waste management [32]. On the contrary, S. Sharmin el al. proposed an intelligent waste management platform which is a cloud-based dynamic system [33]. One of the most effective and dynamic approaches is waste management and monitoring systems using robot [34]. Finally, artificial intelligence (AI) is the most recent and most effective addition to the smart waste management system. There are different fields of waste management like- waste categorizing, efficient sorting, automated identification, estimation of waste generation, etc. and a range of AI models have been deployed to address different issues in each of these fields.

3 Methodology

We followed a systematic literature review approach in order to evaluate and summarize the existing studies regarding this topic. Systematic literature review is meant to be done by following a predefined search strategy and rigorous manner [15]. It is a well-planned strategy to search, identify and evaluate the most relevant studies of any particular research area. By following this scientific method, we tend to understand the scope of the existing literature and map the limitations of them so that those gaps can be explored later on to develop a new agenda. The procedure used in this SLR followed the standard protocol which involves different phases in order to conduct the whole review process.

3.1 Identification of Literature

For conducting SLR, rigorous searching techniques should be followed.

Search Strategy

Firstly, the most relevant keywords which are the core of the research question, should be identified. Boolean operators AND and OR can be used to generate search strings and to use those keywords elaborately. The keywords that are used in this chapter include "Artificial intelligence", "waste management", "smart waste management", "Wastewater management", and "solid waste management".

The strings that we produced in order to perform the identification of the literatures are

- "Artificial intelligence" AND "waste management".
- "Smart waste management".
- "Artificial intelligence" AND "Wastewater management".
- "Artificial intelligence" AND "Solid waste management".

The studies were retrieved from the International digital database SCOPUS. This database was chosen because it offers us a range of facilities. It provides quality research articles; it is authorized and well-known in the research community and the extraction of the information regarding specific studies is easy as well. The searches were done within the article title, abstract, keywords field. The final search was executed on August 30, 2021.

3.2 Screening

Once the results are obtained following search strategy, they need to be measured for relevance. For this, inclusion and exclusion criteria are followed.

3.2.1 Inclusion and Exclusion Criteria

The preliminary search returned 774 results altogether. After that inclusion and exclusion criteria is used to limit them and identify the studies that best explains our research agenda.

Exclusion Criteria

1. Review, Conference review, Book chapter, Notes, Short surveys and Thesis were excluded.
2. Studies that are not entirely based on artificial intelligence techniques for waste management were excluded.
3. Publications other than the English language are excluded.

Inclusion Criteria

1. Articles published on journal and conference proceeding.
2. Full text available on digital database.
3. Articles that involve the subject areas like- Environmental Science, Computer Science, and Engineering.
4. For more precise results these inclusion keywords were identified and selected—"waste management", "artificial intelligence", "machine learning", "deep learning" and "artificial neural networks".
5. Studies published in the English language.

Screening of articles using all these criteria returned 406 articles in total. Quality assessment was performed on those articles next.

3.3 *Quality Assessment*

Here, in this phase, we evaluated each of the papers individually. After the inclusion–exclusion criteria, an excel sheet containing information on 406 articles was extracted. Therefore, 379 articles remained after removing the duplicates. Among them, 273 articles were found which did not match our research area. Some of them were out of our study scope and others were not entirely based on our research area. We analyzed them through the title and abstract field. We identified those papers where AI was used as the major tool for waste management. Each of them was reviewed and based on our research area 40 articles were selected finally.

Final Outcome

After the completion of all criteria, 40 articles were finalized for this systematic literature review. Figure 1 illustrates the whole process.

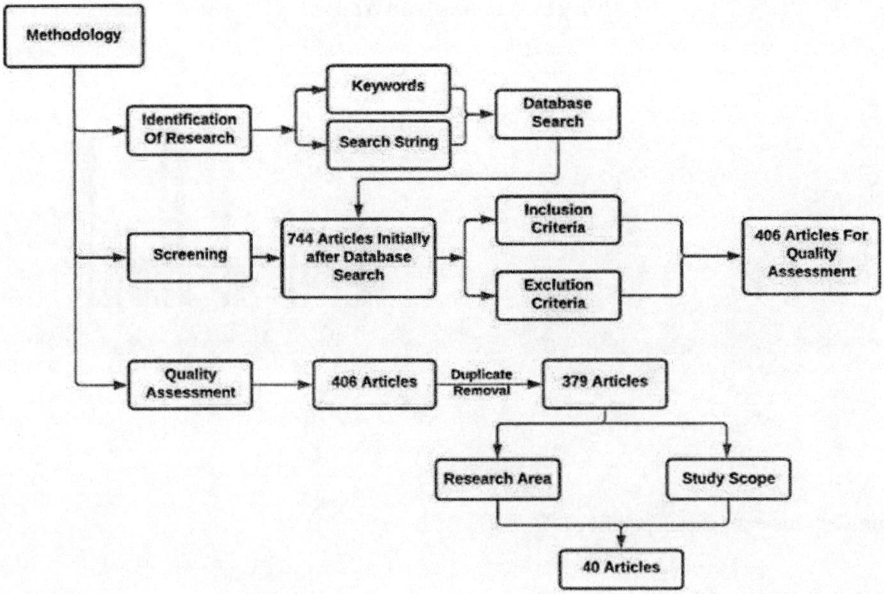

Fig. 1 Methodology of the SLR

4 Result and Discussion

4.1 Research Trend

Observing the research articles published throughout the years is an effective way to visualize research trends and their influence. From the literature search, we observe that articles regarding artificial intelligence first began in 2001. After 2016, the quantity of research increased, with the majority of the studies taking place between 2020 and 2021. Figure 2 depicts data on the number of articles published by various publishers during the previous two decades.

For this systematic literature review, we selected 40 papers. IEEE, Elsevier Ltd, and Springer published the majority of the works. IEEE published 13 papers, Elsevier Ltd published 11, and Springer published 5. Additionally, SAGE Publications Ltd published three articles and MDPI published two. Each of the following organizations published one paper: ACM, ASCE Library, Biomed Library, EDP Science, Frontiers Media S. A., and Hindawi Ltd.

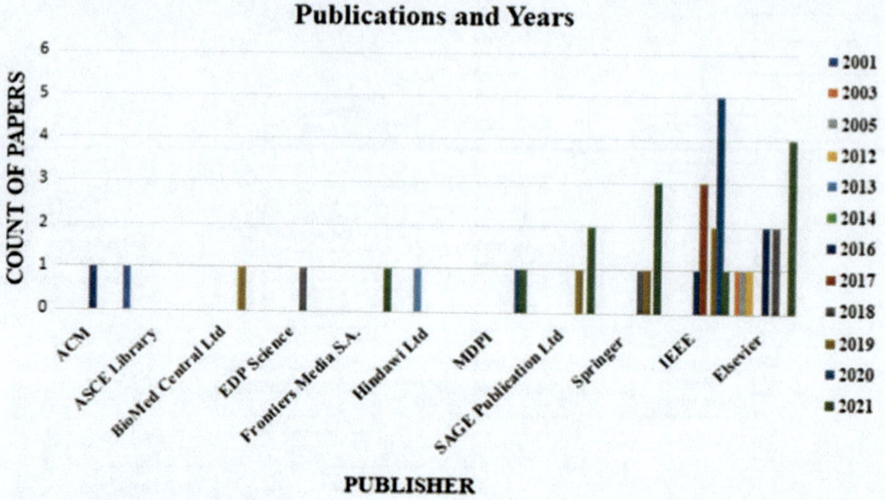

Fig. 2 Number of papers published

4.2 *Addressing RQ1*

The first research question considered for this study is—"What are the waste management fields that involve AI as a major tool?" Fig. 3 shows the number of published articles focusing on different waste management fields.

From Fig. 3, it is quite clear that most of the articles were published focused on the municipal solid waste management field. The least discussed fields were E-waste, construction, and industrial waste management. In order to facilitate the discussion,

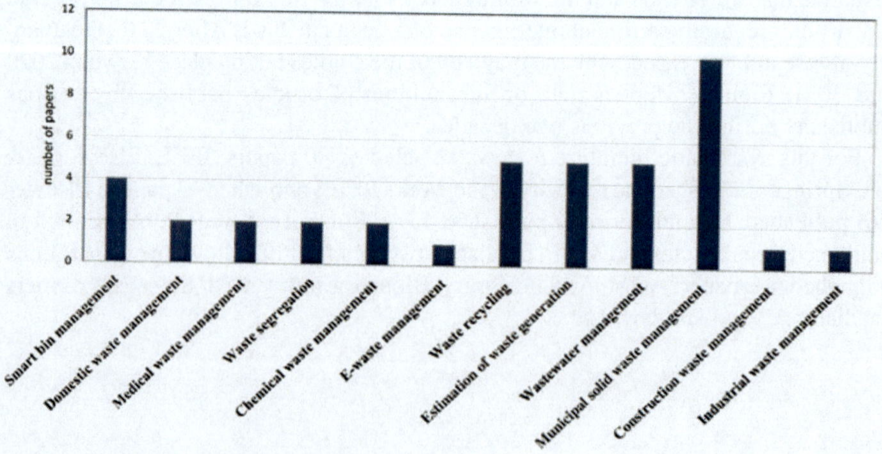

Fig. 3 Number of articles in different fields

the reviewed papers are classified under different waste management fields. The discussion regarding these fields are as follows –

4.2.1 Municipal Solid Waste Management

In this era of urbanization, solid waste management is a growing concern; which needs widespread attention. In developing countries, the challenges of solid waste management are increasing proportionally to the growth of urbanization [35]. One of the most important aspects of solid waste management is the collection and transportation process. In the municipal solid waste management budget, the collection cost takes almost 80–90% and 50–80% in low income and middle-income countries, respectively [36]. Manh Hua et al. discussed an optimized system for waste collection and transportation using K-means algorithm and by applying the vehicle routing problem. Here, K-means algorithm was used to cluster the waste collection centers and VRP was applied to generate an optimized path for the waste collector vehicles [37]. Similarly, Dordevic et al. has presented a model for efficient waste collection, which will eventually save resources. Neural network algorithm (NNA), genetic algorithm (GA) and MATLAB is used for the improvised system [44].

In addition to the transportation issue, unorganized waste dumping is a major concern in metropolitan areas. A mobile application-based smart system has been proposed to raise complaints about it by locals so that the authorities can take action immediately. However, in this case, fake complaints might be a source of concern that must be addressed. Machine learning algorithms like support vector machine (SVM) and convolutional neural network (CNN) have been used in this regard where CNN performed the best with 87% of accuracy [38].

To avoid reckless waste material dumping in urban areas, identifying trash dump yards is a useful approach. In light of this, Ramasami et al. proposed a system that utilizes artificial intelligence algorithms to identify acceptable zones for solid waste dump yards and genetic algorithm (GA) is used in this regard [39]. Detection of different solid waste items is important before conducting waste recycling and sorting. Patel et al. employed a garbage detection system using object detection models which can automatically locate garbage using real-world images as well as videos. EfficientDet-D1, SSD ResNet-50 VA, Faster RCNN ResNet-101 V1, CenterNet ResNet-101 V1, YOLOv5M is used here and YOLOv5M achieved the best result by achieving a mean average precision value of 0.613 [40]. Additionally, a convolutional neural network (CNN)-based prediction model has been proposed to identify waste items and waste mass [43].

Different types of waste can be exploited into energy. To accomplish this, it is necessary to determine the higher heating value (HHV) of municipal solid waste. Machine learning-based systems have been developed in order to forecast the higher heating value which can be used later to transform the waste materials into energy. Multiple regression and genetic programming is used for the implementation. However, genetic programming provides the most precise result [41]. Apart from these, a comparative study of two ML algorithms: Multi-layer perception artificial neural network (MLP-ANN) and support vector regression (SVR) is conducted to analyze performance and accuracy level of different models [42].

4.2.2 Smart Bin Management

AI-driven smart bins are a very innovative way to manage waste materials. Abeygunawardhana et al. proposed a smart bin for classifying waste materials, monitoring filled bins and generating optimal routes for the collectors. This study basically used image processing for the identification of waste materials and convolutional neural network (CNN) is trained to recognize individual trash items in order to facilitate further sorting. To eliminate the manual monitoring, the system used ultrasonic sensors to identify the filled beans and lastly a mobile application was developed for the whole system to produce optimal routes for the waste collectors [45].

Similarly, Jadli et al. offered an architecture in which artificial intelligence techniques are utilized to determine the current status of waste bins where automated surveillance cameras are used to take pictures which are then forwarded to the processing server to detect the fill level. Finally, the architecture produces optimized waste collection schedules for the facility system according to the retrieved data. It employed various algorithms including Inception V3 model for future extraction, then support vector machine (SVM), naive Bayes, and linear regression for classifying the waste materials and lastly convolutional neural network (CNN) for the fill level detection and solar powered cameras were used to take the pictures for the system [46].

In addition to those methods, Ji Sheng et al. demonstrated a smart bin capable of simultaneously detecting and segregating waste materials. The system is configured in such a way that waste products can be separated into distinct compartments within the bin. Tensorflow framework-based deep learning model is used here, which is then exported to the Raspberry Pi mobile microprocessor for waste detection. LoRa communication protocol is used to transmit sensor data, an RFID module is used for bin maintenance, and finally, servo motors controlled by the Raspberry Pi are used to open and close the lids of the bin's individual compartments [47].

On the contrary, M. A. Hannan et al. implemented a method for bin level detection-based on a gray level aura matrix (GLAM). This system can be used to classify the bin level and grade of the solid waste. Both of the information are important features for solid waste collection. A gray level aura matrix (GLAM) is implemented here in order to extract the bin image texture. Then, the extracted image is trained and tested using multi-layer Perceptions (MLP) and K-nearest neighbor classifiers which helps to evaluate the performance of the system. The MLP classifier demonstrated that the bin level and grade classification rates are 98.98 and 90.19%, similarly using K-nearest neighbor classifiers the rates were found 96.91% and 89.14%, respectively [48].

4.2.3 Medical Waste Management

Hospital solid-waste encompasses a variety of infectious, chemical, and radioactive wastes that are extremely hazardous to both humans and the environment. [49]. Golbaz S et al. focused on the development of predicting models using AI techniques in order to find out the HSW rate. Different methods were employed to analyze and forecast the waste generation rate and comparative measurement was also conducted in order to find out the most effective model. In the study multiple linear regression (MLR) and several neuron and kernel-based machine learning methods like ANN, ANFIS, LSSVM, SVM, and FSVM were used. Among all of them kernel-based ML models provided the most satisfactory result [50].

Considering the recent COVID outbreak Kumar N et al. precisely put emphasis on the current COVID situation and addressed an important issue which is COVID related medical waste sorting. During this outbreak a number of infectious medical waste mixed with usual waste types. To solve the problem, the study implemented an automated smart system using AI techniques to sort those wastes before recycling so that the infection can be prevented. For the waste type recognition artificial neural network (ANN), support vector machine (SVM) and K-nearest neighbor (KNN) classifiers are used. Among them, support vector machine performed the best with a 96.5% accuracy [51].

4.2.4 Domestic Waste Management

Nowadays, smart home management incorporates an intelligent approach to waste management. In light of this, Article 50 presents a concept for a smart house management system. Here, advanced technologies are employed to maximize the efficiency of all available resources. For instance, dynamically identifying and resolving issues, optimizing power consumption, and so forth. Machine learning algorithms are employed to reduce waste and carbon emissions. Waste items are categorized into dry, wet, plastic hazardous, etc. Image processing and machine learning techniques are primarily used in this system to ensure an automated home system [52].

Additionally, Papagiannis F. et al. presents an alternative system solution for the European household waste problem where it informs the policymakers on the ambiguous household behaviors. The implementation method is carried out using K-means clustering [53].

4.2.5 Waste-water Management

Gabriel Markovic et al. described the modeling of rainwater and graywater in a school building in Slovakia using artificial intelligence. Fuzzy cognitive maps were used for this management and MATLAB software for the implementation [54]. For wastewater treatment, Chen W. C. et al. stated that by incorporating rough set theory into the neural fuzzy controller, it is possible to achieve superior plant performance in terms of cost effectiveness, control stability, and response time. Neural network model and genetic algorithm was used to meet various control needs. Additionally, they used a hybrid fuzzy control system, which incorporated a variety of artificial intelligence techniques [55].

Similarly, Chen W. et al. presented a three-stage analytical method for advanced fuzzy control. The fuzzy neural controller worked effectively and it achieved required real time control objectives. It can be an efficient and cost effective tool to accommodate the quality control in terms of response time and stability control of the wastewater treatment process. They also utilized neural network model and genetic algorithm [56].

Matheri A. et al. described a model to solve the real-life problem of wastewater treatment using the artificial neural network model and MATLAB. They forecasted the demand for trace metals and chemical oxygen in wastewater treatment [57]. The chemical industry's rise has resulted in an increase of poisonous and dangerous components. Yanbo J. et al. deeply analyzed the wastewater quality and chemical wastewater's biodegradability in a chemical industry park. The time varying and unstable system in the treatment process of toxic and refractory organic wastewater was considered as the research subject. It was analyzed based on the fuzzy control theory of artificial intelligence [58].

4.2.6 Chemical Waste Management

Chemical wastes must be managed appropriately because it can cause severe damage to the environment. Moreover, it can also cause harm to people if they are released or dumped carelessly. Taking this into account, J. M. Aitken et al. developed a reconfigurable rational agent-based robotic system capable of simulating autonomous nuclear waste processing. In this project, random sample consensus algorithm (RANSAC), MSAC and robot operating system (ROS) has been used [59]. Similarly, Fawzy M. et al. conducted a batch biosorption experiment to determine the removal efficiency of Cd(II) ion from aqueous solution by Gossypium Barbadense waste. Artificial neural network (ANN) is used to predict the absorption efficiency of Cd(II) ion removal [60].

4.2.7 Waste Segregation

Along with waste collection and transportation, waste segregation or categorizing is also essential for further processing. Advancement in this sector can accelerate waste recycling. Considering this, few studies have developed. In article 60, an approach was taken in order to classify the waste items into biodegradable and non-biodegradable. This approach was basically developed using deep learning. It is an intelligent system which can learn an update by itself. Boundary algorithm and deep learning framework Caffe was used for the implementation [61].

Similarly in Article 61, a system based on object recognition method is proposed where the waste items are examined through weight sensors. Next a camera is used to capture the pictures and then image segmentation is done for the prediction process. After identifying the items, they were categorized into degradable and non-degradable using AI techniques. Here, TensorFlow library and Raspberry PI Microcontroller was used for the implementation [62].

4.2.8 Estimation of Waste Generation

Estimation of waste generation is an important factor for planning sustainable waste management strategies. In Article 62, four machine learning algorithms were evaluated for their ability to estimate the monthly municipal solid waste generation in Logan city. Support vector machine (SVM), artificial neural network (ANN), K-nearest neighbor (KNN), and adaptive neuro fuzzy inference system (ANFIS) were applied for the estimation process and among them ANFIS performed the best [63].

Another comparative study was done considering Delhi city to find out the monthly MSW generation. Six distinct models were evaluated in this study. There were also a few hybrid models among them. Those models are pure ANN, Pure ANFIS, GA-ANN, DWT-ANN, GA ANFIS, and DWT-ANFIS. Among them GA-ANN performed the best [64]. On the contrary, Coskuner et al. focused on the prediction of domestic, commercial, construction, and demolition waste generation. Here, multi-layer perception artificial neural network (MLP-ANN) is applied to predict the annual waste generation [65]. Similarly, Cha G. W. et al. proposed a RF model to forecast the generation of demolition waste. Here, a small dataset is used for multipurpose demolition waste management. This model is capable of predicting the amount of waste generation based on the type of wastes. In this study, RF-recursive feature elimination (RFE) was used for feature selection and lastly Leave-One-Out Cross-Validation (LOOCV) method was applied to verify the performance of RF models [66].

Finally in Article 66, a model capable of predicting the municipal solid waste generation and diversion of a given region using the demographic and social economic parameters at municipal level is proposed. For the implementation process decision tree and neural network were utilized. Between them neural network performed the best [67].

4.2.9 Waste Recycling

We have encountered the involvement of different machine learning algorithms in the waste recycling process. Ozdemir et al. described several machine learning algorithms along with their workflows, and contributions in the subject of recycling. They employed a variety of AI-based models and machine learning algorithms, including K-nearest neighbor, decision tree, artificial neural network, support vector machine, random forest, and convolutional neural network (CNN) [68].

Huang, Jueru and Dmitry D. Koroteev stated about the development of a machine learning driven predictive analytic framework (MLPAF) for energy and waste management planning. That framework attempted to improve the waste management process by considering energy conservation and material recycling [69]. For an automated recycling system, Nañez Alonso et al. took necessary measures using CNN and image identification to automatically recycle waste material like paper, glass, organic, and plastic. Therefore, (83% of detected paper, 76% of plastic, 97% organic material, 84% glass) using VGG-16 Network, (82% paper, 95% organic material, 60% plastic, 78% glass) using VGG-19 Network, and (88% paper, 98% organic material, 70% plastic, 83% glass) was found using ResNet15V2 Network [70].

D. Rutqvist el al. presented an accurate detection of emptying a recycling container using the measurements from the sensor mounted on the upper part of the container with the help of an automated machine learning approach. They used several machine learning models in this system. For instance, ANN, logistic regression, decision tree, random forest, etc. [71]. A multi-layer hybrid deep learning system was employed in order to automatically sort waste disposed by individuals in the urban public area. This system is constructed with a high-resolution camera for capturing waste image and requires a sensor to detect other necessary features. For extracting the image features this system used a CNN algorithm and MLP to consolidate image features and other related features to identify whether wastes are recyclable or not. Multi-layer hybrid deep learning system (MHS) showed the best result for identifying the recyclable wastes [72].

4.2.10 E-waste Management

Electrical and electronic waste WEEE, or electronic trash, is a top priority in waste management. It has become a significant issue in both developed and developing countries [73]. Król A et al. precisely focused on wastes generated from electrical equipment. To improve WEEE management, the waste transportation structure was developed along with the user interface using artificial intelligence techniques. The system aimed to optimize the route length and number of vehicles so that it results in cost reduction and time efficiency. Additionally, the use of websites and mobile applications was advised to make these services accessible to locals. This concept was primarily articulated in order to facilitate the work of garbage collecting authority. Genetic algorithms (GA) and fuzzy logic were employed in this regard [74].

4.2.11 Industrial Waste Management

According to a study, China produces almost 3.3 billion tons of industrial waste every year. Industrial wastes create a huge problem in the environment, if these wastes are not disposed of within a certain time, it not only hampers the land resource and money but also it affects the ecosystem [75–77]. Liao B. et al. developed an intelligent model for industrial waste planning. They tried to overcome the limitations of traditional ways of industrial waste management and proposed a smart model to upgrade the system. By using the industrial waste images and appropriate detection model, it identifies the target objects and then proceeds for the further recycling process with the help of BP-network prediction model. Here, the faster RCNN algorithm was used for the target detection [78].

4.2.12 Construction Waste Management

The construction industry is likewise experiencing significant difficulties as a result of waste. Construction wastes require huge financial resources to be tackled along with the fact that this waste is a severe threat for the environment. It is estimated in a study that the construction industry produces almost 35% of the overall wastes [78]. The percentage of construction waste in the UK, USA, Australia, Canada, Hong Kong is, respectively, 50%, 33%, 30%, 35%, and 65%. [79, 80]. Application of artificial intelligence is discussed and reviewed through a conceptual framework for an effective waste management system. There T. H. Ali el al. developed an idea of an effective construction waste management system (EMS) by the involvement of AI. Through the EMS the construction practitioners can identify the most effective and economical waste management techniques [81].

4.3 Addressing RQ2

Our second research question is—"What are the AI techniques that are used as a solution for those?" Table 1 represents the usage of different AI models in different papers.

There are 16 AI models that have been identified from our selected research articles. Among them artificial neural network (ANN) was the most used model. This model has been extensively used for estimating the waste generation rates, constructing predictive analytical frameworks and waste type recognition. Apart from that the contribution of support vector machine (SVM), convolutional neural network (CNN), genetic algorithm (GA) were also frequent. Among them CNN is mostly used for categorizing waste items, waste bin fill level detection and object recognition.

Although SVM is another most frequently used model for detection and classifying of waste materials. Usage of GA and fuzzy logic have been found commonly in wastewater treatment and E-waste management. For the waste recycling process, a sheer involvement of KNN is noticed. Few hybrid models for instance; GA-ANN, DWT-ANN, MLP-ANN, DWT-ANN, DWT-ANFIS were mentioned in few studies too. In essence, these models were employed to determine the rates of municipal solid waste generation and they performed comparatively better than the pure machine learning models. The remaining models have also made significant contributions to

Table 1 Percentage of AI technique used in number of papers

AI techniques	Used in no. of papers	Percentage (%)
ANN	9	22.5
CNN	7	17.5
SVM	7	17.5
GA	6	15
KNN	4	10
Fuzzy logic	4	10
ANFIS	3	7.5
Decision tree	3	7.5
Hybrid model	3	7.5
K-means	2	5
RF	2	5
MLR	2	5
Naive Bayes	1	2.5
Linear regression	1	2.5
Logistic regression	1	2.5
BP-network	1	2.5

many domains of smart waste management. Nevertheless, Table 2 outlines the use of many AI models and their contribution to various studies.

4.4 Addressing RQ3

Our third research question is "What are the research gaps in the artificial intelligence and smart waste management domain? In previous sections we provided a categorical representation about the involvement of artificial intelligence in different waste management fields. We reviewed the applications of artificial intelligence, frameworks, and proposed infrastructures for resolving various difficulties in those domains. Following these analytical discoveries, we identified many significant research gaps that can be studied further in the future. The research gaps are as follows:

1. From previous analysis, it is apparent that AI is used frequently in smart waste management systems, and its reach is expanding daily in this sector. The majority of the research we discovered was done on municipal solid waste management (MSW). Apart from these, additional areas of waste management where AI has been extensively applied include- smart bin management, wastewater management, and recycling. However, the rate of studies conducted in these areas is half as compared to the MSW. But wastewater management is a critical component of smart waste management. Due to urbanization, enormous amounts of

Table 2 Usage of different AI models and their contribution

AI models	Articles and references	Applications
ANN	[50, 51, 57, 60], [63, 64, 68, 69], [71]	• MSW generation rate • Hospital solid waste generation rate • Predictive analytical framework • Waste recycling • Wastewater treatment • Waste type recognition
SVM/SVR	[38, 45, 50, 71], [42, 51]	• Classifying waste materials • Identifying spurious complaints about uncollected garbage • Hospital solid waste generation rate
CNN/faster RCNN	[38, 40, 45, 46, 72, 78, 82]	• Waste item recognition • Waste bin fill level detection • Identifying spurious complaints about uncollected garbage • Target object detection • Predicting waste and waste mass
Naive Bayes	[46]	• Classifying waste materials
Linear regression	[46]	• Classifying waste materials
K-means algorithm	[37, 53]	• To group waste collection centers into small clusters
ANFIS	[50, 63, 64]	• Estimating waste generation
RF	[66, 71]	• Waste recycling • Generation of demolition waste
Fuzzy logic	[54–56, 73]	• Cost effective wastewater management • Improved infrastructure for WEEE management
Decision tree	[67, 68, 71]	• Waste recycling • MSW generation
BP-network	[66]	• Industrial waste planning
KNN	[51, 63, 68, 71]	• Estimation of waste generation • Waste type recognition • Waste recycling
MLR	[41, 50]	• Higher heating value prediction of waste materials • HSW generation rate
Logistic regression	[71]	• Waste recycling
GA	[39, 41, 56, 74], [55]	• WEEE management • Location prediction for waste dump yard • Higher heating value prediction of waste materials • Wastewater treatment • Optimizing waste collection

(continued)

Table 2 (continued)

AI models	Articles and references	Applications
Hybrid model	[42, 64, 65]	• Estimation of waste generation

wastewater are generated every day, and the majority of it is just discharged carelessly into surrounding canals and rivers, causing severe environmental harm, yet only five studies [54–58] on wastewater treatment and processing were identified. As a result, wastewater management could be a promising field of research for further developing the notion of smart waste management.

2. Second, we observed that there is very limited research on E-waste management. Due to urbanization, electrical and electronic equipment-related wastes are also increasing at an alarming rate, yet relatively few research have examined this topic. The involvement of artificial intelligence in this area is remarkably less than other fields despite the fact that this is a critical issue in the modern era. Only one article [74] was found regarding AI-based WEEE management which aimed to optimize the transportation of E-wastes so that cost efficiency can be achieved. Extensive research and effective frameworks should be very impactful and highly recommended.
3. Another comparatively least discussed topic is industrial and construction waste management. We can notice an emerging industrial revolution in our world. As a result, a large number of wastes are generated on a daily basis in the sector. Most importantly, the waste generated in this industry can be extremely dangerous due to the fact that the majority of it is chemical waste, which is extremely detrimental to the environment. But we found only one study regarding this topic where a smart model for object detection was proposed. However, sophisticated frameworks can be established to mitigate the risks associated with this subject. On the other hand, in construction and demolition waste management AI has been rarely discussed. Further research in this area can be very helpful to come up with advanced solutions for construction waste management systems.
4. We have noticed the existence of a good number of papers for smart bin management [45–48]; where the bins are monitored through surveillance cameras and weight sensors and the waste materials are also identified automatically once they are dumped into the bin. However, in addition to identifying waste items, efficient sorting is required. Automated systems are required for this. We discovered only one publication in which waste items were automatically detected and sorted [48], implying that such a smart device was installed to enable waste products to be segregated without human intervention. This is a very novel concept, but this concept needs to be explored more so that more efficient and reliable automated systems can be accomplished. Therefore, it can be a potential research area for the future.
5. Finally, the efficiency and accuracy level of AI models needs to be increased. Different kinds of machine learning and deep learning algorithms were used in different studies to train the model. Additionally, few comparative studies

[38, 40, 42, 50, 63, 64] were also done to find out the most efficient models. Therefore, further research for improving these models is highly recommended.

However, we have noticed that different models have different accuracy levels under different circumstances. Table 3 summarizes the comparative studies of different models.

5 Contribution

We have identified the following implication after analyzing the selected articles. The implications are outlined as follows:

1. The first and foremost contribution of this work is to the best of our knowledge there is no previous systematic literature review on artificial intelligence and smart waste management. Although there are a few research studies that concentrated on a specific field of waste management and conducted a systematic review of the literature. [12–14] However, this literature review considers several fields of waste management. We have discussed about a number of waste management fields in this chapter, where AI techniques have been used extensively. Additionally, we offered an overview of the rate of work in each field. This literature review will aid the researchers in identifying the less discussed fields. In this way they will be able to contribute to those fields.
2. Secondly, this chapter provides a comprehensive summary of the AI models that have been used for solving different problems at one place. Researchers can gain a thorough understanding of the most often addressed issues in this field. Additionally, they can visualize current issues that require additional attention.
3. Thirdly, we have discussed the most frequently used machine learning and deep learning algorithms along with their applications. Moreover, we have provided the findings about the comparative studies of different models. We have summarized the findings in such a way so that the researchers can have a clear conception about the accuracy level and functionality of each model.
4. We have analyzed and provided a very detailed overview of different research fields of waste management that have been taken into consideration until now. For example, municipal solid waste management, wastewater management, smart bin management are the most explored domains. A few other areas, for instance, E-waste management, hazardous waste management, construction, and industrial waste management are the less discussed areas. Our observation from this analysis will help the potential researchers to gaze their site onto those areas. As a result, those unexplored areas will be developed.
5. Finally, the article can be a guideline for learning more about artificial intelligence and its application to smart waste management. Our analysis revealed that waste management is a wide topic with numerous potential areas for progress and innovative tactics to be implemented. Artificial intelligence may be a viable option in this instance. The findings from this study can assist

Table 3 Comparative studies of different models

Articles	Objective	Key findings
[63]	Comparison of different AI techniques for estimating monthly Municipal waste generation to find out the most reliable method	R2 = coefficient of determination • ANFIS performed better where train value of R2 was 0.98 • SVM, ANN, KNN were performed, respectively, with train value of R2 0.71, 0.46 and 0.51, respectively
[64]	Comparison of different AI techniques for estimating monthly municipal waste generation to find out the most reliable method	• Pure ANN models performed with a R2 value of 0.72 • DWT-ANN performed with a R2 value of 0.67 • GA-ANN, Pure ANFIS, DWT-ANFIS, GA-ANSIS showed R2 value of 0.87, 0.36, 0.73, 0.56, respectively. Therefore, GA-ANN was considered as the most effective model
[38]	Identification of uncollected garbage in urban areas so that necessary steps can be taken immediately for processing the waste items	• Accuracy of CNN hit a peak of 83% to 85% and the intermittent spikes was up to 87%
[40]	The main objective of this research is to conduct a comparative study between different object detection model based on their performance on their dataset	mAP@0.5 = mean average precision is calculated by thresholding intersection over union (IoU) at 0.5 • The mAP@0.5 value for EfficientDet-D1 was 0.360 • SSD ResNet-50 FPN (RetinaNet50), Faster RCNN ResNet-101, CenterNet ResNet-101 FPN and YOLOv5M's mAP@0.5 value were, respectively, 0.511, 0.586, 0.595, and 0.613 • YOLOv5M performed better result in mAP@0.5
[42]	This chapter compared two machine learning techniques one was multi-layer perceptron artificial neural network (MLP-ANN) and another one was SVR to predict annual municipal solid waste in Bahrain using MSW dataset (1997–2019)	• Test data of R2 value of ZSN was 0.98 and MMN was 0.99 • Entire data value of R2 of ZSN was 0.93 and MMN was 0.94
[50]	Development of predicting models for hospital solid waste (HSW) management system	• In ANN modeling the testing value of R2 was 0.73, 0.65 and 0.70, respectively • Same as for ANFIS the value of R2 was 0.66, 0.71, and 0.59, respectively • For SVM the value of R2 was 0.79, 0.98, and 0.90, respectively • For LSSVM the value of R2 was from 0.70 to 0.77 • In FSVM model it was 0.79–0.92 Hence, kernel-based models provided the most satisfactory result

researchers, enthusiasts, and other stakeholders to evolve the concept of smart waste management.

6 Conclusion

The purpose of this systematic literature review is to assess prior research on artificial intelligence smart waste management. We acquired the relevant studies from Scopus which is a well-known database and conducted a thorough analysis of each one. We examined the number of publications published each year on this subject and discovered that the scope of this field of study has expanded in recent years. This systematic review examined the role of various AI models in the waste management system by examining 40 publications published between 2001 and 2021. A comprehensive analysis was conducted on several AI models and their applications in the waste management industry. According to the findings of this SLR study, different types of AI models have been used to anticipate, simulate, and enhance waste management systems, including pure machine learning and deep learning models, as well as hybrid AI models. Despite increased study in this field, AI systems are still mostly in the research and development phase. This SLR demonstrates unequivocally which domains have received the greatest attention and which domains have received the least. The most significant drawback of this SLR is that we only searched one database, Scopus. To discover other publications, we employed citation chaining. Despite this, we may have overlooked some essential articles that should have been included in this SLR. This constraint can be alleviated in the future by increasing the number of databases accessible for article searching.

References

1. Low, N., Gleeson, B., Green, R., Radovic, D.: The Green City: Sustainable Homes, Sustainable Suburbs. Routledge (2016)
2. Roo, M.D., Kuypers, V.H.M., Lenzholzer, S.: The Green City Guidelines: Techniques For a Healthy Liveable City. The Green City (2011)
3. Hoornweg, D., Bhada-Tata, P., Kennedy, C.: Environment: waste production must peak this century. https://doi.org/10.1038/502615a
4. Besen, G., Fracalanza, A.: Challenges for the sustainable management of municipal solid waste in Brazil. disP—Plan. Rev. **52**(2), 45–52 (2016)
5. World Bank. What a waste 2.0: a global snapshot of solid waste management to 2050. Int. Bank Reconstr. Develop. (2018)
6. Triassi, M., Alfano, R., Illario, M., Nardone, A., Caporale, O., Montuori, P.: Environmental pollution from illegal waste disposal and health effects: a review on the "triangle of death". Int. J. Environ. Res. Public Health **12**, 1216–1236 (2015)
7. Pardini, K., Rodrigues, J.J.P.C., Kozlov, S.A., Kumar, N., Furtado, V.: IoT-based solid waste management solutions: a survey. J. Sens. Actuator Netw. **8**, 5 (2019)
8. Vijay, S., Raju, S., Nitish Kuma, P., S, V.: Smart waste management system using ARDUINO. Int. J. Eng. Res. Technol. (IJERT) (2019). Available at: https://www.ijert.org/>. Accessed 8 Dec 2020

9. Islam, M., Arebey, M., Hannan, M., Basri, H.: Overview for solid waste bin monitoring and collection system. In: 2012 International Conference on Innovation Management and Technology Research (2012)
10. Abdulla Al Mamun, M., Hannan, M., Hussain, A., Basri, H.: Integrated sensing systems and algorithms for solid waste bin state management automation. IEEE Sens. J. **15**(1), 561–567 (2015)
11. Yetilmezsoy, K., Ozkaya, B., Cakmakci, M.: Artificial intelligence-based prediction models for environmental engineering. Neural Netw. World. **21**(3), 193–218 (2011)
12. Basri, H., Stentiford, E.: Expert systems in solid waste management. Waste Manag. Res. J. Sustain. Circular Econ. **13**(1), 67–89 (1995)
13. Jayawardhana, L., de Alwis, A., Pilapitiya, S., Ransinghe, M.: Bestcity: developing clean cities. In: Li, D., Wang, B. (eds) Artificial Intelligence Applications and Innovations. AIAI 2005. IFIP—The International Federation for Information Processing, vol. 187. Springer, Boston. https://doi.org/10.1007/0-387-29295-0_6(2005)
14. Xia, W., Jiang, Y., Chen, X., Zhao, R.: Application of machine learning algorithms in municipal solid waste management: a mini review. Waste Manag. Res. J. Sustain. Circular Econ. 0734242X2110337 (2021)
15. Agarwal, V., Goyal, S., Goel, S.: Artificial intelligence in waste electronic and electrical equipment treatment: opportunities and challenges. Int. Conf. Intell. Eng. Manage. (ICIEM) **2020**, 526–529 (2020). https://doi.org/10.1109/ICIEM48762.2020.9160065
16. Abdallah, M., Abu Talib, M., Feroz, S., Nasir, Q., Abdalla, H., Mahfood, B.: Artificial intelligence applications in solid waste management: a systematic research review. Waste Manag. **109**, 231–246 (2020)
17. Kitchenham, B., Charters, S.: Guidelines for Performing Systematic Literature Reviews in Software Engineering. Accessed 27 Dec 2020. Available: http://citeseerx.ist.psu.edu/viewdoc/summary?doi=10.1.1.117.471
18. Haenlein, M., Kaplan, A.: A brief history of artificial intelligence: on the past, present, and future of artificial intelligence. Calif. Manag. Rev. **61**(4), 5–14 (2019)
19. McCarthy, J.: What is AI? [Internet]. Jmc.stanford.edu. 2021 [cited 22 October 2021]. Available from: http://jmc.stanford.edu/articles/whatisai.html
20. Ning, S., Yan, M.: Discussion on research and development of artificial intelligence. In: 2010 IEEE International Conference on Advanced Management Science (ICAMS 2010), pp. 110–112. doi:https://doi.org/10.1109/ICAMS.2010.5553039 (2010)
21. Wirtz, B., Weyerer, J., Geyer, C.: Artificial intelligence and the public sector—applications and challenges. Int. J. Public Adm. **42**(7), 596–615 (2018)
22. European Commission, "Waste": http://ec.europa.eu/environment/waste/. Accessed 23 May 2018
23. The World Counts: Theworldcounts.com. Available: https://www.theworldcounts.com/challenges/planet-earth/state-of-the-planet/world-waste-facts/story (2021)
24. "Wang, Z., Zhang, B., Guan, D.: Take responsibility for electronic-waste disposal. Nat. Cell Biol. **536**, 23–25 (2016)
25. Trends in Solid Waste Management: Datatopics.worldbank.org. Available: https://datatopics.worldbank.org/what-a-waste/trends_in_solid_waste_management.html (2021)
26. The World Counts: Theworldcounts.com. Available: https://www.theworldcounts.com/challenges/planet-earth/waste/hazardous-waste-statistics/story (2021)
27. Liao, B., Wang, T.: Research on industrial waste recovery network optimization: opportunities brought by artificial intelligence. Math. Prob. Eng. **2020**, 1–11. Available: https://doi.org/10.1155/2020/3618424 (2020)
28. Yang, H., Xia, J., Thompson, J., Flower, R.: Urban construction and demolition waste and landfill failure in Shenzhen, China. Waste Manag. **63**, 393–396 (2017). Available: https://doi.org/10.1016/j.wasman.2017.01.026
29. Malinauskaite, J., et al.: Municipal solid waste management and waste-to-energy in the context of a circular economy and energy recycling in Europe. Energy, **141**, 2013–2044 (2017). Available: https://doi.org/10.1016/j.energy.2017.11.128

30. Smart Waste Management by Sensoneo: Sensoneo. Available: https://sensoneo.com/ (2021)
31. What is Smart Waste Management?: IoT For All. Available: https://www.iotforall.com/smart-waste-management (2021)
32. Big belly-Smart city solution: http://bigbelly.com/. Accessed 27 May 2018. Available :http://bigbelly.com/
33. Commscope.com: https://www.commscope.com/globalassets/digizuite/1440-1270-sb-bigbelly.pdf?utm_source=ruckus&utm_medium=redirect (2021)
34. Chowdhury, B., Chowdhury, M.U.: RFID based real-time smart waste management system. In: 2007 Australasian Telecommunication Networks and Application Conference, pp.175–180 (2007)
35. Sharmin, S., Al-Amin, S.T.: Cloud-based dynamic waste management system for smart cities. In: Proceeding of the 7th Annual Symposium on Computing for Development, ser. ACM DEC 16
36. Purushotham Vijay Naidu, V., Dhikhi, T.: Smart Garbage Management Systems. Int. J. Pharm. Technol. **8**(4) (2016)
37. Ahmed, S., Ali, M.: Partnerships for solid waste management in developing countries: linking theories to realities. Habitat Int. **28**(3), 467–479 (2004)
38. Aremu, A.: In-town tour optimization of conventional mode for municipal solid waste collection. Ajol.info. 2021 [cited 22 October 2021]. Available from: https://www.ajol.info/index.php/njt/article/view/123619
39. Towards a decision support system for municipal waste collection by integrating geographical information system map, smart devices and agent-based model. In: Proceedings of the Seventh Symposium on Information and Communication Technology [Internet]. Dl.acm.org. 2021 [cited 22 October 2021]. Available from: https://dl.acm.org/doi/https://doi.org/10.1145/3011077.3011129
40. Identifying uncollected garbage in urban areas using crowdsourcing and machine learning [Internet]. Ieeexplore.ieee.org. 2021 [cited 22 October 2021]. Available from: https://ieeexplore.ieee.org/document/8070078
41. Location prediction for solid waste management—A Genetic algorithmic approach [Internet]. Ieeexplore.ieee.org. 2021 [cited 22 October 2021]. Available from: https://ieeexplore.ieee.org/abstract/document/7919609
42. Garbage Detection using Advanced Object Detection Techniques [Internet]. Ieeexplore.ieee.org. 2021 [cited 22 October 2021]. Available from: https://ieeexplore.ieee.org/abstract/document/9395916
43. Boumanchar, I., Chhiti, Y., M'hamdi Alaoui, F.E., et al.: Municipal solid waste higher heating value prediction from ultimate analysis using multiple regression and genetic programming techniques. Waste Manag. Res. 2019;37(6):578–589. doi:https://doi.org/10.1177/0734242X18816797
44. Jassim, M.S., Coskuner, G., Zontul, M.: Comparative performance analysis of support vector regression and artificial neural network for prediction of municipal solid waste generation. Waste Manag. Res. Apr 4:734242X211008526. doi:https://doi.org/10.1177/0734242X211008526. Epub ahead of print. PMID: 33818220 (2021)
45. Rutqvist, D.K., Blomstedt, F.: An Automated Machine Learning Approach for Smart Waste Management Systems. In: IEEE Transactions on Industrial Informatics, vol. 16, no. 1, pp. 384–392 (2020). doi:https://doi.org/10.1109/TII.2019.2915572
46. Miєić, M.., Dordevic, A., Arsić, A.K.: The optimization of vehicle routing of communal waste in an urban environment using a nearest neighbirs' algorithm and genetic algorithm: Communal waste vehicle routing optimization in urban areas. In: 2017 Ninth International Conference on Advanced Computational Intelligence (ICACI), pp. 264–271 (2017). doi:https://doi.org/10.1109/ICACI.2017.7974519
47. Abeygunawardhana, A.G.D.T., Shalinda, R.M.M.M., Bandara, W.H.M.D., Anesta, W.D.S., Kasthurirathna, D., Abeysiri, L.: AI—driven smart bin for waste management. In: 2020 2nd International Conference on Advancements in Computing (ICAC), 2020, pp. 482–487, doi: https://doi.org/10.1109/ICAC51239.2020.9357151.

48. Jadli, A., Hain, M.: Toward a deep smart waste management system based on pattern recognition and transfer learning. In: 2020 3rd International Conference on Advanced Communication Technologies and Networking (CommNet), pp. 1–5 (2020). doi: https://doi.org/10.1109/CommNet49926.2020.9199615
49. Sheng, T.J., et al.: An internet of things based smart waste management system using LoRa and tensorflow deep learning model. IEEE Access **8**, 148793–148811 (2020). https://doi.org/10.1109/ACCESS.2020.3016255
50. Hannan, M., Arebey, M., Begum, R., Basri, H.: An automated solid waste bin level detection system using a gray level aura matrix. Waste Manag. **32**(12), 2229–2238 (2012)
51. Sabour, M.R., Mohamedifard, A., Kamalan, H.: A mathematical model to predict the composition and generation of hospital wastes in Iran. Waste Manag. **27**(4), 584–587 (2007). https://doi.org/10.1016/j.wasman.2006.05.010. Epub 2007 Jan 18 PMID: 17239577
52. Golbaz, S., Nabizadeh, R., Sajadi, H.S.: Comparative study of predicting hospital solid waste generation using multiple linear regression and artificial intelligence. J. Environ. Health Sci. Eng. **17**(1), 41–51 (2019). https://doi.org/10.1007/s40201-018-00324-z.PMID:31297201; PMCID:PMC6582046
53. Kumar, N., Mohammed, M., Abdulkareem, K., Damasevicius, R., Mostafa, S., Maashi, M., et al.: Artificial intelligence-based solution for sorting COVID related medical waste streams and supporting data-driven decisions for smart circular economy practice (2021)
54. Naveen Ananda Kumar, J., Chimmani, S.: Proposal of smart home resource management for waste reduction and sustainability using AI and ML. In: 2019 International Conference on Communication and Electronics Systems (ICCES), pp. 992–998 (2019). doi:https://doi.org/10.1109/ICCES45898.2019.9002031
55. Papagiannis, F., Gazzola, P., Burak, O., Pokutsa, I.: A European household waste management approach: Intelligently clean Ukraine. J. Environ. Manag. **294**, 113015 (2021). doi:https://doi.org/10.1016/j.jenvman.2021.113015. Epub 2021 Jun 10. PMID: 34119987
56. Wastewater management using artificial intelligence. Wastewater management using artificial intelligence [Internet]. 2018 [cited 22 October 2021]. p. 8. Available from: https://doi.org/10.1051/e3sconf/20184500050
57. Chen, W.C., Chang, N.B., Chen, J.C.: Rough set-based hybrid fuzzy-neural controller design for industrial wastewater treatment. Water Res. **37**(1), 95–107 (2003). https://doi.org/10.1016/s0043-1354(02)00255-5. PMID: 12465791
58. Chen, W., Chang, N., Shieh, W.: Advanced hybrid fuzzy-neural controller for industrial wastewater treatment. J. Environ. Eng. **127**(11), 1048–1059 (2001)
59. Matheri, A., Ntuli, F., Ngila, J., Seodigeng, T., Zvinowanda, C.: Performance prediction of trace metals and cod in wastewater treatment using artificial neural network. Com. Chem. Eng. **149**, 107308 (2021)
60. Yanbo, J., Jianyi, J., Xiandong, W., Wei, L., Lincheng, J.: Bioaugmentation technology for treatment of toxic and refractory organic waste water based on artificial intelligence. Front Bioeng Biotechnol, **9**, 696166 (2021). Published 2 July 2021. doi:https://doi.org/10.3389/fbioe.2021.696166
61. Aitken, J.M., et al.: Autonomous nuclear waste management. IEEE Intell. Syst. **33**(6), 47–55 (2018). doi:https://doi.org/10.1109/MIS.2018.111144814
62. Fawzy, M., Nasr, M., Nagy, H., Helmi, S.: Artificial intelligence and regression analysis for Cd(II) ion biosorption from aqueous solution by Gossypium barbadense waste. Environ. Sci. Pollut. Res. Int. **25**(6), 5875–5888 (2018). https://doi.org/10.1007/s11356-017-0922-1. Epub 2017 Dec 12 PMID: 29235028
63. Sudha, S., Vidhyalakshmi, M., Pavithra, K., Sangeetha, K., Swaathi, V.: An automatic classification method for environment: friendly waste segregation using deep learning. IEEE Technological Innovations in ICT for Agriculture and Rural Development (TIAR) **2016**, 65–70 (2016). https://doi.org/10.1109/TIAR.2016.7801215
64. A. R., R. B. A. R.: An improvised smart bin management system using an object recognition method. In: 2020 International Conference on System, Computation, Automation and Networking (ICSCAN), pp. 1–6. doi: https://doi.org/10.1109/ICSCAN49426.2020.9262333 (2020)

65. Abbasi, M., El Hanandeh, A.: Forecasting Municipal Solid Waste Generation Using Artificial Intelligence Modelling Approaches (2021)
66. Soni, U., Roy, A., Verma, A., Jain, V.: Forecasting Municipal Solid Waste Generation Using Artificial Intelligence Models—A Case Study In India (2021)
67. Coskuner, G., et al.: Application of artificial intelligence neural network modeling to predict the generation of domestic, commercial and construction wastes. Waste Manag. Res. **39**, 499–507 (2020)
68. Cha, G.-W., Moon, H.J., Kim, Y.-M., Hong, W.-H., Hwang, J.-H., Park, W.-J., Kim, Y.-C.: Development of a prediction model for demolition waste generation using a random forest algorithm based on small datasets. Int. J. Environ. Res. Public Health. **17**(19), 6997 (2020). https://doi.org/10.3390/ijerph17196997
69. Kannangara, M., Dua, R., Ahmadi, L., Bensebaa, F.: Modeling and prediction of regional municipal solid waste generation and diversion in Canada using machine learning approaches. Waste Manag. **74**, 3–15 (2018)
70. Erkinay Ozdemir, M., Ali, Z., Subeshan, B. et al.: Applying machine learning approach in recycling. J. Mater. Cycles. Waste. Manag. **23**, 855–871 (2021). https://doi.org/10.1007/s10163-021-01182-y
71. Huang, J., Dmitry D.K.: Artificial intelligence for planning of energy and waste management. Sustain. Energy Technol. Assess. **47**, 101426 (2021)
72. Nañez Alonso, S.L., Reier Forradellas, R.F., Pi Morell, O., Jorge-Vazquez, J.: Digitalization, circular economy and environmental sustainability: the application of artificial intelligence in the efficient self-management of waste. Sustainability **13**, 2092 (2021). https://doi.org/10.3390/su13042092
73. Rutqvist, D., Kleyko, D., Blomstedt, F.: An automated machine learning approach for smart waste management systems. IEEE Trans. Industr. Inf. **16**(1), 384–392 (2020). https://doi.org/10.1109/TII.2019.2915572
74. Chu, Y., Huang, C., Xie, X., Tan, B., Kamal, S., Xiong, X.: Multilayer Hybrid Deep-Learning Method for Waste Classification And Recycling (2021)
75. Ongondo, F.O., Williams, I.D., Cherrett, T.J.: How are WEEE doing? A global review of the management of electrical and electronic wastes. Waste Manag. **31**(4), 714–730 (2011). https://doi.org/10.1016/j.wasman.2010.10.023. Epub 2010 Dec 13 PMID: 21146974
76. Król, A., Nowakowski, P., Mrówczyńska, B.: How to improve WEEE management? Novel approach in mobile collection with application of artificial intelligence. Waste Manag. **50**, 222–233 (2016). https://doi.org/10.1016/j.wasman.2016.02.033. Epub 2016 Mar 2 PMID: 26944864
77. Lu, Z., Zhang, J., Lu, S., et al.: Pollution characteristics and evaluation of heavy metal elements in the soil around the municipal solid waste incineration power plant and in the plant area. Environ. Sci. **40**(5), 483–2492 (2019)
78. Peng, Y., Lei, L., Peng, X., Yang, P., Zhao, X., Ma, W.: /edevelopment of domestic waste classification in China? Obstacles and countermeasures. China Environ. Sci. **38**(10), 3874–3879 (2018)
79. Dan, L., Chen, G., Ma, W., Duan, N.: Characteristics and treatment status of domestic garbage in villages and towns in China. China Environ. Sci. **38**(11), 4187–4197 (2018)
80. Liao, B., Wang, T.: Research on Industrial Waste Recovery Network Optimization: Opportunities Brought by Artificial Intelligence (2021)
81. Huang, X.S., Xu, X.: Legal regulation perspective of eco efficiency construction waste reduction and utilization. Urban Dev. Stud. **9**, 90–94 (2011)

82. Malinauskaite, J., Jouhara, H., Czajczynska, D., Stanchev, P., Katsou, E., Rostkowski, P., Thorne, R.J., Colón, J., Ponsá, S., Al-Mansour, F., Anguilano, L.: Municipal solid waste management and waste-toenergy in the context of a circular economy and energy recycling in Europe. Energy **141**, 2013–2044 (2017). https://doi.org/10.1016/j.energy.2017.11.128
83. Ali, T.H., Akhund, M.A., Memon, N.A., Memon, A.H., Imad, H.U., Khahro, S.H.: Application of artifical intelligence in construction waste management. In: 2019 8th International Conference on Industrial Technology and Management (ICITM), pp. 50–55. doi:https://doi.org/10.1109/ICITM.2019.8710680 (2019)
84. Jude, A.B., Singh, D., Islam, S., et al.: An artificial intelligence based predictive approach for smart waste management. Wireless Pers. Commun. (2021). https://doi.org/10.1007/s11277-021-08803-7

Machine Learning and Green Transportation

Traffic Sign Detection for Green Smart Public Transportation Vehicles Based on Light Neural Network Model

Riadh Ayachi, Mouna Afif, Yahia Said, and Abdessalem Ben Abdelali

Abstract Aiming to rise the security degree and the safety level of drivers, vehicles, and pedestrians, a traffic sign detection system is proposed in this work based on deep learning technologies. By developing the proposed assisting system, we contribute to build a new public smart transportation system used in smart cities and smart environments. Traffic sign detection presents one of the most important parts in an ADAS system due to its safety reasons. Detecting road sign can widely prevent people from accidents by respecting the traffic rules. Ensuring a reliable implementation on edge devices as field programmable gate arrays (FPGAs) presents an increasing challenge. To address this problem, we propose to build in this paper a new traffic sign detection system based on deep convolutional neural networks (DCNNs). The proposed detection system has been built based on YOLO as an objects detectors model in combination with SqueezeNet model which was used as lightweight backbone for features extraction. The use of SqueezeNet has been proposed to ensure a lightweight implementation on FPGA. In order to ensure the model implementation on FPGA, different optimizations techniques have been proposed. The proposed lightweight implementation of the traffic sign detection system has been performed on the pynq z1 platform. Training and testing experiments have been performed using the Chinese traffic sign detection (CTSD) dataset. Based on the experiments results, the proposed detection system achieved very interesting results in terms of detection accuracy and processing time. It achieves 96% mAP as a detection accuracy with 16 FPS as a processing time.

Keywords Traffic sign detection · Advanced driver assistance system · Edge implementation · Deep learning · Pruning and quantization

R. Ayachi (✉) · M. Afif · A. B. Abdelali
Laboratory of Electronics and Microelectronics, Faculty of Sciences of Monastir, University of Monastir, Monastir, Tunisia
e-mail: riadh.ayachi@fsm.rnu.tn

Y. Said
Electrical Engineering Department, College of Engineering, Northern Border University, Arar, Saudi Arabia

M. Lahby et al. (eds.), *Computational Intelligence Techniques for Green Smart Cities*, Green Energy and Technology, https://doi.org/10.1007/978-3-030-96429-0_4

1 Introduction

Recently, building new systems used for advanced driver assistance system presents an increasing need to increase the safety degree for the driver control of the vehicles. In the last few years, deep learning-based module has been used to solve different computers vision issues such as indoor objects detection [1, 2], indoor wayfinding assistance [3], scene recognition [4], and pedestrian detection [5].

Different well-known car producers are implementing the newest technologies in their products. Audi [6] integrated a traffic light detection system in its new cars, while VOLO company [7] integrated an autopilot in its tracks aiming to take control of the highway. When building an ADAS system, two main concerns must be addressed: detection accuracy and processing time. In order to build a reliable ADAS system, different concerns have to be fully addressed to ensure its lightweight implementation on low-end devices with limited memory consumption. Also, power consumption concern must be addressed as the system runs on mow battery. Incorporating new technologies into new intelligent transportation systems as environment sustainability, smart green cities become a major concern to contribute for intelligent public transport vehicles. Intelligent autonomous vehicles have to fully address different components and contributes for better traffic conditions such as less accidents number, less congestion, short trip time, more public safety, better urban network, and extremely less pollution contributing for better effects on the environment and improve the life quality. All these concerns contribute for the green smart cities. The right to a quality environment is important. The transportation plan in the few coming years must combine different major concerns as mobility and environment quality.

The latest advances in electronics technologies as embedded sensors and automated vehicles technologies contribute to the emergence of connected automated vehicles [8]. Building a reliable traffic sign detector must ensure a balance between model size, memory consumption, and real-time implementation. Smart transportation presents one from the most important challenging field as it collects data from sensors and process the information in a real-time way. Industry 4.0 presents a new component to contribute to smart transportation system by ensuring powerful hardware platform combined with software algorithms. Developing new co-design devices contributes for higher software–hardware performances which can widely improve the system results when deploying ADAS systems on edge devices.

In order to ensure a low-end implementation of the new technologies, various manufacturers as Xilinx and Intel develop new generation of FPGAs which provides very interesting performances while using low memory and energy consumption. Recently, deep learning-based architecture has gained a huge attention thanks to their powerful architectures. These architectures have contributed to solve various computer vision and artificial intelligence problems. Deep learning-based technique can be divided into three main categories: convolutional neural networks (CNNs), recurrent neural networks (RNNs), and deep belief networks (DBNs). All these architectures present a great ability to learn directly from the input data. We note that in image processing field, the most used architectures are the CNNs. CNN architectures are heavy and need great devices performances to be implemented. To address this problem, different optimization techniques have been proposed [9] to ensure a reliable implementation on edge devices. Various optimizations techniques can be applied at different stages and can be divided into three main categories: model optimization stage, training optimization stage, and inference optimization stage. Generally, the most used technique is the model optimization stage. Various neural networks have been proposed to reduce applications and energy consumption and ensure better performances.

SqueezeNet [10] has been proposed for the above reasons. This model includes the use of fire module which is proposed to replace the convolution layers which require huge memory consumption to process data as they present huge number of parameters. The use of fire module is to reduce the parameters number without accuracy drop. Also, MobileNet [11] has been proposed for lightweight implementation. In this architecture, the convolution layers are replaced by separable depthwise convolution blocks. These blocks are used to process the data in a faster way and to ensure a lightweight implementation of the model on low-end devices. EfficientNet [12] model presents a combination of optimizations and parameters scaling. In order to decrease the network computation complexity, we propose in this work to build a new traffic sign detection system based on YOLO [13] and SqueezeNet [10]. The proposed work that was used in this work was named YOLO-SqueezeNet. YOLO network has been used to solve the detection problem as a regression problem and designed for real-time implementations. SqueezeNet network has been used as a backbone network used for features extraction and highly recommended for FPGA implementations. Aiming to ensure a lightweight implementation, we applied different types of optimizations. We applied pruning and quantization techniques in order to compress the model size and to fit into the hardware memory.

The traffic sign detection system was trained and tested using CTSD dataset [14]. The proposed traffic sign system is robust enough as it was trained and evaluated on real-world images. The proposed traffic sign detection system achieved a detection accuracy of 96% mAP and 16 FPS as a processing time. After pruning and quantization application, the developed model presents a size of 2.5 MB which can be implemented on edge devices. In order to implement the proposed work, we used pynq platform.

The main contributions of this work are the following:

- Edge implementation of the proposed traffic sign detection system.
- Using a combination between YOLO and SqueezNet for traffic sign detection
- Using different optimization techniques as pruning and quantization.
- The proposed system achieved very interesting results in terms of detection accuracy and processing speed.

The reminder of this chapter is the following: Section 2 is reserved for presenting related works. The proposed approach is described and detailed in Sect. 3. In Sect. 4, experiments and results are presented and discussed. Conclusions are provided in Sect. 5.

2 Related Works

Various works have been proposed in the literature regarding the traffic sign detection task. In [15], authors proposed a CNN-based system used for traffic sign detection that is reliable for FPGA implementations. The system uses 32×32 image size resolution for the input. The proposed model was named ResCoNN. The model presents a small number of weights compared to state-of-the-art models which makes it suitable for edge implementations.

In [16], authors benchmarked the different deep learning frameworks and investigated the deployment of traffic sign detection on FPGA devices. An evaluation of the training speed and inference accuracy on GPU is performed using the GTSRB dataset [17]. Famous deep learning frameworks have been evaluated for FPGA implementations such as MXnet, Pytorch, TensorFlow, and CNTK. It also evaluates various neural networks such as VGG [18], ResNet [19], and SSD [20].

YOLO v3 model [21] has been proposed for object detection purpose. It presents a compressed version named YOLO v3-tiny which was used for FPGA implementations. The model presents the use of quantization techniques which were performed by changing 32 floating point with 8 bits fixed-point representation.

Shabarinath et al. [21] proposed a CNN model for traffic sign recognition suitable for edge devices. The proposed CNN model is composed of two convolution layers, two pooling layers, three fully connected layers, and an output layer. Each convolution layer was followed by a nonlinear activation layer. An input image size of 32×32 was used. The input images were preprocessed before feeding them to the CNN model. Many preprocessing techniques were applied such as gray scaling, histogram equalization, and image normalization. Those techniques all reduce the processing time of features extraction by the CNN model. The pruning [22] and post-training quantization [23] were applied to compress the model size and speed up the inference. The proposed model was evaluated on the GTSRB dataset [17].

After a careful study of state-of-the-art works, we discover that all works focus on traffic sign recognition on low-resolution images (e.g., 32×23) that contain only the traffic sign and none of them is investigating traffic sign detection in real scenes. In this work, we proposed a traffic sign detection system in a real scene with complex background. Besides, the proposed system was implemented on an edge device suitable for integration in real applications. More details on the proposed approach are presented in the next section.

3 Proposed Approach

Generally, the existing works focus on building new traffic recognition systems. But, implementing these systems on FPGAs has not been well studied in the literature. In this paper, we propose to build a traffic sign detection system using real-world images with real-world conditions.

The proposed traffic sign detection system is developed based on you only look once (YOLO) network with SqueezeNet. YOLO architecture has been designed for real-time objects detection. In the proposed work, the detection problem is considered as a regression problem. YOLO architecture divides the input data after passing through the backbone network to S×S grid. Every cell in the grid is able to detect just one object. In order to ensure a best localization for the object in the input image, YOLO uses a fixed set of anchors to predict a fixed number of bounding boxes. Each predicted bounding box is characterized by five parameters (coordinates x, y, width: w, height: h, and confidence score). The confidence score defines how likely the predicted bounding box contain an object. The bounding box width and height are fixed after the input image normalization. And x, y are the offset coordinates. The confidence score is predicted for each cell. YOLO architecture divides the output feature map into a 7×7 grid. Figure 1 illustrates the main architecture of YOLO framework.

The confidence score of the object class is calculated by applying a multiplication between the bounding box confidence score and the conditional class score c. As the YOLO treats the object detection problem as a regression problem, it allows to simultaneously predict bounding box class and score. Based on this fact, YOLO model is very fast and generally meet the real-time requirements. To optimize the network performances more, a combination of two loss function was proposed by combining a localization and a classification loss. To optimize equally the error for small and wide predicted bounding boxes, YOLO introduces the use of square root prediction of height and width. To obtain a more precise bounding box precision, the total loss function is multiplied by λ_{coord} which is fixed to 5. If various predicted bounding boxes do not contain objects, this fact leads to class imbalance problem. To address this problem, the total loss function is multiplied by λ_{noobj} term. Its value is fixed to 0.5. The YOLO loss function is calculated as the following equation.

$$\text{loss} = \lambda_{\text{coord}} \sum_{i=1}^{s^2} \sum_{j=1}^{k} q_{ij}^{\text{obj}} (x_i - \hat{x}_i)^2 + (y_i - \hat{y}_i)^2$$

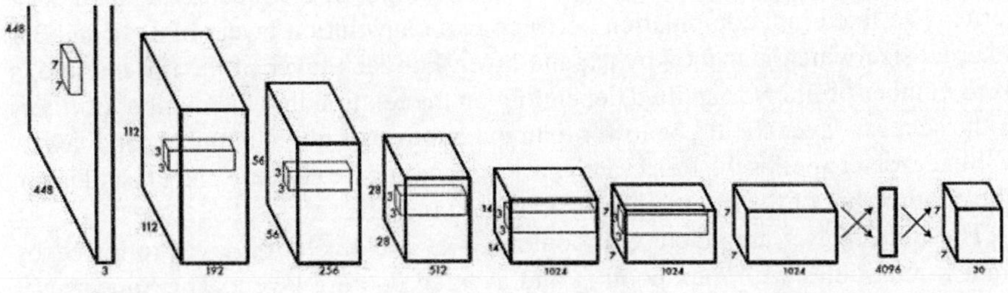

Fig. 1 YOLO architecture

$$+ \lambda_{\text{coord}} \sum_{i=1}^{s^2} \sum_{j=1}^{k} q_{ij}^{\text{obj}} (\sqrt{w_i} - \sqrt{\hat{w}_i})^2 + (\sqrt{h_i} - \sqrt{\hat{h}_i})^2$$

$$+ \sum_{i=1}^{s^2} \sum_{j=1}^{k} q_{ij}^{\text{obj}} (c_i - \hat{c}_i)^2 + \lambda_{\text{noobj}} \sum_{i=1}^{s^2} \sum_{j=1}^{k} q_{ij}^{\text{noobj}} (c_i - \hat{c}_i)^2 \quad (1)$$

where

(x_i, y_i)	the *ith* center coordinate of the ground truth bounding box.
$(\hat{x}_i, \hat{y}_i)$	the *ith* center coordinate of the predicted bounding box.
w_i	the width of the *ith* ground truth bounding box.
$\hat{w}_i$	the width of the *ith* predicted bounding box.
$\hat{h}_i$	the height of the *ith* predicted bounding box.
h_i	the height of the *ith* ground truth bounding box.
$\hat{c}_i$	the confidence score of the bounding box j in the cell *i.*
c_i	the target confidence score.
s	the grid size.
k	the number of predicted bounding boxes.

After regression and detection process, a duplicated number of predictions for the same objects is detected; to eliminate the duplicated number, YOLO applies the non-maximum suppression technique. In order to ensure a lightweight implementation for the proposed traffic sign detection system, we changed the YOLO backbone, and we used SqueezeNet to extract features from the input data. It was designed especially for FPGA implementations and devices with low computation requirements. SqueezeNet model has proposed three types of strategies. In order to reduce the model complexity without accuracy degradation. The first optimization is by using 1×1 convolution instead of using 3×3 convolution. This strategy allows to reduce the parameters number of the network. The second strategy is to reduce the number of input channels of 3×3 convolution kernels. While the third strategy is to perform a down-sampling process to the last stage. This fact led to obtain better accuracy.

To perform the above strategies, SqueezeNet replaces the regular convolution layers by the fire module. The fire module is composed of Squeeze layer of 1×1 kernel size fixed in a combination between two convolution layers of 1×1 and 3× 3 kernel size which is named by expand layer. For each layer of the fire module, a fixed number of filters is applied depending on the relation that the number of filters in the squeeze layer should be lower than the number of filters in the expand layer.

SqueezeNet applies the ReLU activation layer after the squeeze and expand layer. The architecture of the fire module is presented in Fig. 2.

The SqueezeNet architecture is composed of a convolution layer followed by eight fire modules with max pooling and average pooling layers. The SqueezeNet architecture is presented in Fig. 3.

Max pooling has been applied in the SqueezeNet architecture to compress the features map. Authors in [25] mentioned that by using strided convolution instead of max pooling has shown better results for hardware implementations. In the proposed

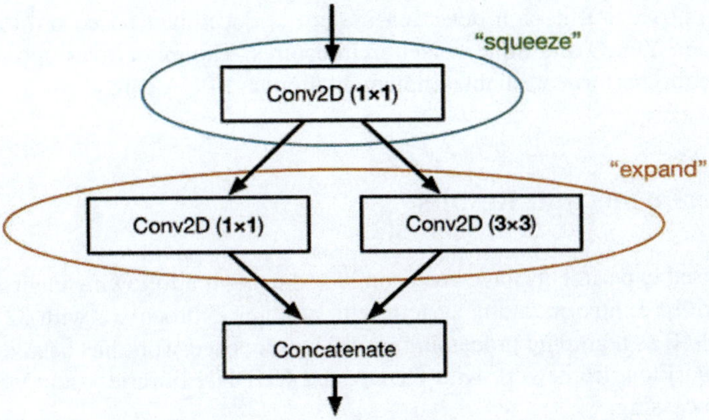

Fig. 2 Fire module architecture

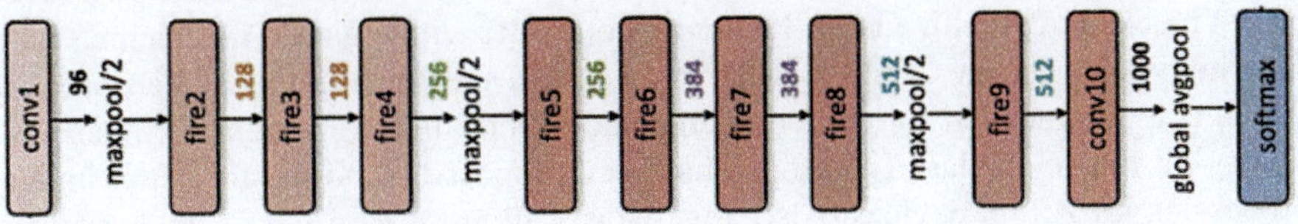

Fig. 3 SqueezeNet original architecture

work, we converted the network to a fully convolutional network, and we removed the average pooling layer. The model size cannot be fit into the FPGA memory. To reduce the model size of the traffic sign system, we applied the channel pruning technique after the model training. To apply the pruning technique, two main steps have to be fully addressed. First, a selection of channels based on the regression loss is performed. This technique is applied to remove the weak channels. Also, the filters that generate the redundant channels are removed. Secondly, features maps are reconstructed using last square method. After all these steps, the pruned model was retrained to recover the accuracy.

In the proposed work, we applied the post-training transform quantization [24] in order to reduce the model size. This method allows to transform the model using a decorrelation method. This technique allocates a bit-depth and transform it to a required bit-depth. This technique is applied to maximize the performance of the quantization method. The quantization technique has been applied after the training process, and the model was fine-tuned to recover the accuracy. The single-bit representation was performed using a negative and positive bit. The weights bits transformation can be presented as 2.

$$\text{sign}(w) : \left\{ \begin{matrix} +1 & \text{if } w \geq 0 \\ -1 & \text{otherwise} \end{matrix} \right\} \tag{2}$$

The proposed traffic sign detection system is developed based on a combination between YOLO and SqueezeNet architectures. The model was optimized and compressed to perform a reliable implementation on FPGA cards.

4 Experiments and Results

The proposed experiments have been conducted using the following environment: a desktop with a Linux operating system with an Intel i7 processor with 32 of RAM and GTX 960 as a graphic processing unit. The proposed work has been developed using TensorFlow framework with CUDA and CUDNN libraries and OpenCV for images processing.

To implement the proposed work, we used pynq z1 engine. The pynq z1 is equipped with a CPU ARM cortex A9 and ZYNQ XC7Z020-1CLG400C FPGA based on the Artix-7 programmable logics family. Pynq z1 device provides a 630 KB of fast blocks RAM with a DDR3 memory controller with eight DMA channels and four high-performance AXI3 slave part. We used the lite version of TensorFlow which is recommended for edge implementation. For this, we used the xfopen CV library for image display. The xfopen has been proposed by Xilinx for better image processing on Xilinx devices. Also, this dataset allows a partition of the hardware and software co-designs automatically.

Training and testing experiments have been conducted using the CTSD dataset [14]. CTSD dataset provides 10,000 images taken under real-world conditions. The data was divided into 8000 images for raining and 2000 for model testing. The dataset provides three classes: prohibited signs, danger signs, and mandatory signs. During training and test processes, the input images has been resized into 128 × 128 pixels per image. We used ADAM as a network optimizer during the back propagation process. We used 0.002 as an initial learning rate with 0.005 as a weight decay. Data augmentation has been applied using translation, mirror, and rotation. Mean average precision (mAP) has been used as an evaluation metric. The per-class average precision is presented in Table 1.

As mentioned in Table 1, we achieved very encouraging precision for the three classes present in the CTSD dataset. YOLO-SqueezeNet model has achieved 96% mAP. After model compression, we achieved 93.6% as a mean average precision. Table 2 provides the results obtained before and after model compression.

The proposed model presents 2.5 MB after model compression. The compression of the model allows to reduce the model size by 1.3 MB. The parameters number

Table 1 Per-class average precision

Class	AP (%)
Mandatory	96.2
Danger	95.3
Prohibited	96.5

Table 2 Mean average precision obtained before and after model compression

Model	mAP (%)	Speed (FPS)	Model size (MB)	Parameters (million)
Yolo-SqueezeNet	96	38	2.57	1.24
Yolo-SqueezeNet compressed	93.4	46	1.32	0.42

has been reduced by 0.42 million. Generally, FPGA present less than 10 MB of on-chip memory. So, the model can be implemented on FPGA due to its small size. The traffic sign detection system is robust enough as it was trained and tested under various challenging conditions: weather conditions, heavy occlusion, lighting conditions, distance, and viewpoint. Table 3 reports the obtained results under various challenging conditions.

The inference part of YOLO-SqueezeNet model has been implemented on the pynq z1 device. We achieved 16 FPS as an inference speed which is considered as real-time result. The xfopen CV library was used to load images and display the output. This library is designed for Xilinx FPGAs and SoCs. We used xfDNN library to distribute the model through the platform. Convolutional neural networks depend mainly on matrix multiplication and require huge parallelism to achieve good performances, we implemented the convolution layers on the hardware device. We note that the convolution layer parameters were stored into the on-chip memory. For the output layer, it was implemented on the software part as it need a lot of storage memory. Figure 4 provides an illustration of the proposed model detection.

Table 3 Obtained average precision under different conditions

Conditions		AP (%)
Weather	Sun	96.5
	Cloud	95.9
	Rain	95.2
	Snow	95.6
	fog	94.4
Light	Good	96.7
	Bad	95.4
Occlusion	Yes	95.1
	No	96.8
Distance	Far	96.1
	Close	96.8
Color fading	Yes	96.2
	No	96.7
Camera facing	Yes	96.6
	No	95.8

Fig. 4 Model detection example

5 Conclusions

Traffic sign detection presents one of the most important tasks for an ADAS system and to ensure more safety in the urban network. We propose in this paper to build a new traffic sign detection system using deep learning techniques. A combination between YOLO and SqueezeNet networks has been used to develop the proposed system. We also propose a lightweight implementation of the proposed work on FPGA devices. Different optimization techniques have been proposed to reduce the model and to ensure an edge implementation of the developed traffic sign detection system. Experiments have proved the model efficiency in terms of accuracy as well as processing time achieved. YOLO-SqueezeNet model allowed to achieve 96% mAP before model compression and 93.4% mAP after the model compression and 16 FPS as a processing time on Pynq z1 device. The obtained results proved the performance of the proposed work and the applied optimization techniques.

References

1. Afif, M., Ayachi, R., Atri, M.: Indoor objects detection system implementation using multi-graphic processing units. Cluster Comput. (2021). https://doi.org/10.1007/s10586-021-03419-9
2. Afif, M., Ayachi, R., Pissaloux, E., et al.: Indoor objects detection and recognition for an ICT mobility assistance of visually impaired people. Multimed. Tools Appl. **79**, 31645–31662 (2020). https://doi.org/10.1007/s11042-020-09662-3
3. Afif, M., Ayachi, R., Said, Y., et al.: Deep learning-based application for indoor way finding assistance navigation. Multimed. Tools Appl. **80**, 27115–27130 (2021). https://doi.org/10.1007/s11042-021-10999-6
4. Afif, M., Ayachi, R., Said, Y., et al.: Deep learning based application for indoor scene recognition. Neural Process Lett. **51**, 2827–2837 (2020). https://doi.org/10.1007/s11063-020-10231-w
5. Ayachi, R., Afif, M., Said, Y., Abdelaali, A.B.: Pedestrian detection for advanced driving assisting system: a transfer learning approach. In: 2020 5th International Conference on Advanced Technologies for Signal and Image Processing (ATSIP), pp. 1–5. IEEE (2020)
6. Audi is advancing the tech that teaches cars to talk to traffic lights: Available at: https://www.digitaltrends.com/cars/audi-traffic-light-recognition-v2i-technology-gains-new-features/. Accessed 1 Jul 2021
7. Driver Support services: Available at: https://www.volvotrucks.com/en-en/services/driver-support.html. Accessed 01 Jul 2021
8. Giuffrè, T., Canale, A., Severino, A., Trubia, S.: Automated vehicles: a review of road safety implications as a driver of change. In: Proceedings of the 27th CARSP Conference, vol. 16 (2017)
9. Ayachi, R., Said, Y., Abdelali, A.B.: Optimizing neural networks for efficient FPGA implementation: A survey. Arch. Comput. Methods Eng. 1–11 (2021)
10. How Howard, A.G., Zhu, M., Chen, B., Kalenichenko, D., Wang, W., Weyand, T., Adam, H.: Mobilenets: Efficient convolutional neural networks for mobile vision applications (2017). arXiv preprint arXiv:1704.04861
11. Iandola, F.N., Han, S., Moskewicz, M.W., Ashraf, K., Dally, W.J., Keutzer, K.: SqueezeNet: AlexNet-level accuracy with 50x fewer parameters and <0.5 MB model size (2016). arXiv preprint arXiv:1602.07360
12. Tan, M., Le, Q.: Efficientnet: Rethinking model scaling for convolutional neural networks. In: International Conference on Machine Learning, pp. 6105–6114 (2019). PMLR
13. Redmon, J., Divvala, S., Girshick, R., Farhadi, A.: You only look once: Unified, real-time object detection. In: Proceedings of the IEEE Conference on Computer Vision and Pattern Recognition, pp. 779–788 (2016)
14. Zhang, Y., Wang, Z., Qi, Y., Liu, J., Yang, J.: Ctsd: A dataset for traffic sign recognition in complex real-world images. In 2018 IEEE Visual Communications and Image Processing (VCIP), pp. 1–4. IEEE (2018)
15. Lechner, M., Jantsch, A., Dinakarrao, S.M.P.: ResCoNN: Resource-efficient FPGA-accelerated CNN for traffic sign classification. In: 2019 Tenth International Green and Sustainable Computing Conference (IGSC), pp. 1–6. IEEE (2019)
16. Lin, Z., Yih, M., Ota, J.M., Owens, J.D., Muyan-Özçelik, P.: Benchmarking deep learning frameworks and investigating FPGA deployment for traffic sign classification and detection. IEEE Trans. Intell. Veh. **4**(3), 385–395 (2019)
17. Stallkamp, J., Schlipsing, M., Salmen, J., Igel, C.: The German traffic sign recognition benchmark: a multi-class classification competition. In The 2011 international joint conference on neural networks, pp. 1453–1460. IEEE (2011)
18. Simonyan, K., Zisserman, A.: Very deep convolutional networks for large-scale image recognition. arXiv preprint arXiv:1409.1556 (2014)
19. He, K., Zhang, X., Ren, S., Sun, J.: Deep residual learning for image recognition. In Proceedings of the IEEE conference on computer vision and pattern recognition. pp. 770–778 (2016)

20. Liu, W., Anguelov, D., Erhan, D., Szegedy, C., Reed, S., Fu, C.Y., Berg, A.C.: Ssd: Single shot multibox detector. In: European conference on computer vision, pp. 21–37. Springer, Cham (2016)
21. Shabarinath, B. B., Muralidhar, P.: Convolutional neural network based traffic-sign classifier optimized for edge inference. In: 2020 IEEE region 10 conference (TENCON), pp. 420–425. IEEE (2020)
22. Yeom, S. K., Seegerer, P., Lapuschkin, S., Binder, A., Wiedemann, S., Müller, K.R., Samek, W.: Pruning by explaining: a novel criterion for deep neural network pruning. arXiv preprint arXiv:1912.08881 (2019). Yeom, S. K., Seegerer, P., Lapuschkin, S., Binder, A., Wiedemann, S., Müller, K.R., Samek, W.: Pruning by explaining: A novel criterion for deep neural network pruning. Patt. Recogn. **115**, 107899 (2021)
23. Nahshan, Y., Chmiel, B., Baskin, C., Zheltonozhskii, E., Banner, R., Bronstein, A.M., Mendelson, A.: Loss aware post-training quantization. Mach. Learn. 1–18 (2021)
24. Young, S., Wang, Z., Taubman, D., Girod, B.: Transform quantization for CNN compression. In: IEEE Transactions on Pattern Analysis and Machine Intelligence (2021)
25. Ayachi, R., Afif, M., Said, Y., Atri, M.: Strided convolution instead of max pooling for memory efficiency of convolutional neural networks. In International conference on the Sciences of Electronics, Technologies of Information and Telecommunications, pp. 234–243. Springer, Cham (2018s)

Green Transportation Balanced Scorecard Model: A Fuzzy-Delphi Approach During COVID-19

Badr Bentalha

Abstract Affected by the COVID-19 epidemic, logistics has received much attention by practitioners and researchers. The paper develops an integrated model for green transportation performance evaluation using the Balanced Scorecard. We propose a set of indicators to evaluate green transportation. These indicators are selected based on Fuzzy-Delphi method following different quantitative judgments of experts. The various criteria are related to the green transportation imperatives especially sustainability, stakeholder engagement and integration, continuous process improvement and green innovation. The proposed Balanced Scorecard is based on progressive methodological steps integrating the precision of the performance levers, the balancing of the components of this performance, the selection of the sustainability factors and the possibility of continuous improvement and adaptation of the selected factors.

Keywords Green transport · Green supply chain · Balanced Scorecard · Fuzzy-Delphi · Green transport balanced scorecard

1 Introduction

As globalization intensifies rapidly and continuously, supply chains have become much more interconnected and global. Coronavirus (COVID-19) has shown, once again that the rise of an endemic disease can cause serious problems and disrupt supply chains (SC).

This pandemic seems to be disastrous and long. For [1] it is the largest and longest global crisis since World War II. According to Johns Hopkins University and Medicine, the number of deaths from COVID-19 will outnumber 5,300,000, in 2021. In addition, the pandemic has had unprecedented social consequences. A large part of the world's population has been affected in one way or another by restrictions on their movement. It also triggered and sustained a global economic crisis and

B. Bentalha (✉)
National School of Business and Management, Sidi Mohammed Ben Abdellah University, Fez, Morocco

M. Lahby et al. (eds.), *Computational Intelligence Techniques for Green Smart Cities*, Green Energy and Technology, https://doi.org/10.1007/978-3-030-96429-0_5

national airspace and border closures. COVID-19 has also changed the economic and managerial world. Indeed, whatever their size or stage of development, the erection of these barriers has put a strain on supply chains. A part of our social and commercial exchanges, our professional relations and our organizational structures are already significantly modified by this pandemic [2].

Given the economic and social consequences of COVID-19, several environmental and ethical considerations arise for supply chains. These chains appear to be very active during the epidemic and continue to trigger ethical and environmental concerns. A green supply chain is described as the management of the facilities, operations, transportation, and environmental impacts of all components of the supply chain. The constraint of environmental protection is added to other economic and social constraints, in order to improve competitiveness while taking into consideration environmental effects [3]. Green transport is a component of this green supply chain. It refers to a specific form of transport that makes it possible to link economic objectives with the requirements of sustainable development [4]. It aims to accommodate current needs for the transportation of goods and people without harming or disadvantaging the ability to sustain this mode of transportation for future generations. Current green transportation objectives focus on continued optimization of transportation costs, reduction of negative externalities of transportation, and matching logistics choices with legal prerogatives.

In this work, we propose an empirical analysis of the possible indicators of green transport during the COVID-19 pandemic. This examination will allow us to identify the main opportunities of logistic chains and the steering mechanisms of these chains in order to respect the prerogatives of sustainable development. First, a presentation of the conceptual and theoretical framework through the concepts of supply chains and green transport will be made. Then, an analysis of the chosen methodology will be conducted by discussing the possibilities offered by the Balanced Scorecard model (BSC) and the Fuzzy-Delphi method. These two methods combined allow for a purification, ranking and synthesis of green transportation indicators. Finally, the results will be commented and discussed in order to present an integrated and global framework of the green transport balanced scorecard.

2 Conceptual and Theoretical Framework

2.1 Supply Chains and Green Supply Chain Management

Lambert et al. [5] explain the supply chain as a sequence of combinations of firms that provide products and services to the market. The member companies of a supply chain are connected by three types of flows. These are physical flows of goods, materials or merchandise, information and data flows and finally, monetary or financial flows. The information flows represent all the transfers or exchanges of data between the different actors in the supply chain. In addition, monetary and financial flows materialized by receipts, disbursements and financial asset transactions. The continuous circulation of these flows aims at bringing value to the different stakeholders [6].

The current evolution of markets has changed organizational structures toward more flexibility and responsiveness ([7, 8]). Thus, in this era marked by the development of information technologies, the multiplication of transport channels and the globalization of markets, the management of supply chains aims primarily at creating value. Indeed, a well-managed supply chain minimizes risks and increases profits ([9–11]). This perspective includes the desire to marry economic objectives with environmental concerns.

Green supply chain management is a new managerial direction that aims to integrate environmental concerns into the economic and shareholder management of the supply chain [12]. The term "green" has become increasingly abundant. It evokes several meanings depending on the perceptions of its users, but all of which refer to the environment [13]. We can trace the origin of the term green supply chain back to 1996. Indeed, it is the Manufacturing Research Consortium of the University of Michigan that initiated this term in order to analyze the ecological consequences of supply chains [14]. In a broad sense, a green supply chain is described as the management of facilities, operations, transportation and environmental impacts of all facilities in the supply chain. The environmental protection constraint is added to other economic constraints in order to improve competitiveness while taking environmental effects into consideration. Beamon [15] have proposed that the green supply chain should be considered a particular extension of the traditional supply chain. Indeed, this green chain integrates a concern for the reduction of environmental impacts. This continuous learning environmental process aims to reduce the negative externalities of transportation and logistics throughout the life cycles [16].

For [17], the green supply chain refers to a focal firm that engages sustainability in its relationships with different stakeholders with the objective of integrating ecological performance with the economic performance of the logistics network. Green supply chain design involves adding to the traditional goal of cost minimization, a second goal of minimizing the environmental impacts of the companies' products and processes.

A green chain appears to integrate responsible sourcing, green and sustainable industry, and green delivery and returns management [18]. The application of these principles aims at optimizing costs, complying with regulatory frameworks, seeking competitive advantage and aligning chain processes with stakeholder requirements ([19, 20]). Green supply chain management (GSCM) is therefore a new managerial vision that stimulates the integration of environmental objectives with the search for long-term economic performance. It is therefore a management method that coordinates and organizes a multitude of logistics practices upstream and downstream of supply chains ([21, 22]). It is a homogeneous and orderly integration of various ecological considerations into inter-organizational logistics management practices ([23, 24]).

Green supply chain management is a topic that is currently the subject of much debate among researchers and practitioners. It has been the subject of a multitude of publications due to pressure from different stakeholders ([25–27]). These pressures are oriented toward taking into account the negative impact of transport on the environment and society. This implies that companies that have embarked on this path must now rethink their business strategies in order to reach a more appropriate compromise between the usual economic and environmental targets. Thus, in the various issues related to the green chain, transportation is of primary importance. For this reason, the analysis of green transportation seems to be a very important topic.

2.2 Green Transport and Green Transport Indicators

Going back to the origin of the word, green transportation (GT) seems to have a close relationship with the idea of sustainability evoked by the Brundtland Report [28]. This is motivated by the juxtaposition of three components, economic, social and environmental performance, in most definitions of green transportation [29]. It is therefore transport that is carried out with respect to environmental and ethical considerations between the different partners in the logistics chain. The adoption of green transportation has a multitude of stated or latent objectives. These include cost optimization, compliance with legal frameworks and responsible image of supply chains [30]. In addition to these considerations, we can add the desire to reduce the negative externalities of transport, the optimization of routes and routes and the integration of renewable energies in the logistics management mechanisms.

Achieving these goals over the long-term requires a green transport evaluation system. Indeed, the evaluation of the performance of this mode of transport and the continuous modifications of the adopted system seems to be a major condition for the sustainability and suitability of green transport. Indeed, GT is an opportunity-oriented solution. It is about federating the supply chain partners around the possibility of obtaining a competitive advantage of the supply chain, the commercial importance of sustainability toward customers and consumers and finally, the possible savings of this mode of transport.

Thus, quantitative measurement of logistics performance is a process of continuous evaluation and quantification of the effectiveness and efficiency of the supply chain [31]. The measurement of logistics performance aims at verifying the adequacy of achievements with the objectives set, especially in the field of green transport. This instantaneous comparison is based on a set of linked and homogeneous indicators.

3 Research Methodology

There are currently several tools and methods that aim to improve environmental performance. The most widely discussed tool is the BSC. Indeed, it offers the advantage of simplicity of use and the integration of different performance variables in

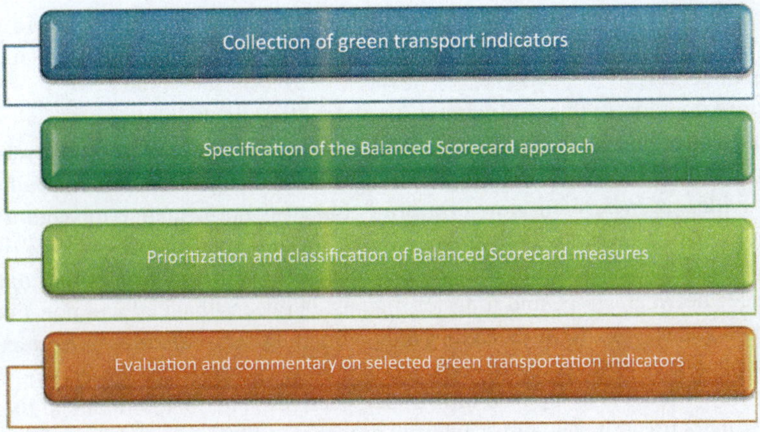

Fig. 1 Research methodology adopted

the same framework [32]. It also includes financial and non-financial, internal and external components and a juxtaposition of short and long-term indicators. For these reasons, we have chosen to use the BSC as a lever for identifying green transport indicators.

The starting point of our approach was a literature review on our main topic, namely, green transport. We also interviewed experts and practitioners in order to refine a preliminary list of green transport criteria. This allowed us to identify the basic factors of green transport from a BSC perspective. Next, a proprietary approach to the various criteria is taken. The objective of this step is to distinguish the criteria most related to our theme. We called upon the expertise and knowledge of 16 experts in order to classify and refine the criteria retained by the literature review. This review is based on the Fuzzy-Delphi Method (FDM). In this approach, triangular fuzzy numbers [33] are used to specify the preferences of one criterion over another. The proposed model globalizes a multitude of green transport indicators. The objective is to select and prioritize green transportation indicators (Fig. 1).

3.1 The Balanced Scorecard Model BSC

Focusing exclusively on traditional financial accounting measures, such as return on investment and payback period, has been criticized as recondite and elementary. By focusing on short-term financial performance measures, managers tend to sacrifice current profitability for actions such as new product development, process improvement, human resource development, information technology, and customer and market development that can bring long-term benefits, thereby limiting future investments and growth opportunities. Such actions by managers result in unbalanced performance measurement systems that focus only on short-term financial performance [34].

In order to address this problem by supplementing financial measures with additional measures that can help assess the long-term performance of a company, Kaplan and Norton introduced the BSC (Fig. 2). It is a performance measurement framework that provides an integrated view of a firm's business performance through a set of measures, which includes both financial and non-financial metrics ([35–39]). The name BSC is in the intention to keep the score of a set of measures that maintain a balance "between short- and long-term goals, between financial and non-financial metrics, between lagging and leading indicators, and between internal and external performance perspectives" ([40, 41]).

The BSC model is composed of four perspectives. The first is financial in nature and the others track the customer, internal process, and learning and growth perspectives. These three perspectives are non-financial and complementary. We can briefly introduce the different perspectives of the BSC [37]:

- Financial: This is the standard and traditional vision of performance. This vision, although criticized, remains important and basic. It is the measurement of financial performance via profit, return on investment or cash-flow.
- Customer: It is the understanding and satisfaction of customer needs. These variables depend on the company's business strategy, market research and operational

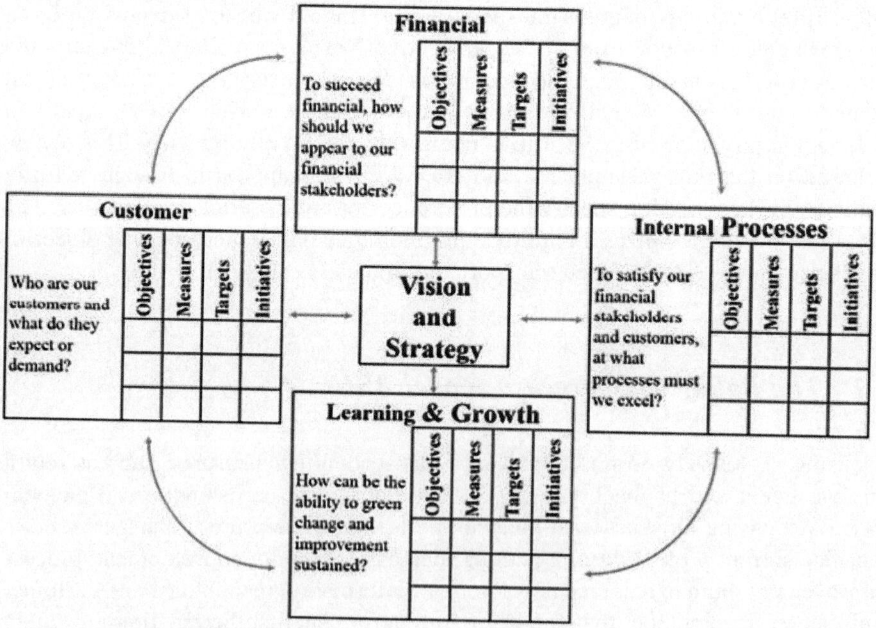

Fig. 2 The BSC model [35]

marketing elements. The main indicators of this perspective are customer retention and loyalty, market share or complaints.

- Internal Processes: It is about analyzing the company's internal processes and the elements of its value chain in order to satisfy the various stakeholders.
- Learning and Growth: This perspective aims at the complementarity of all the components of the BSC. The goal is to create and sustain long-term growth and improvement.

The BSC is then presented as a set of indicators. These are both strategic and operational indicators. They are directly linked to the strategy and provide the opportunity to monitor the determinants of performance on a daily basis [42]. In addition, they are global and integrated, as they are broken down into action and result variables. Thus, it is necessary to choose and classify the indicators retained in the BSC. For this reason, the Fuzzy-Delphi method is proposed to purify and prioritize the green transport indicators.

3.2 *Fuzzy-Delphi Method*

The Fuzzy-Delphi method is a combination of a qualitative methodology of Delphi with a quantitative approach of prioritization of elements via the triangular numbers or Fuzzy approach. It was initiated by [43] and seems to be today a vision allowing to have a consensus of the experts via a mixed approach. Its purpose is to have, after cycles of discussions, a consensus of the various stakeholders ([45–47]). The combination of these two methods has the advantage of taking the time of the study and sharing several expert opinions ([48, 49]). However, it requires several attempts at consensus and series of questionnaires and expert feedback to reach consensus [50].

In order to converge the opinions of the experts, a small group can be formed. There is no specific size for this method but the minimum number of experts should not be less than 10 [51]. A group of 16 experts was formed to finalize the previously identified green transport indicators. The questionnaire was coded based on the 5-level Likert metric scales in order to identify a homogeneous and ranked set of green transport indicators (Table 1). A score is obtained for each criterion coded in Ri.

The methodology of the work is summarized as follows [52]:

Step 1 Compute the index $O_i = (L_i, M_i, U_i)$ for each indicator coded by i. This index represents the triangular number of each indicator. L_i is the coding chosen for the minimum value. M_i is the coding chosen for the geometric mean of each indicator ($M_i = (R_{i1} * R_{i2} * \ldots * R_{ik})1/k$) with k is the index of each expert. Finally, U_i is the maximum value of each indicator (Fig. 3).

Step 2 When the triangular fuzzy numbers (O_i) are fixed for each of the indicators, a center of area (COA) coded in G_i is calculated with [53]:

$$G_i = [(U_i -- L_i) + (M_i -- L_i)]/3 + L_i$$

Table 1 List of green transportation indicators proposed to experts

	Indicators	Authors
Financial dimension	Customer inquiry time	[56]
	Presentation of new products according to customer needs	[56]
	Customer satisfaction	[56]
	Reduction of external costs	[57]
	Respect of deadlines	[58]
	Quality of service	[59]
Stakeholder performance dimension	Cost of transportation and inventory	[56]
	New investment for stakeholders	[56]
	Budget efficiency used for environmental improvement	[56]
	Green partnership platform with all stakeholders	[60]
	Logistics service providers that have implemented a green policy	[60]
	Supplier agreement to share their warehouses	[60]
	Collaborative delivery planning with your suppliers/customers	[60]
	Intensity of cooperation	[61]
	Level of conflict	[61]
	Level of transparency	[61]
Internal process dimension	Efficiency of the reverse logistics system	[56]
	Flexibility of service systems	[56]
	Efficiency improvement of business processes	[56]
	Implementation of environmental standards (e.g., ISO 14001)	[56]
	Short distance to key suppliers	[60]
	Eco-efficient network optimization	[60]
	Sustainable waste management	[57]
	Total energy usage	[58]
	Maximum utilization of transportation capacity	[60]
	Effective tracking system for green transportation indicators	[60]
Learning, growth and innovation dimension	Staff training and empowerment	[56]
	Development of R and D activities	[56]

(continued)

Table 1 (continued)

	Indicators	Authors
	Knowledge sharing among supply chain members	[56]
	Development of green technologies in partnership with supply chain members	[56]
	Use of alternative fuels and engines	[60]
	Sophisticated software for route optimization	[60]
	Use of telematics systems for efficient transportation operations	[60]
	Light-duty vehicles	[60]
	Eco-efficient driver incentive system	[60]
	Accessibility, connectivity and travel time	[57]
	Promotion of green vehicles	[57]
	Green training for company logistics staff	[60]
	Awareness, education and transition	[57]

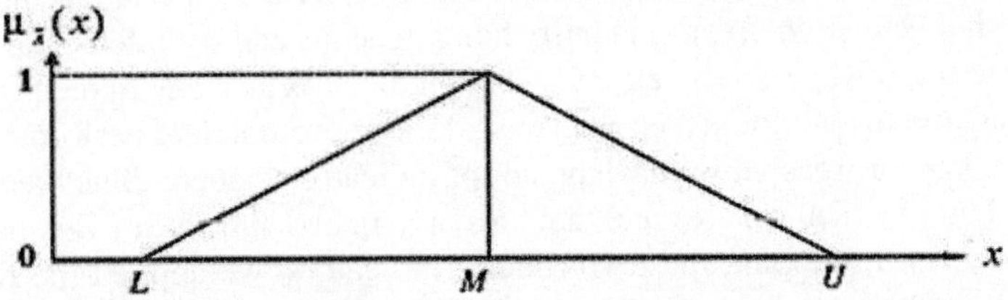

Fig. 3 Fuzzy triangular numbers [53]

Step 3 The factors are analyzed based on the defuzzification threshold named α. This threshold is obtained via the value of G_i. This defuzzification value is 3.50 (Fig. 4) and is the average of the mean value of the "important" variable (0.7) and the maximum value of the "normal" language variable (0.7) [54]. The threshold value thus serves as the sole criterion for selecting or rejecting an item. Thus, all indicators with a score below this threshold are eliminated.

Linguistic variable	Rating	Corresponding TFN
Extremely unimportant	1	(0.1, 0.1, 0.3)
Unimportant	2	(0.1, 0.3, 0.5)
Normal	3	(0.3, 0.5, 0.7)
Important	4	(0.5, 0.7, 0.9)
Extremely important	5	(0.7, 0.9, 0.9)

Fig. 4 Delphi fuzzy language scale for 5-level scale [54]

4 Results and Discussion

4.1 Proposed Green Transportation Indicators

The choice of the indicators proposed to the green transport experts is made through a literature review. Indeed, we have synthesized and summarized the articles dealing with this topic in order to identify homogeneous and complementary criteria according to the BSC perspectives [35]. These indicators also have a reciprocal relationship and aim to overcome the criticisms addressed to standard performance indicators [55]. For this reason, we have opted for indicators that combine the strategic, operational and tactical spheres and that have a long and short-term vision.

Therefore, in developing the performance indicators, we applied the following principles:

(1) Each indicator selected must be consistent with its category;
(2) Stakeholders must be able to use or adapt the indicator;
(3) Each indicator must be measurable and quantified;
(4) Indicators must have causal properties, i.e., they must be linked by BSC perspectives
(5) The overall model selected must have balance and consistency among the indicators.

The list of different green transport criteria was distributed to 16 experts (Table 1). This list includes 39 proposed indicators for green transportation that were derived from the scientific literature in the field and also from discussions with the experts.

4.2 *Selected Green Transportation Model With a Balanced Scorecard Approach*

Referring to the accepted standards of acceptance or rejection of indicators, 30 green transport indicators are selected and accepted in the designed model given the met threshold value of 3.50 and this reflects a relative consensus of experts on these indicators. Nevertheless, 9 indicators are rejected and excluded from the final model (Table 2).

The FDM results show that from a financial point of view, external cost reduction, customer satisfaction and timeliness are the top three indicators. We can also mention an average ranking of this financial dimension among the overall green transport criteria (Table 3).

Regarding the stakeholder dimension, the level of transparency, the cost of transport and the platform between partners are the three most important indicators of this dimension.

These three indicators reflect the strong presence of the relational aspect of green transport (Table 4).

The internal process perspective integrates the various supply chain processes into the different performance components. It traces a global homogeneity of the supply chain in order to achieve positive results in green transport. Indeed, it considers the links in a supply chain as a single, connected component and involves complementary internal processes (Table 5).

Sustainable waste management, eco-efficient network optimization and total energy used are the three main indicators.

Finally, from the perspective of the learning, growth and innovation dimension, the use of alternative fuels and engines, the sharing of knowledge and best practices among chain members, and the development of R&D activities are the main criteria selected (Table 6). These different indicators reflect a commitment to the ability of green transportation to adapt and improve. This is a perspective of continuous improvement and organizational learning.

4.3 *Discussion of the Criteria Adopted*

We followed several methodological rules by combining the BSC approach with the Fuzzy-Delphi approach. The result is a global and integrated model of green transport with the capacity to reflect all the facets of the GT and also with a hierarchy and modularity in the indicators selected.

A deeper analysis of our results shows that improving customer relations and stakeholder satisfaction and reducing costs can have an impact on improving sustainable financial goals. This impact can be direct or indirect, through influencing other aspects of the BSC and especially the perspectives of stakeholders. Therefore, green and ecological transportation appears as a way to improve the performance of the

Table 2 List of selected and rejected green transportation indicators

	Indicators	L_i (Min)	M_i (Mean)	U_i (Max)	G_i (COA)	Screening
Financial dimension	Customer inquiry time	1	4,75	5	3,583	Accepted
	Presentation of new products according to customer needs	1	4,18	5	3,394	Rejected
	Customer satisfaction	3	10,37	5	6,124	Accepted
	Reduction of external costs	4	10,99	5	6,662	Accepted
	Respect of deadlines	2	8,24	5	5,079	Accepted
	Quality of service	1	3,84	5	3,279	Rejected
Stakeholder performance dimension	Cost of transportation and inventory	2	8,44	5	5,148	Accepted
	New investment for stakeholders	2	5,50	5	4,166	Accepted
	Budget efficiency used for environmental improvement	1	4,10	4	3,032	Rejected
	Green partnership platform with all stakeholders	2	7,10	5	4,699	Accepted
	Logistics service providers that have implemented a green policy	1	4,19	5	3,396	Rejected
	Supplier agreement to share their warehouses	2	5,96	5	4,320	Accepted
	Collaborative delivery planning with your suppliers/customers	2	5,82	5	4,274	Accepted
	Intensity of cooperation	1	3,27	3	2,423	Rejected
	Level of conflict	1	4,17	5	3,391	Rejected
	Level of transparency	3	9,98	5	5,994	Accepted
Internal process dimension	Efficiency of the reverse logistics system	2	6,84	5	4,613	Accepted

(continued)

Table 2 (continued)

	Indicators	L_i (Min)	M_i (Mean)	U_i (Max)	G_i (COA)	Screening
	Flexibility of service systems	1	3,63	5	3,211	Rejected
	Efficiency improvement of business processes	1	4,17	5	3,391	Rejected
	Implementation of environmental standards (e.g., ISO 14001)	3	8,65	5	5,548	Accepted
	Short distance to key suppliers	2	7,18	5	4,727	Accepted
	Eco-efficient network optimization	4	12,01	5	7,004	Accepted
	Sustainable waste management	4	12,28	5	7,094	Accepted
	Total energy usage	3	8,98	5	5,661	Accepted
	Maximum utilization of transportation capacity	2	8,06	5	5,019	Accepted
	Effective tracking system for green transportation indicators	2	7,53	5	4,842	Accepted
Learning, growth and innovation dimension	Staff training and empowerment	2	7,15	5	4,717	Accepted
	Development of R and D activities	4	12,01	5	7,004	Accepted
	Knowledge sharing among supply chain members	4	12,28	5	7,094	Accepted
	Development of green technologies in partnership with supply chain members	4	11,75	5	6,915	Accepted
	Use of alternative fuels and engines	4	12,56	5	7,186	Accepted
	Sophisticated software for route optimization	3	9,18	5	5,728	Accepted

(continued)

Table 2 (continued)

Indicators	L_i (Min)	M_i (Mean)	U_i (Max)	G_i (COA)	Screening
Use of telematics systems for efficient transportation operations	2	6,55	5	4,517	Accepted
Light-duty vehicles	2	7,02	5	4,674	Accepted
Eco-efficient driver incentive system	4	11,49	5	6,829	Accepted
Accessibility, connectivity and travel time	1	3,54	5	3,181	Rejected
Promotion of green vehicles	2	6,82	5	4,608	Accepted
Green training for company logistics staff	2	7,03	5	4,677	Accepted
Awareness, education and transition	1	5,56	5	3,853	Accepted

Table 3 Indicators of the financial dimension of green transportation

	Indicators	Classification by dimension	Ranking Total
Financial dimension	Reduction of external costs	1	8
	Customer satisfaction	2	9
	Respect of deadlines	3	15
	Customer's inquiry time	4	30

company's operations. Indeed, green transport mode aims to unite the satisfaction of the company's economic objectives with the internal requirements and in terms of the different stakeholders. For this reason, we can see that the selected indicators seem to be linked with the different facets of customer satisfaction, other stakeholders and internal processes. In fact, there is a strong presence of relational and behavioral variables in the evaluation and analysis of the selected green transport indicators.

This finding is all the more remarkable given that transparency and the platform with partners seem to be two prevalent means in the perspective of green transport stakeholders. Indeed, the exchange of information, collaborative planning of deliveries with suppliers, customers, investments for stakeholders, and agreements by suppliers to share their warehouses are relevant indicators from the stakeholder perspective. These indicators trace the importance of relationship and trust in the

Table 4 Indicators of the green transportation stakeholder dimension

	Indicators	Classification by dimension	Ranking total
Stakeholder performance dimension	Level of transparency	1	10
	Cost of transportation and inventory	2	14
	Green partnership platform with all stakeholders	3	20
	Supplier agreement to share their warehouses	4	26
	Collaborative delivery planning with your suppliers/customers	5	27
	New investment for stakeholders	6	28

Table 5 Indicators of the green transportation internal process dimension

	Indicators	Classification by dimension	Ranking total
Internal process dimension	Sustainable waste management	1	2
	Eco-efficient network optimization	2	4
	Total energy usage	3	12
	Implementation of environmental standards (e.g., ISO 14001)	4	13
	Maximum utilization of transportation capacity	5	16
	Effective tracking system for green transportation indicators	6	17
	Short distance to key suppliers	7	18
	Efficiency of reverse logistics system	8	23

prospects of green transport. Indeed, the green transport approach is an integrated and comprehensive one requiring consensus among the supply chain.

From an internal perspective, sustainable waste management is a key indicator of green transportation. Indeed, environmental impacts are related to the eco-efficiency of the network, the energy used and environmental standards. There is also the

Table 6 Indicators of the green transportation stakeholder dimension

	Indicators	Classification by dimension	Ranking total
Learning, growth and innovation dimension	Use of alternative fuels and engines	1	1
	Knowledge sharing among supply chain members	2	2
	Development of R&D activities	3	4
	Development of green technologies in partnership with supply chain members	4	6
	Eco-efficient driver incentive system	5	7
	Sophisticated software for route optimization	6	11
	Staff training and empowerment	7	19
	Green training for the company's logistics staff	8	21
	Light vehicles	9	22
	Promotion of green vehicles	10	24
	Use of telematics systems for efficient transportation operations	11	25
	Awareness, education and transition	12	29

maximum use of transport capacity, the monitoring systems of green transport indicators, the distances of suppliers and the efficiency of the reverse logistics system. We can see that these indicators are linked and global. The use of a set of relevant and multi-level indicators will allow to see the green transport as a whole and avoid relatively inefficient local or partial optimizations.

The precise and detailed knowledge of the relationships between the variables of the learning, growth and innovation axis shows the reciprocal interrelations between these elements. Indeed, these parameters seem to be interdependent, complementary and homogeneous. For example, the stated technological innovations require knowledge sharing and human resource involvement. Therefore, these measures cannot be taken individually but rather globally and in a complementary way. For this reason, it is vital to take these indicators together. The most illustrative criteria for this are the use of alternative fuels and engines, knowledge sharing between members of the supply chain and the development of R and D activities.

Improvement in these three criteria must be considered to achieve sustainability. In addition, ensuring a safe working environment and healthy employees in the company has a plausible impact on business relations and the quality of ecological transport. This includes staff training and empowerment, green training for the company's logistics staff and awareness, education and transition.

The IT and information component is also important and reported by two main criteria. These are the use of telematics systems for efficient transport operations and sophisticated software for route optimization. Finally, the relationships that emerge from the different criteria selected show that the efficiency of operations and the use and development of renewable resources translate into social efficiency. Indeed, the variables related to personnel are omnipresent in the indicators selected from this perspective.

Thus, the proposed model is rich by several varied and related indicators. Figure 5 represents the final proposed model with the selected indicators.

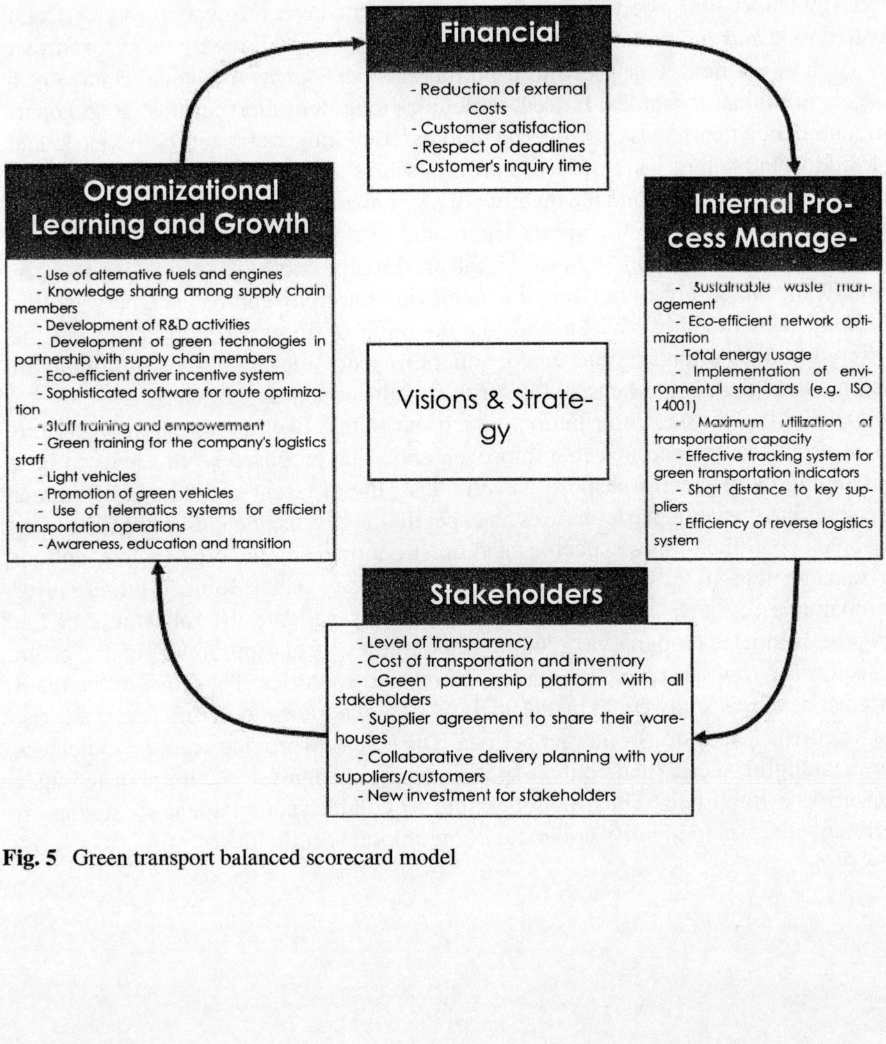

Fig. 5 Green transport balanced scorecard model

5 Conclusion

The objective of our research is to provide a comprehensive process for evaluating green transportation in relation to the various sustainability criteria. The study sought to design a framework linking several green transport indicators using the principles of the BSC. The rationale for choosing this tool is its comprehensiveness, simplicity, stakeholder integration, and flexibility. The selection and ranking of the BSC green transport indicators is conducted via a Fuzzy-Delphi approach that links qualitative aspects and quantitative models to rank the different green transport criteria. The end result is a specific and innovative model for green transportation that provides a diverse and consistent set of indicators useful for practitioners and policy makers.

Several practical, theoretical and managerial contributions can be enumerated by our study. First, in this study, the three sustainability objectives of green transportation are linked to create a balanced and homogeneous framework. It covers some weaknesses and gaps in research in this scientific field. Indeed, much of the work on green transportation is partial and fragmented. In this sense, this work provides a relatively broad and comprehensive view that can serve as a preliminary basis for further research in the field. Our green transportation model seems to take advantage of a variety of indicator profiles. Indeed, it includes long- and short-term, strategic operational and tactical measures, internal and external indicators, and both metric and qualitative indicators. This variety of profiles seems to be a way of showing the globalism of green transport and the inability of a fragmented view to capture all the facets of this specific mode of transport. The model also incorporates several relational, social and environmental indicators that are largely overlooked in other models. Finally, the work carried out aimed at achieving harmony and balance between the different facets of the BSC. Indeed, the adoption of an expert consensus and the different rounds of discussions and negotiations seem relevant to the multiplicity of opinions and the comprehensiveness of the results obtained.

Despite the various contributions noted our study, like any research work, has some shortcomings and possible improvements. The proposed work seems to be a unique model of green transport. Nevertheless, the different sectors and economic branches have several particularities and specificities. Thus, the indicators have to be adapted according to these specifics and other contingency factors like size, culture or management style. Due to the relative immaturity of the subject, empirical or comparative research is desirable to quantitatively validate the robustness of the proposed model. Comparative or longitudinal studies are also possible. Finally, green transport is a topic that can vary across sectors and countries. Thus, one of the main extensions of this study could be the analysis of green transport in different contexts, such as crisis, expansion or merger periods. The goal is to provide accurate indicators for meaningful, sector- and context-specific empirical analysis. Cultural differences can also be important. Thus, it is possible to conduct complementary studies in different contexts to identify contextual or temporal specificities.

1. References

1. Boccaletti, S., Ditto, W., Mindlin, G., Atangana, A.: Modeling and forecasting of epidemic spreading: the case of Covid-19 and beyond. Chaos Solitons Fractals. **135**(1), (2020)
2. Trzebiński, J., Cabański, M., Czarnecka, J.Z.: Reaction to the COVID-19 pandemic: the influence of meaning in life, life satisfaction, and assumptions on world orderliness and positivity. J. Loss Trauma. **25**(6–7), 544–557 (2020)
3. Wong, C.Y., Wong, C.W., Boon-itt, S.: Effects of green supply chain integration and green innovation on environmental and cost performance. Int. J. Prod. Res. **58**(15), 4589–4609 (2020)
4. Manoj, J.K., Ogallo, H.G., Owolabi, O.: A Quantitative analysis of sustainability and green transportation initiatives in highway design and maintenance. Proc. Soc. Behav. Sci. **111**(1), 1185–1194 (2014). doi:https://doi.org/10.1016/j.sbspro.2014.01.153
5. Lambert, D.M., Cooper, M.C., Pagh, J.D.: Supply chain management: implementation issues and research opportunities. Int. J. Logistics Manag. **9**(2), 1–20 (1998)
6. Rota-Frantz, K., Thierry, C., Bel, G.: Gestion des flux dans les chaînes logistiques (supply chain management), pp. 103–128. Performances Industrielles et Gestion Des Flux, Hermes Science-Lavoisier Paris (2001)
7. Camman, C.: Quelles potentialites de/pour la critique en supply chain management? Logistique Manag. **27**(1), 55–67 (2019)
8. Hmioui, A., Bentalha, B., Alla, L.: Service supply chain: a prospective analysis of sustainable management for global performance. In: 2020 IEEE 13th International Colloquium of Logistics and Supply Chain Management (LOGISTIQUA), 1–7 (2020). doi:https://doi.org/10.1109/LOGISTIQUA49782.2020.9353940
9. Fawcett, S.E., Magnan, G.M., McCarter, M.W.: Benefits, barriers, and bridges to effective supply chain management. Supply Chain Manag. Int. J. (2008)
10. Yanya, M., Mahamat, N.: The impact of supply chain management practices on competitive advantages: moderation role of total quality management. Polish J. Manag. Stud. **21**(1), (2020)
11. Bentalha, B., Hmioui, A., Alla L.: Last mile logistics applied to the distribution of COVID-19 vaccines: a prospection of good practices. Altern. Managé. Economiq. **3**(3), 41–61 (2021). doi:https://doi.org/10.48374/IMIST.PRSM/ame-v3i3.27423
12. Tseng, M.L., Islam, M.S., Karia, N., Fauzi, F.A., Afrin, S.: A literature review on green supply chain management: trends and future challenges. Resour. Conserv. Recycl. **141**, 145–162 (2019)
13. Çankaya, S.Y., Sezen, B.: Effects of green supply chain management practices on sustainability performance. J. Manufact. Technol. Manag. (2019)
14. Handfield, R.B., Walton, S.V., Melnyk, S.A.: Green supply chain: best practices from the furniture industry, USA. In: Proceedings of the Annual Meeting of the Decision Science Institute, pp. 1295–1297 (1996)
15. Beamon, B.M.: Designing the green supply chain. Logist. Inf. Manag. **12**(4), 332–342 (1999)
16. Wang, H.F., Gupta, S.M.: Green supply chain management: product life cycle approach. McGraw-Hill Education (2011)
17. Ahi, P., Searcy, C.: A comparative literature analysis of definitions for green and sustainable supply chain management. J. Clean. Prod. **52**(1), 329–341 (2013)
18. Hervani, A.A., Helms, M.M., Sarkis, J.: Performance measurement for green supply chain management. Benchmarking Int. J. **12**(4), 330–335 (2005)
19. Kumar, S., Teichman, S., Timpernagel, T.: A green supply chain is a requirement for profitability. Int. J. Prod. Res. **50**(5), 1278–1296 (2012)
20. de Oliveira, U.R., Espindola, L.S., da Silva, I.R., da Silva, I.N., Rocha, H.M.: A systematic literature review on green supply chain management: Research implications and future perspectives. J. Clean. Prod. **187**, 537–561 (2018)
21. Zhu, Q., Sarkis, J.: Relationships between operational practices and performance among early adopters of green supply chain management practices in Chinese manufacturing enterprises. J. Oper. Manag. **22**(3), 265–289 (2004)
22. Srivastava, S.K.: Green supply-chain management: a state-of-the-art literature review. Int. J. Manag. Rev. **9**(1), 53–80 (2007)

23. Sarkis, J., Zhu, Q., Lai, K.H.: An organizational theoretic review of green supply chain management literature. Int. J. Prod. Econ. **130**(1), 1–15 (2011)
24. Dubey, R., Gunasekaran, A., Papadopoulos, T.: Green supply chain management: theoretical framework and further research directions. Benchmarking Int. J. **24**(1), 184–218 (2017)
25. Pinto, L.: Green supply chain practices and company performance in Portuguese manufacturing sector. Bus. Strateg. Environ. **29**(5), 1832–1849 (2020)
26. Upadhyay, A.: Antecedents of green supply chain practices in developing economies. Manag. Environ. Qual. Int. J. (2020)
27. Khan, S.A.R., Yu, Z., Golpira, H., Sharif, A., Mardani, A.: A state-of-the-art review and meta-analysis on sustainable supply chain management: future research directions. J. Clean. Prod. **278**(1), (2021)
28. WCED (World Commission on Environment and Development). Our Common Future: Report of the World Commission on the Environment and Development [Bruntlandt Report]. General Assembly, United Nations, Forty-second Session, Supplement No. 25, A/42/25. Also published as Our Common Future. Oxford and New York: Oxford University Press (1987)
29. CEMT (The European Council of Ministers of Transport) Les services réguliers interurbains d'autocars en Europe, Table Ronde 114, Centre de Recherche Économique, Conférence Européenne des Ministres des Transports, Paris (2001)
30. Bongardt, D., Creutzig, F., Hüging, H., Sakamoto, K., Bakker, S., Gota, S., Böhler-Baedeker, S.: Low-Carbon Land Transport: Policy Handbook. Routledge (2013)
31. Zhu, Q., Sarkis, J., Lai, K.H.: Confirmation of a measurement model for green supply chain management practices implementation. Int. J. Prod. Econ. **111**(2), 261–273 (2008)
32. Northcott, D., Taulapapa, T.M.A.: Using the balanced scorecard to manage performance in public sector organizations: Issues and challenges. Int. J. Public Sect. Manag. **25**(3), 166–191 (2012)
33. Ishikawa, A.: The new fuzzy Delphi methods: economization of GDS (group decision support). In: Proceedings of the Twenty-sixth Hawaii International Conference on System Sciences, vol. 4, no. 1, pp. 255–264 (1993)
34. Bentalha, B., Hmioui, A., Alla, L.: The global performance of a service supply chain: a simulation-optimization under arena. In: The Proceedings of the Third International Conference on Smart City Applications. Springer, Cham, pp. 489–502 (2020). doi:https://doi.org/10.1007/978-3-030-66840-2_37
35. Kaplan, R.S., Norton, D.P.: The balanced scorecard—measures that drive performance. Harv. Bus. Rev. 71–79 (1992)
36. Kaplan, R.S., Norton, D.P.: Using the balanced scorecard as a strategic management system. Harv. Bus. Rev. (1996)
37. Kaplan, R.S., Norton, D.P.: Linking the balanced scorecard to strategy. California Manag. Rev. (1996)
38. Kaplan, R.S., Norton, D.P.: The balanced scorecard: translating the strategy action. Harv. Bus. Sch. Publ. (1996)
39. Kaplan, R.S., Norton, D.P.: Building a strategy-focused organization. Harv. Bus. Sch. Publ. (1999)
40. Kaplan, R.S., Norton, D.P.: Having trouble with your strategy? Then map it. Harv. Bus. Rev. (2000)
41. Kaplan, R.S.: The balanced scorecard: comments on balanced scorecard commentaries. J. Acc. Organ. Change. (2012)
42. Asiaei, K., Bontis, N.: Translating knowledge management into performance: the role of performance measurement systems. Manag. Res. Rev. (2019)
43. Kaufmann, A., Gupta, M.M.: Fuzzy mathematical models in engineering and management science. New York, NY: Elsevier Science Inc., p. 362 (1988)
44. Spinelli, T.: The Delphi decision-making process. J. Psychol. **113**(1), 73–80 (1983)
45. Baumfield, V.M., Conroy, J.C., Davis, R.A., Lundie, D.C.: The Delphi method: gathering expert opinion in religious education. Br. J. Relig. Educ. **34**(1), 5–19 (2012)

46. Ishikawa, A., Amagasa, M., Shiga, T., Tomizawa, G., Tatsuta, R., Mieno, H.: The max-min Delphi method and fuzzy Delphi method via fuzzy integration. **55**(3), 241–253 (1993)
47. Noorderhaven, N.G.: Strategic decision making. Addison-Wesley, Wokingham (1995)
48. Bui, T.D., Tsai, F.M., Tseng, M.-L., Ali, M.H.: Identifying sustainable solid waste management barriers in practice using the fuzzy Delphi method. Resour. Conserv. Recycl. **154**(1), (2020)
49. Lee, C.H., Wu, K.-J., Tseng, M.-L.: Resource management practice through eco-innovation toward sustainable development using qualitative information and quantitative data. J. Clean. Prod. **202**(1), 120–129 (2018)
50. Saido, G.A.M., Siraj, S., DeWitt, D., Al-Amedy, O.S.: Development of an instructional model for higher order thinking in science among secondary school students: a fuzzy Delphi approach. Int. J. Sci. Educ. **40**(8), 847–866 (2018)
51. Chao, C.-C., Lirn, T.-C., Lin, H.-C.: Indicators and evaluation model for analyzing environmental protection performance of airports. J. Air Transp. Manag. **63**(1), 61–70 (2017)
52. Singh, P.K., Sarkar, P.A: framework based on fuzzy Delphi and DEMATEL for sustainable product development: a case of Indian automotive industry. J. Clean. Prod. **246**(1), (2020)
53. Hsieh, T.Y., Lu, S.-T., Tzeng, G.-H.: Fuzzy MCDM approach for planning and design tenders selection in public office buildings. Int. J. Project Manag. **22**(1), 573–584 (2004)
54. Kumar, A., Dash, M.K.: Using Fuzzy Delphi and Generalized Fuzzy TOPSIS to evaluate technological service flexibility dimensions of internet malls. Glob. J. Flex. Syst. Manag. **18**(1), 153–161 (2017)
55. Nouri, F.A., Nikabadi, M.S., Olfat, L.: Developing the framework of sustainable service supply chain balanced scorecard (SSSC BSC). Int. J. Product. Perform. Manag. **68**(1), 148–170 (2019)
56. Mohaghar, A., Janatifar, H., Dehghan, M.: Performance evaluation of green supply chain based on LFPP and balanced scorecard approach. Glob. J. Manag. Stud. Res. **1**(3), 158–163 (2014)
57. Rahman, M.H., Chin, H.C.: A balanced scorecard for performance evaluation of sustainable urban transport. Int. J. Dev. Sci. **2**(3), 1671–1702 (2013)
58. Teslya, A., Gutman, S.: Forming and developing a green transport corridor in the Arctic. IOP Conf. Ser. Earth Environ. Sci. **434**(1), (2020)
59. Péra, T.G., Bartholomeu, D.B., Su, C.T., Caixeta Filho, J.V.: Evaluation of green transport corridors of Brazilian soybean exports to China. Br. J. Oper. Prod. Manag. **16**(3), 398–412 (2019)
60. Staš, D., Lenort, R., Wicher, P., Holman, D.: Conceptual framework for assessing the green transport level in industrial companies and supply chains. Appl. Mech. Mater. **708**(1), 87–92 (2014)
61. Prause, G.: A green corridor balanced scorecard. Transp. Telecommun. **15**(4), (2014)

Green Smart City Intelligent and Cyber-Security-Based IoT Transportation Solutions for Combating the Pandemic COVID-19

Salma Ait Oussous, Fatima Zahra Hamza, Siham Beloualid, Abdelhadi El Allali, Abderrahim Bajit, and Ahmed Tamtaoui

Abstract This chapter presents the fruit of our research by merging smart transportation and smart health to provide IoT transportation solutions based on green smart city intelligence and safety to fight against the COVID-19 pandemic. For this, we have realized a model that allows transporting citizens via a means of transportation based on electric mobility to reduce energy consumption, reduce CO_2 emission and cost by searching the optimal path that this vehicle will be used. And to transport people, we need a system that allows us to check the health situation of citizens to avoid and prevent the spread of the COVID-19 pandemic. In order to find solutions to this work, we have proposed approaches to calculate the most optimal path that meets our needs, as well as to propose scenarios that allow checking the situation of the citizens. And to complete this work with minimum consumption of memory and time, we made a comparative study on the nodes used for the different IoT network topologies to choose the best one for our platform. Concerning the communication, we chose to use the CoAP protocol to ensure the communication between the nodes, and we used the AES-SHA256 encryption algorithm to compare it with RSA-SHA256 to ensure the elements of security and protection the data from any intrusion.

Keywords Smart transportation · Smart health · Green smart city · IoT network topologies · CoAP protocol · AES-SHA256/RSA-SHA256

S. Ait Oussous (✉) · F. Z. Hamza · S. Beloualid · A. El Allali · A. Bajit
Laboratory of Advanced Systems Engineering (ISA), National School of Applied Sciences, Ibn Tofail University, Kenitra, Morocco
e-mail: salma.aitoussous@uit.ac.ma

F. Z. Hamza
e-mail: fatimazahra.hamza@uit.ac.ma

A. Tamtaoui
Laboratory of Advanced Systems National Institute of Posts and Telecommunications, Department of SC, Mohammed V University, Rabat, Morocco
e-mail: tamtaoui@inpt.ac.ma

M. Lahby et al. (eds.), *Computational Intelligence Techniques for Green Smart Cities*, Green Energy and Technology, https://doi.org/10.1007/978-3-030-96429-0_6

1 Introduction

The fraction of the world's population that lives in cities has increased dramatically. And to meet the demands of inhabitants, the growth into urban areas necessitates the optimization of available resources such as power, water, housing, healthcare, and transportation. As a result, new technologies like artificial intelligence, cognitive technology, the Internet of Things, machine learning, and cloud computing are now being employed extensively to transform cities into "smart cities." Smart city building relies heavily on energy-efficient green communication and seamless networking to connect the various vital aspects of smart cities [1]. Currently, the transformation of digital cities into smart cities is a key source of concern for many countries. Intelligent and digital cities are technology-driven, but a city becomes "smart" when it is well-organized, interconnected, self-repairing, self-decision, and healthy.

In the literature, various models and explanations for smart city descriptions have been proposed [2]. A smart city architecture model consists of seven layers, each of which performs a critical role. The third layer plays an important role (i.e., the communication and networking layer). Infrastructure is the first layer, which includes roads, buildings, the electrical grid, and parking. From a personal area network (PAN) to a wide area network, the communication and networking layer facilitates and supports many forms of communication such as voice, text, data, video, multimedia, and real-time services (WAN). A question that arises is how communication and networking may help a city become smarter. The explanation is that this architecture's midway layer connects the top information and core framework to the infrastructure layer, and it is in charge of transferring services and features across one or several networks. This layer makes it easier for interconnected objects to communicate with one another. However, to apply easily integrated engineering to the design of smart cities, adequate communication must be established. Within the last decade, the smart city has become one of the most popular research subjects. The focus of research is currently on how green communication may be used to promote urban development and government administration.

Many applications have been urbanized for smart cities that use information and communication technology, among the most important applications are smart health and smart transportation. For the first, the IoT has a significant impact in the sector of health care. Patient data monitoring, detection, and sensing are some of the appliances. Some daily health needs, such as blood, ambulance, and so on, may be readily tracked using smart health technology. It can also be used to help prevent human errors in data collection [3]. However, smart transportation aids in the reduction of some traffic issues such as traffic congestion, the design of parking areas, and the use of traffic data for determining the arrival time at destinations [4].

The term "green" is associated with a number of goals, including long-term sustainability, adherence to a circular economy (recycle, reuse, reduce), and environmental protection, preservation, and recovery. Green technologies are a set of techniques or practices that limit the technical impact on the environment while also taking into account the recycling of many of the materials used in these processes, allowing people to profit from them.

Green transportation is defined as a mode of transportation that has no negative impact on the ecosystem or the environment. It makes life easier, reduces traffic, and reduces reliance on automobiles and foreign fuels. It is both safer and less expensive, and it would benefit the environment as a whole. People all across the world should shift to smarter and greener transportation due to environmental deterioration, depletion of natural resources, air pollution, greenhouse gas emissions, rapid depletion, and rising oil prices. The following are some of the major factors: air pollution, rising oil prices, depletion and extraction of natural resources, and reliance on oil for energy. People are currently working on a variety of green transportation systems that will have a low environmental impact by integrating sustainable transportation with green technology, such as electric-powered automobiles, hybrid electric vehicles, and compressed air vehicles [5].

In this work, we integrate smart transportation with smart health (Fig. 1) to come out with a good solution. The main goal is to encourage people to use a common smart transportation model based on electric mobility to reduce CO_2 emission, time, energy consumption, and the cost of transportation, by proposing an IoT platform for autonomous electric vehicles which will be informed by the optimal path that must select according to take a maximum people with a minimum stops in route and reduced costs. And considering that the several waves of pandemics of COVID-19 that we live and that changed our lives and our habits, we have reached a point where the reopening of public and private spaces is mandatory, but it is necessary to respect the sanitary rules and reduce the spread of the virus.

For this, we have proposed another IoT platform which will continue the first one and that will allow managing the access of the citizens to the autonomous electric vehicle. Access to this transport requires that people have a QR code obtained after the second dose of vaccination, or a negative PCR test dating less than 48 h. Without these conditions, the citizen is not allowed to access. The platform is divided into three tests that we will discuss in the proposed approach section.

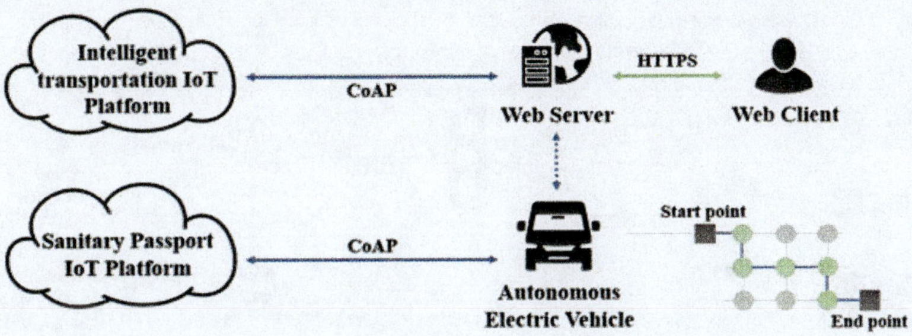

Fig. 1 IoT transportation platform for combatting COVID-19 architecture

Finally, for connecting the node devices to the internet, we devoted ourselves to the implementation of a CoAP protocol with an authentication server in the IoT. The intervention of encryption algorithms to solve the above-mentioned problems by encrypting and decrypting messages. We have therefore implemented a secure version of the CoAP protocol using the AES encryption algorithm.

2 Supervising IoT Transportation Platform for Combating Pandemic COVID-19

In our work, IoT is used to facilitate the collection of people in optimal conditions, focusing on face detection to detect the number of people's faces in each station (50 people maximum) to choose the optimal path.

Figure 2 shows the architecture of the intelligent transportation IoT platform. Firstly, we will use the autonomous vehicle to transport people, and this vehicle will search for the optimal path to save time and energy by collecting the number of faces in each station using nodes. This nodes will monitor the condition of the road if it is empty or full. The data collected will be sent to the Web server via the Proxy or Broker who plays as an intermediary.

Secondly, in order for citizens to access, an intelligent, secure, and synchronized sanitary passport, IoT platform (Fig. 3) was created to predict and combat the virus. So, first, we collect the data about the citizen to process its analysis and to act appropriately on the citizen's access rights. The system can examine a person's facial shape and compare it to information in a database before recognizing the person. At the space's entry, a PIR sensor detects the presence of a person. An RFID identification node used to identify data, and a node to monitor the citizen's temperature. We also use a camera to scan the QR code on the immunization pass and to identify the individual.

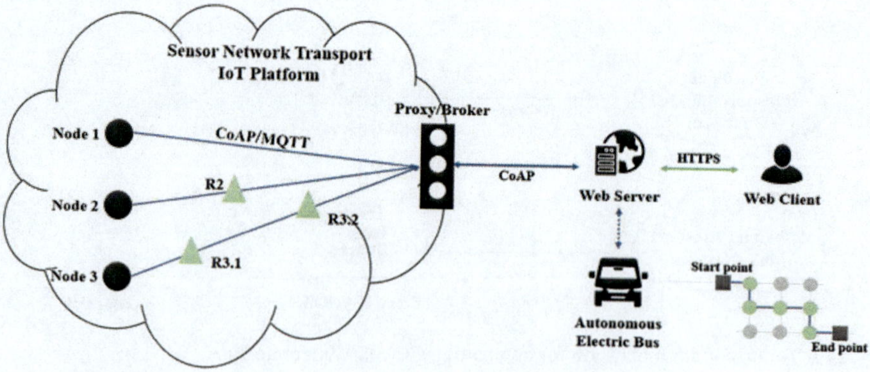

Fig. 2 Intelligent transportation IoT platform architecture

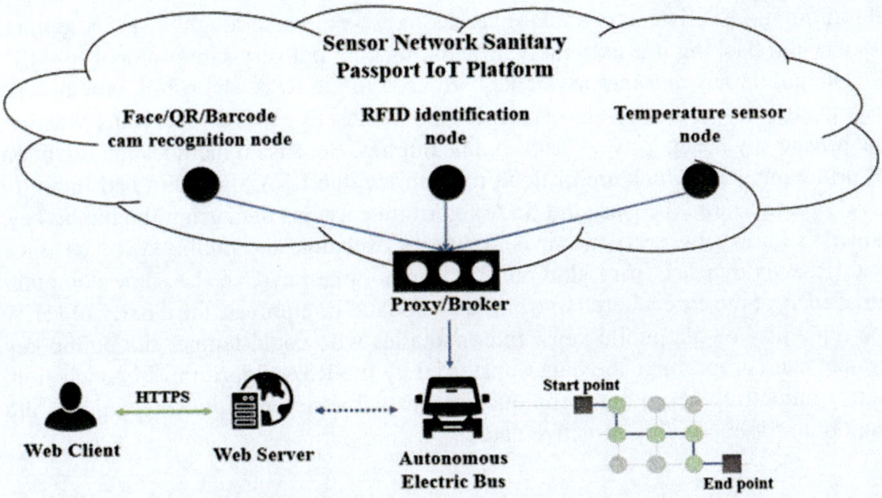

Fig. 3 Sanitary passport IoT platform architecture

Finally, to secure our data for both platforms, we used the AES-SHA256 encryption algorithms and the CoAP communication protocol to see its efficiency in real-time. And for networking, we applied a comparative study of different topologies to decide which is best for our work.

3 Methodology

In our work, we employed the CoAP protocol to ensure communication between nodes, and we used AES-SHA256/RSA-SHA256 data encryption algorithms to secure them.

3.1 Encryption Algorithms

The Advanced Encryption Standard (AES) is used to encrypt data and protect it from illegal access. This is accomplished using the cryptography process's variable-length key. AES encryption can be used to verify that data is authentic and originates from the location claimed. It also allows users to verify that the content has not been changed while being transmitted. AES is a widely used security technique that employs bigger key sizes for encryption, such as 128, 192, and 256 bits. It is used in a range of applications including wireless communication, financial transactions, e-commerce, encrypted data storage, and more. As for 128 bits, this makes the AES algorithm more resistant to cracking; breaking it would take roughly 2128 attempts. As a result, hacking it is extremely difficult, making it a very safe protocol [6–8].

The public key and the private key are used in the RSA algorithm, which is an asymmetric cryptography algorithm. The public key is given to everyone, whereas the private key is kept private, as the name implies. Because it includes factorization of prime integers, which are difficult to factorize, the RSA algorithm is difficult to crack. Furthermore, because the RSA algorithm encrypts data using the public key, which is known by everyone, it is simple to exchange the public key. The issue with RSA is that it is very slow in that it uses large keys, so the same computer must encrypt the case where large amounts of data. It requires a third party to verify the reliability of the public keys. Intermediaries who could temper the public key system can compromise the data transferred by the RSA algorithm. In conclusion, both symmetric encryption techniques and asymmetric encryption techniques are important for encrypting sensitive data [9].

3.2 CoAP Protocol

The CoAP protocol is used to overcome HTTP's limitations. CoAP, like HTTP, is based on a client–server approach, and it supports Http's REST APIs for retrieving data from sensors. CoAP messages are in binary format and are sent through UDP datagrams by default, which provides extra security [10].

CoAP is made up of two layers: Messages and request/response, with four different sorts of messages: confirmation, unconfirmable message, acknowledgment (ACK), and reset (RST). The first layer is in charge of UDP and messages, while the second is in charge of the request/response interaction based on those messages [11]. The confirmable message is a trustworthy message that is dependent on the retransmission timeout. With this form of message, the client can be confident that it will be sent to the intended recipient. The server sends a confirmation message with the same identification number as the confirmable message after it gets the data (CON). CoAP sends a non-confirmable message (NON) with a unique identification in the event of an unreliable transmission. The server does not require to acknowledge NON-messages [12, 13]. The CoAP protocol is used in IoT communication because it can send data in a limited network and allows connection-oriented transmission.

3.3 IoT Network Topologies

During our project, we applied different types of IoT topologies for networking, in order to determine which network is suitable for our application; for this, we focused on understanding the benefits and drawbacks of each.

The star topology is a topology when all nodes are separately connected to the central connection which might be a hub, router, or switch. As a result, every computer is indirectly connected to every other node via the hub. This topology has the advantage of being easy to install and discovering errors because the links are plainly recognized. However, it is quite expensive to use, and if a network connection or the hub fails, all of the nodes connected are disabled [14].

Tree topology, as the name implies, is a topology in which the connected elements are arranged in the form of tree branches. A tree topology combines two previous topologies: Bus topology, which connects all nodes to a single wire, and star topology, which connects nodes to a central hub device. There are two types of nodes in a tree topology: parent and child nodes, often known as network leaves. A single parent node can connect to two child nodes. It is employed in the network's expansion.

The benefits of this topology include the ability to search and sort a huge amount of data in the network, the use of less cable, and the fact that if one node is disconnected from the network, the others continue to function normally. The disadvantage of a tree topology is that if the main bus fails or becomes detached, the network ceases to function. This topology is also difficult to set up, and the cable size, as well as the number of nodes employed, are limited [15].

When we talk about the cluster, it is similar to tree topology as stated in the 802.15.4 standards, because it is built on trees and has a simple father–child relationship-based hierarchical [16]. A cluster head (father) can be assigned to each cluster and communicate with the coordinator. In addition, when the network is created using association requests and replies, the clusters' parents operate as intermediate routers [17].

For mesh, it is used to ensure network communication dependability; there is no central connection point in this topology, and each node is connected to numerous other nodes; in fact, each node can send and receive messages from other nodes. The nodes act as relays, transmitting messages to their final destinations; the strength of this topology is the speed with which messages are sent; there are multiple connections, which means that each node can transmit and receive data from multiple nodes at the same time; and new nodes can be added without disrupting or interfering with existing nodes.

Mesh networking, because of its broad reach, can be used physically at various physical layers and logically at various logical layers. For physically applicable, it can be used in optical backbone networks, MHNs, access networks, fixed and wireless networks, or cognitive radio networks. Moreover, Mesh can be used rationally in peer-to-peer networks, dynamic virtual circuit networks, and overlay networks employing MPLs [18, 19].

4 Related Works

With the help of several distribution centers, the authors of this study [20] aimed to tackle the problem of optimizing the electric vehicle EV distribution path. The EV distribution path issue must then be optimized with various distribution channels and charging facilities in mind in order to develop EV distribution paths with high resilience, low susceptibility to uncertainty factors, and exact route-by-route schemes. To reduce transportation time, a robust EV distribution path optimization model with variable robustness is built based on Bertsimas' theory of robust discrete optimization. Using the three-segment mixed coding and decoding approach, the model is also solved using an improved three-segment genetic algorithm, such that the optimal distribution scheme initially includes all of the route-by-route data. This approach evaluated three challenges for the EV distribution path problem with several distribution locations.

Moreover, in this paper [21], the authors have tried to regularize the problem of pollution and time loss by using electric mobility. The IoT platform is used to detect the optimal path that the minibus will take to reduce CO_2, energy consumption, time, and cost of transportation, by detecting the faces of people in each station and applying approaches.

Then, in this work [22], the authors have tried to find a solution to access public and private spaces in view of the situation we are living with the pandemic of COVID-19. The IoT platform is used to manage the access of citizens in a public or private area to avoid contamination and to check the sanitary situation of people by applying tests and scenarios.

Based on these two topics of smart transportation and smart health, and the results obtained, we tried to combine these two topics into one to regulate the problems of the environment in order to have a green smart city by offering an autonomous vehicle while respecting the health situation of people to fight the COVID-19 virus.

5 The Proposed Approach

To solve the problems related to congestion, energy consumption, and CO_2 emissions, we worked to regulate them by creating a model of transportation, with a green energy source powering the charging stations, to get the optimal path by taking a maximum number of people avoiding the maximum stops. So, we applied two approaches. For the first algorithm approach, after identifying the number of people in each station, the bus will move ahead to each station one by one, stopping at each one except the one with no people, until the capacity (50 people) is reached, then it will proceed to the nearest tram station, allowing us to find the most efficient route.

Algorithm 1 The first approach algorithm for getting the optimal path

Require: $STATIONS \leftarrow [St_1, St_2, ..., St_n]$
  $size \leftarrow 50$
  $sum \leftarrow \sum_1^n St_n$
  $K \leftarrow length(STATIONS)$
  for $i \leftarrow 1$ `to` k **do**
    if $sum > n_{max}$ **then**
      if $St_i \neq 0 \wedge St_{i+1} \leq n_{max} \wedge size \leq n_{max} \wedge size \leq size - St_i$ **then**
        Move the bus to St_i
        Stay 20s in St_i
      end if
    else
      if $St_i \neq 0 \wedge St_{i+1} \neq 0$ **then**
        Move the bus to St_i
        Stay 20s in St_i
      end if
    end if
    Move the bus to St_{i+1} (5s)
  end for

To better understand, Fig. 4, illustrates the two examples of getting the optimal path for this approach. The bus drives from one station to the next, and when station 9 is full, it transfers to the nearest station tram, which is s.tram 2. For the second example, the bus transports passengers from one station to another. And when it is full in station 13, it moves to s.tram 3.

Moreover, for the second algorithm approach, the Web server identifies the number of persons in each station, and we move to the three stations at minimum or to five stations at maximum.

Algorithm 2 The second approach algorithm for getting the optimal path

Require: $STATIONS \leftarrow [St_1, St_2, ..., St_n]$
  $n_{max} \leftarrow 50$
  $size \leftarrow 50$
  $sum \leftarrow \sum_1^n St_n$
  $Max \leftarrow MaxNumb(St, 5)$
  $SUM \leftarrow sum(Max_{1..3})$
  if $sum > size$ **then**
    Move the bus to St_{Max1}, St_{Max2}, St_{Max3}
  else
    if $sum(SUM, Max_4) > size$ **then**
      Move the bus to St_{Max1}, St_{Max2}, St_{Max3}
    else
      if $sum(SUM, Max_5) > size$ **then**
        Move the bus to St_{Max1}, St_{Max2}, St_{Max3}, St_{Max4}, St_{Max5}
      end if
    end if
  end if

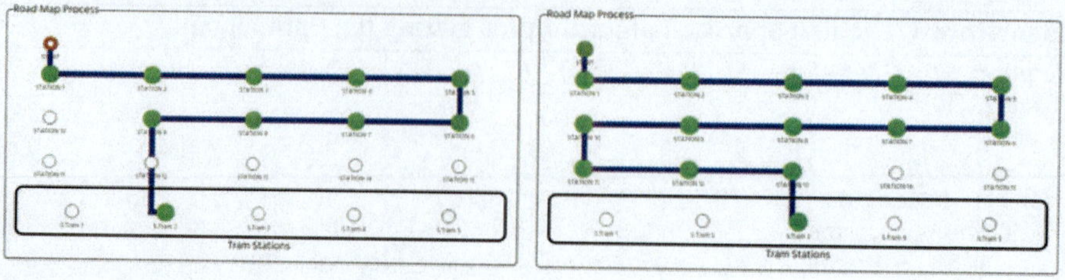

Fig. 4 Optimal path examples for the first approach

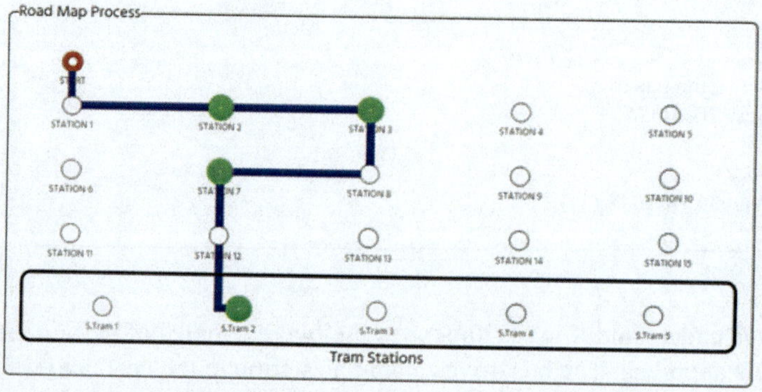

Fig. 5 Optimal path for the second approach with 3 stations maximum

The Web server will first classify the five largest numbers before calculating the sum of the three largest numbers. If the sum is less than the capacity, the sum of the four largest will be calculated, in this case, If the sum is greater than the capacity, it will move to the station with the three largest number of people, otherwise it will calculate the sum of the five largest; if the sum is greater than the capacity, it will move to the station with the four largest number of people, otherwise it will calculate the sum of the five largest; if the sum is smaller or equal to the capacity, it will move to the station with the five largest number; if the sum equals the capacity, it will be assigned to the station with the three largest numbers of people.

To make things clearer, we have an example in Fig. 5 which illustrates the first case. So as shown in the figure the number of personal data in each station was collected, then we calculate the optimal path, so we found the best route is from the starting position, passing from stations 2, 3, and 7.

The second case in Fig. 6 is for a number of people bigger than 50 for four stations maximum. So, we found the optimal path is from the start point, moving to stations 2, 4, 9, and 7.

The last case in Fig. 7 is for a number of people bigger than 50 for five stations maximum. The same treatment was done, and we get the best route is from the start point to station 1, selecting stations 4, 9, 10, and 11.

On the other hand, in order to control and check the people getting on the vehicle, we have proposed three scenarios to reduce the contamination of COVID-19.

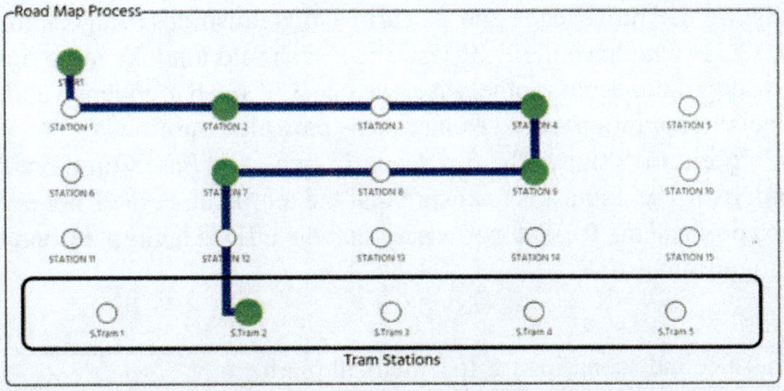

Fig. 6 Optimal path for the second approach with four stations maximum

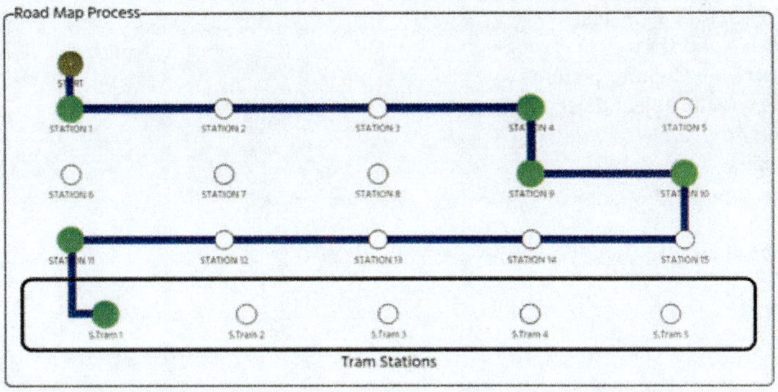

Fig. 7 Optimal path for the second approach with 5 stations maximum

Our algorithm's scenarios begin with the verification of people who have been vaccinated using two tests as shown in algorithm 3, scanning a QR code and using a facial recognition test. If these data are correct, we agree to give them authorized access; if one of these tests is invalid, for example, the citizen does not have a health pass, the access is denied.

Algorithm 3 First scenarios for IoT medical platform

```
authorized ← False
object_detection ← Get Iot object detection
if object_detection is detected then
    Qr_code ← Get Iot object QR Code
    if Qr_code is valid then
        picture ← Capture picture
        if picture is object detected then
            authorized ← True
        end if
    end if
end if
return authorized
```

Secondly, we will move to the 2nd scenario which illustrates in algorithm 4, which is for the citizens who have the PCR test; the test should not take more than 48 h, if it does, we deny them access; otherwise, we check if his bar code is valid, then we go to the facial recognition test to ensure these data are confirmed.

Finally, algorithm 5 shows the 3rd scenario, which is based on three tasks that must accomplish. The 1st test is to ensure that the temperature does not exceed 37o, the 2nd is to produce his RFID tag to verify that his UID is legitimate, and the 3rd is a facial recognition test.

Algorithm 4 Second scenarios for IoT medical platform

```
authorized ← False
object_detection ← Get Iot object detection
if object_detection is detected then
    PCR ← Get the PCR test
    if PCR is valid then
        picture ← Capture picture
        if picture is object detected then
            authorized ← True
        end if
    end if
end if
return authorized
```

Algorithm 5 Third scenarios for IoT medical platform

```
authorized ← False
object_detection ← Get Iot object detection
if object_detection is detected then
    Temperature ← Get the Temperature
    if Temperature is taken then
        RFID ← Get RFID
        if RFID is identified then
            picture ← Capture picture
            if picture is object detected then
                authorized ← True
            end if
        end if
    end if
end if
return authorized
```

6 Discussion and Results

The goal of this chapter is to apply the AES-SHA256 and RSA-SHA256 cryptographic algorithms to our work, as well as to develop it according to different topologies in order to analyze the impact of each one on the platforms, taking into consideration that CoAP is a high constraint protocol, so it is critical to preserve the power of IoT nodes while also reducing memory usage and execution time.

The comparative outcomes of the three scenarios offered by our platform are depicted in the following graphs. In the first scenario, we scan the QR code that is used in the vaccine pass, as well as using facial recognition to identify the person. A second scenario is one in which the PCR test is combined with facial recognition. A third scenario is dedicated to examining the body temperature of citizens after their recovery from COVID-19, as well as the UID stored in the RFID tag and facial recognition.

When comparing the security mechanisms of the three scenarios in the results Figs. 8, 9 and 10, it is definite that RSA is more costly in terms of memory occupation and execution time, as it takes time to execute and consumes more power and memory, compared to AES, which is more functional in our case as it has almost

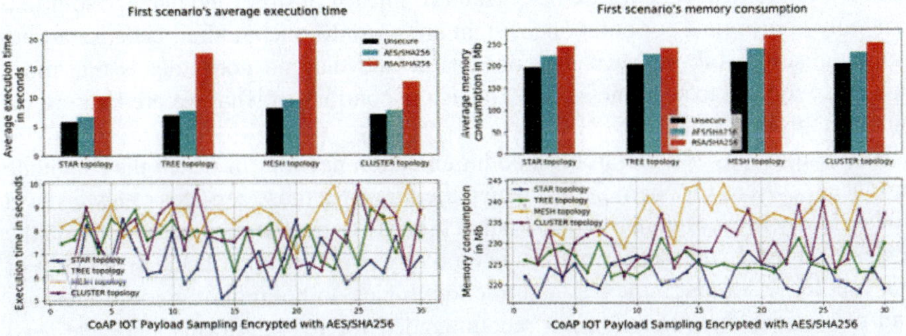

Fig. 8 The average execution time and memory consumption during the first scenario

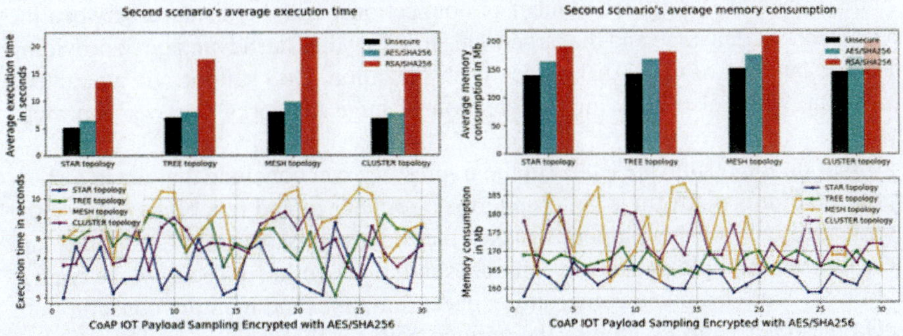

Fig. 9 The average execution time and memory consumption during the second scenario

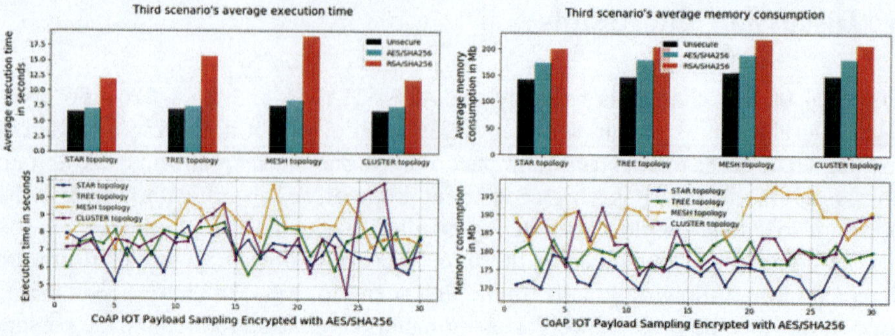

Fig. 10 The average execution time and memory consumption during the third scenario

no effect on platform performance while also being a powerful encryption method. Furthermore, AES is energy, time, and memory-efficient, processing data swiftly, providing excellent performance, encrypting and decrypting huge amounts of data quickly, and requiring little resources and memory.

When the results of each topology are examined, it is clear that the star topology has the best memory and time outcomes when compared to the other topologies. Although star is a dependable, efficient, and simple to construct and manage solution, it is only appropriate for components that are close to one another, because a long communication link between the coordinator and the end node means that more power is required to send messages. This is the contrary of what we are looking for on our platform.

A tree topology is a hierarchical communication network in which the root node produces services for client nodes, which then transfer those services to lower-level nodes. Every layer of nodes in this architecture can construct a star network with the nodes it serves. In this scenario, the tree topology structure inherits the drawbacks of star topology. Although the star and tree topologies do not require a lot of memory and processing time, they are not recommended for IoT platforms since we need real-time and rapid responses without losing data in this field, and if we opt between mesh and cluster tree. We will choose mesh because it is based on characteristics like a node's residual energy, the number of connections, and the distance between the nodes' devices or a node and the target. This topology has the advantage of providing all of the benefits of the signal mode synchronization, QoS support via guaranteed time slots, while also allowing the creation of huge networks to cover enormous areas.

Based on the results presented in the figures, we can conclude that using a mesh topology for this platform is better in terms of efficiency and reliability because the interconnectivity of the endpoints makes them extremely resistant to failures, and there may be multiple paths between nodes in the network; for security, the configuration is secure from any compromise, execution time, and memory consumption.

On the other hand, to choose the optimal path for the vehicle, we have studied the impact of different topologies in terms of execution and memory. Based on the

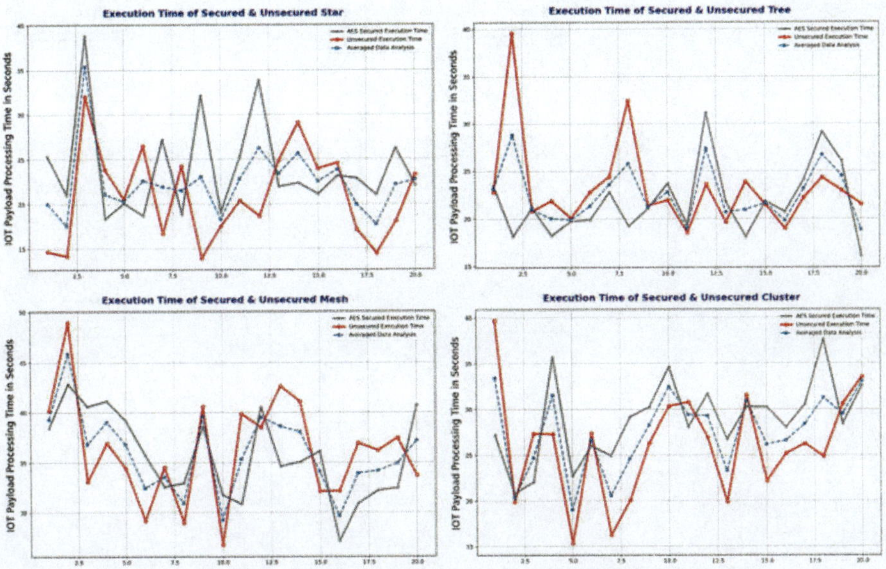

Fig. 11 Execution time by topology and security layer for the intelligent transportation IoT platform

results obtained above, and according to the results of those articles [23, 24], we found that AES-SHA256 is better than RSA-SH256, since it consumes less memory and time. So we opted to secure the data with AES-SHA256 for the second platform.

From Figs. 11 and 12, we found that topologies take time to execute and memory in secured mode compared to unsecure mode which is normal. But we note that the mesh topology takes longer to execute and consumes more memory compared to the other topologies.

In addition, from Fig. 13, we can see that the average execution time of the mesh topology is higher than the other also. However, for the average memory consumption, we noticed that there is not much difference between the topologies. The increase of time and memory of the mesh topology can be explained by its strong structure that looks like a backbone where all the nodes are connected to each other which allows them to communicate and transmit a large amount of data, as well as the low loss of it in case of failure of one of the nodes, unlike the others that are very fast in transmitting the data, consume less memory and less time, but they are easy to fail. For this reason, we can suggest the mesh topology for this work.

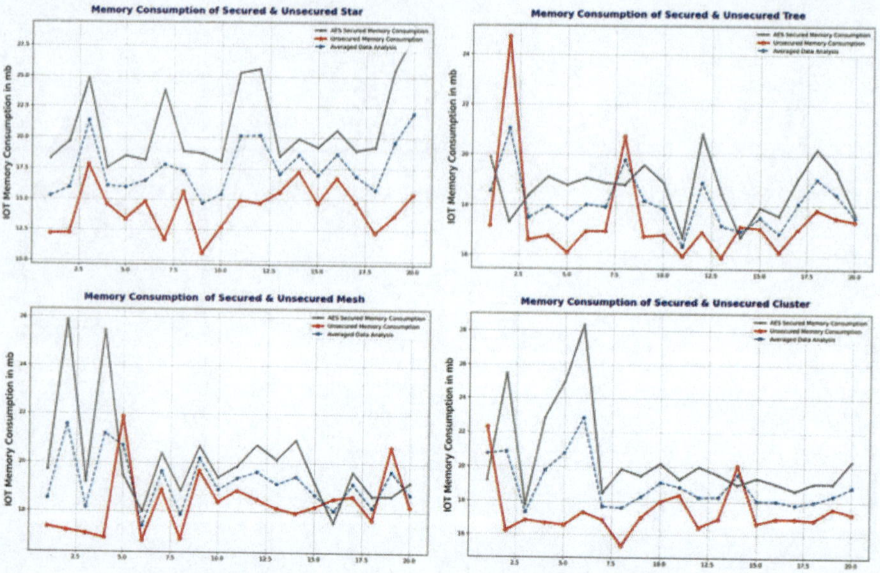

Fig. 12 Memory consumption by topology and security layer for the intelligent transportation IoT platfrom

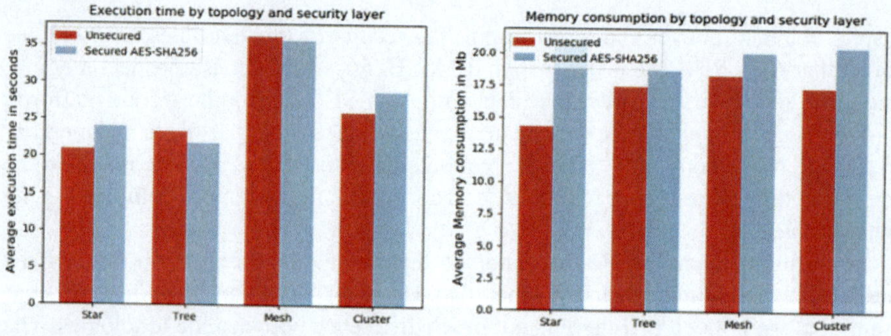

Fig. 13 Average execution time and memory consumption by topology and security layer for the intelligent transportation IoT platfrom

7 Conclusion

In this work, we have been able to combine smart transportation and smart health to come up with a solution that is based on the green smart city model and cyber-security while respecting health regulations to combat COVID-19. We were able to create two platforms, one for transportation and the other to check the health status of the people riding in the minibus. We used real IoT nodes to have a real-time response, we chose the CoAP protocol to ensure the communication between the nodes, we used the mesh topology to ensure the reliability of this communication and for the security of the data, and we opted for the AES-SHA256.

In future work, we will implement this work using the 6LoWPAN protocol as well as deploying another security method namely ECC, in order to propose the most efficient, secure, and optimal protocol for this platform. Not forgetting to look for other feasible solutions.

References

1. Tomar, P., Kaur, G.: Green and Smart Technologies for Smart Cities, 1st edn., p. 381. Springer International Publishing (2019)
2. Wenge, R., Zhang, X., Dave, C., Chao, L., Hao, S.: Smart city architecture: a technology guide for implementation and design challenges. China Commun. IEEE **11**(3), 56–69 (2014)
3. Sharma, R., Mishra, M., Nayak, J., Naik, B., Pelusi, D.: Green Technology for Smart City and Society: Proceedings of GTSCS 2020, pp. 597/604. Springer International Publishing (2020)
4. Shafie-Khah, M., et al.: Optimal trading of plug-in electric vehicle aggregation agents in a market environment for sustainability. Appl. Energy **162**, 601–612 (2016)
5. Srivastava, A., Gupta, M.S., Kaur, G.: Green smart cities, 1st edn. In: Green and Smart Technologies for Smart Cities, pp. 18–200 (2019)
6. Barodi, A., Bajit, A., Benbrahim, M., Tamtaoui, A.: Improving the transfer learning performances in the classification of the automotive traffic roads signs. E3S Web Conf. Proc. (2020)
7. Zhou, J., Wang, Y., Ota, K., Dong, M.: AAIoT: accelerating artificial intelligence in IoT systems. IEEE Wirel. Commun. Lett. **8**(3), 825–828 (2019)
8. Lu, C.-C., Tseng, S.-Y.: Integrated design of AES (advanced Encryp-tion Standard) encrypter and decrypter. In: Proceedings IEEE International Conference on Application—Specific Systems, Architectures, and Processors, San Jose, CA, USA, 2002, pp. 277–285. https://doi.org/10.1109/ASAP.2002.1030726
9. El Aidi, S., Bajit, A., Barodi, A., Chaoui, H., Tamtaoui, A.: An optimized security vehicular internet of things -IoT-application layer protocols MQTT and COAP based on cryptographic elliptic-curve. In: 2020 IEEE 2nd International Conference on Electronics, p. 9314579. Optimization and Computer Science, ICECOCS, Control (2020)
10. Kothmayr, T.: A security architecture for wireless sensor networks based on DTLS. Master's Thesis in the Software Engineering Elite Graduate Program at the University of Augsburg, 2011
11. Giuseppe, N., Maria Carla, C.: Security of IoT Application Layer Protocols: Challenges and Findings. MDPI (2020)
12. Chanwit, S., Chatchai, K.: Congestion Control in CoAP Observe Group Communication. MDPI (2019)
13. Naik, E.N.: Choice of effective messaging protocols for IoT systems: MQTT, CoAP, AMQP and HTTP. In: 2017 IEEE International Systems Engineering Symposium (ISSE), 2017, pp. 1–7. https://doi.org/10.1109/SysEng.2017.8088251
14. Star Topology: Advantages and Disadvantages, Penna Sparrow. https://www.ianswer4u.com/2011/05/star-topology-advantages-and.html
15. Estrada, R., Tomasi, C., Schmidler, S.C., Farsiu, S.: Tree topology estimation. IEEE Trans. Pattern Anal. Mach. Intell. **37**(8), 1688–1701 (2015)
16. Ouadou, M., Zytoune, O., Aboutajdine, D., El Hillali, Y., Menhaj-Rivenq, A.: Improved Cluster-tree Topology Adapted for Indoor environement in Zigbee Sensor Network, 2016, pp. 272–279. https://doi.org/10.1016/j.procs.2016.08.041

17. Chatterjee, A., Mukhopadhyay, A.K., Mukherjee, D.: A transport protocol for congestion avoidance in wireless sensor networks using cluster-based single-hop-tree topology. Third Int. Conf. Emerg. Appl. Inf. Technol. **2012**, 389–393 (2012). https://doi.org/10.1109/EAIT.2012.6407982
18. Network topologies, protocols and layers. https://www.bbc.co.uk/bitesize/guides/zr3yb82/revision/1
19. Plageras, A.P., Psannis, K.E., Ishibashi, Y., Kim, B.: IoT-based surveillance system for ubiquitous healthcare. In: IECON 2016—42nd Annual Conference of the IEEE Industrial Electronics Society, 2016, pp. 6226–6230. https://doi.org/10.1109/IECON.2016.7793281
20. Ma, C., Hao, W., He, R., Jia, X., Pan, F., Fan, J., et al.: Distribution path robust optimization of an electric vehicle with multiple distribution centers. PLoS One **13**(3), e0193789 (2018)
21. Yachou, M., Ait Oussous, S., El Harrouti, T., Beloualid, S., El Aidi, S., EL Allali, A., Bajit, A., Tamtaoui, A.: Applying advanced IOT network topologies to enhance intelligent city transportation cost based on a constrained and secured applicative IOT CoAP protocol. In: The International Conference on Information, Communication and Cybersecurity (ICI2C'21) (2021)
22. Hamza, F., El Aidi, S., Beloualid, S., El Allali, A., Bajit, A., Chaoui, H., Tamtaoui, A.: Applying advanced IOT network topologies to optimize COVID-19 sanitary passport multi-scenarios platform based on a constrained and secured applicative CoAP protocol. In: The International Conference on Information, Communication and Cybersecurity (ICI2C'21) (2021)
23. El Aidi, S., Bajit, A., Barodi, A., Chaoui, H., Tamtaoui, A.: An Advanced Encryption Cryptographically-Based Securing Applicative Protocols MQTT and CoAP to Optimize Medical-IOT Supervising Platforms. Lecture Notes on Data Engineering and Communications Technologies **72**, 111–121 (2021)
24. El Aidi, S., Bajit, A., Chaoui, H., Tamtaoui, A.: An advanced, synchronized and cryptographically AES-RSA-SHA encryptionBased protocols MQTT-CoAP applied to secure medical IoT platforms. Adv. Data Sci. Adapt. Anal. (ADSAA) J. Proc

Machine Learning and Green Health

Deep Learning-Based Convolutional Neural Network with Cuckoo Search Optimization for MRI Brain Tumour Segmentation

Kalimuthu Sivanantham

Abstract In recent scenario there is a huge requirement of image processing in various applications, namely, pattern recognition, Image compression, multimedia computing, remote sensing, secured image data communication, biomedical imaging and content-based image restoration. The medical image processing is the process and technique where the human body images are created for the purpose of medical field to examine, reveal or diagnose the diseases. The internal anatomy of the human body is visualized by the medical imaging technique without opening the body. Proposed research consist of various steps preprocessing used to remove noise, lung MRI images that are diagnosed by a radiologist are segmented using basic thresholding and morphological operations to extract the lung parenchyma. Next the ROIs of pleural effusion are extracted followed by the extraction of the ROIs of pneumothorax. Ten shape and texture features area, convex area, equivalent diameter, mean, eccentricity, solidity, perimeter, entropy, smoothness and standard deviation are extracted from the ROIs. The CNN is trained to identify the feature vectors belonging to the four class's pleural effusion, pneumothorax, normal lung and chest CT slices affected by other diseases. When the query MRI slice is applied, based on the training received, the classifies the query slice into the two classes for pneumonia or not. The classified result parameter optimized using Cuckoo search optimization algorithm (CSO). CSO algorithm a non-greedy local heuristic approach is used to solve optimization issues. The optimization results exhibit an accuracy of 94.18%.

Keywords Computed tomography · Convolutional neural network · Cuckoo search optimization · Segmentation algorithm · Feature extraction · MRI image · Application of medical imaging

K. Sivanantham (✉)
Crapersoft, Coimbatore, Tamilnadu, India
e-mail: sivanantham.kalimuthu@crapersoft.com

M. Lahby et al. (eds.), *Computational Intelligence Techniques for Green Smart Cities*, Green Energy and Technology, https://doi.org/10.1007/978-3-030-96429-0_7

1 Introduction

Images are processed by using basic techniques or steps either to remove unwanted data or structures such as noises, or to improve the quality of images for human visualisation. One of the important areas in IP is image segmentation. Segmentation of MRI brain image involves dividing or partitioning the image space into different cluster region with the same intensity image values [7]. Medical image mostly varies and are corrupted due to the presence of overlapping intensity values of different tissue regions. In the simplest form most people use the camera in their mobile phones for capturing images as a record of places and people they are associated with, in their time-line of their life. The level and purpose of acquiring images are higher in scientific, engineering and medical fields. Images when systematically analysed with appropriate algorithms reveal many information that are not apparently seen in a naked eye. Digital image processing covers the entire spectrum of methods and techniques used to exploit the information carried by an image. To deal with all such techniques will be beyond the scope of the thesis. We restrict our discussion to segmentation in medical images. By using computers, the digital images are manipulated and hence it is called as digital image processing [8].

When an organ of a human body is injured, or deformity occurs or afflicted by a disease, a physician in the first instance would like to study the present shape and structure of the organ. A physician can inspect it externally, if the organ like eye, finger, nose, skin lies on the surface of the body. When the organ is inside the body, then a method is to be devised to know about its condition. One of the most commonly used method is to get an image of the organ and study it. Imaging technique is a non-invasive and non-destructive method. Imaging of a human organ is done by utilizing various techniques involving mechanical, electrical, magnetic and electromagnetic waves and fields. There are various medical imaging modalities such as ultrasound scan (US), X-ray, computed tomography (CT), single photon emission computed tomography (SPECT), photon emission computed tomography (PECT), magnetic resonance imaging (MRI) [17]. Each of the imaging modalities has its own merits and demerits. Imaging can be done by any one of the imaging modalities by the radiographer depending on the nature of the tissue to be investigated. A radiographer or an expert view the images and produce an impression about the tissues from the images and report to the doctor. The doctor, using that report, proceeds to diagnose the diseases.

The existence of a brain tumour is the greatest cause of cancer-related death in children and young people. The majority of primary CNS cancers are brain tumours, accounting for 85–90% of all cases. This year, an estimated 23,800 adults in the United States (13,450 men and 10,350 women) would be diagnosed with primary malignant tumours of the brain and spinal cord. In 2017, an estimated 26,070 primary malignant and 53,200 non-malignant cancers will be diagnosed in the United States. Females (24.46 per 100,000) have a higher rate than males (20.10 per 100,000) [30]. Every year, 40,000–50,000 people in India are diagnosed with a brain tumour. Children account for 20% of the total. Until a year ago, the percentage was at around

5%. More than 600,000 people worldwide are currently afflicted with a brain tumour. Although there are various causes for brain tumour, early detection has major impact in diagnosis [13]. The applications of image processing are as follows,

- Remote sensing
- Morphological image processing
- Microscope image processing
- Medical image processing
- Non-photorealistic rendering
- Lane departure warning system.

In image processing applications, the medical image processing application is used in our work. The ultrasound is a key medical imaging technique which is used for the imaging of soft tissues and organs. The ultrasound technique is a radiation free, non-ionizing and non-invasive technique. There are four major areas in medical image processing [14]. They are,

- Image creation
- Image visualization
- Image examination
- Image management.

The image formation can form a digital image matrix by capturing the image steps. The image analysis contains all the processing steps that are used for the interpretation abstract of biomedical images for quantitative measurements. The image management combines all the techniques which supply image data retrieval (access), arching, transmission, communication and effective storage. In medical field, the accuracy of the disease diagnosis plays vital role as it leads to further treatment of the patient. So the prime objective dissertation is to improve the diagnosing efficiency of the medical expert system by,

- Employing feature optimization techniques to select most significant feature subset in the medical data [16].
- Constructing various classifier models (two-class) to train and test the clinical data [20]
- Optimizing classifier parameters and fuzzy rules by using single and hybrid optimization techniques [21].

Consequently, the various medical MRI image classifications to be implemented and produced better efficiency algorithms are support vector machine [19], decision tree [22], K-nearest neighbor algorithm [23], naive Bayes [33] and artificial neural network algorithm [26].

1.1 Magnetic Resonance Imaging (MRI)

Magnetic Resonance Imaging (MRI) is an imaging technique that produces internal views of images in our bodies using magnetic and radio frequency pulses. MRI produces high-resolution pictures with a clear view of soft tissue and anatomic structures in the brain, such as grey and white matter. It is used for a variety of human tissue investigations, including brain malignancies, spine inflammation and measuring blood flow and heart function. MRI scans are used for obtaining high contrast and resolution three-dimensional (3-D) images of the human brain and other organs in our body. MRI does not emit any ionizing radiation and also does not require any contrast agents. In virtually every direction, MRI scanners can produce direct pictures of the human body. MRI is commonly used to scan brain tissues because it gives images with higher resolution and contrast than other imaging techniques. A collection of two-dimensional (2-D) images of the human head make up an MRI. A MRI volume is a collection of these 2-D slices [24]. A schematic view of an MRI scanner is shown in Fig. 1.

In this research work, Matlab with digital image processing tools are used for simulating the results and an efficient machine learning algorithms are used for processing of MRI image datas. Artificial Intelligence-based classification is a long-standing field showing continuous and vigorous growth. It reduces the dimension of the data and incorporates intelligent behaviour into machines and software. It is an interdisciplinary field which includes a number of sciences, professions and specialized areas of technology. Artificial Intelligence assists in the decision making process by performing data collection, treatment, processing, presentation, testing

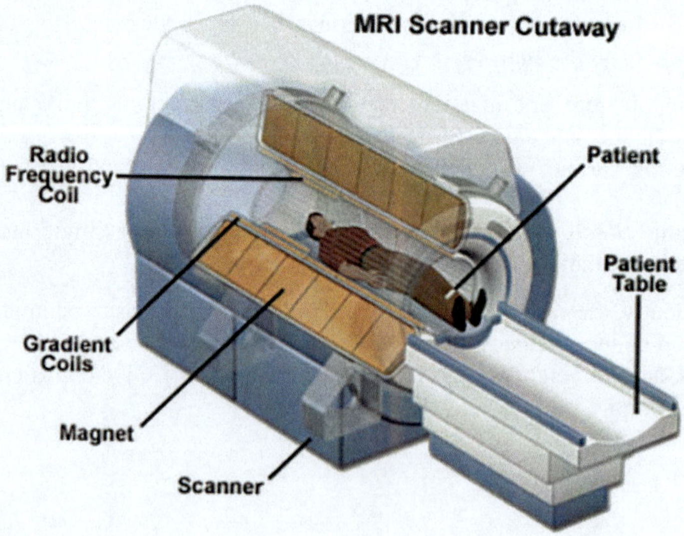

Fig. 1 A schematic view of a MRI scanner

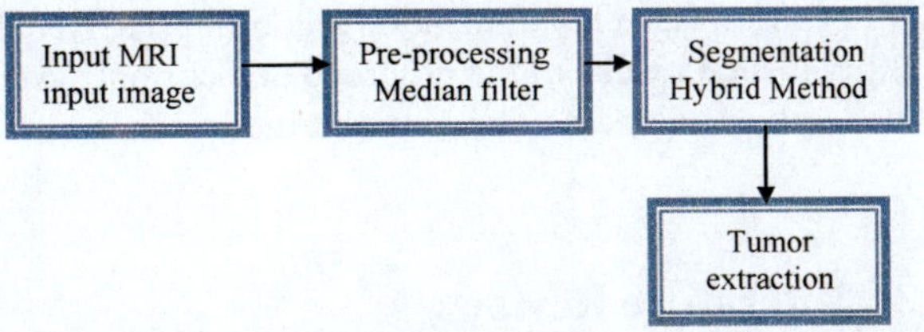

Fig. 2 General MRI image prediction for machine learning approach

and simulating new treatments, scenarios and devices. During training process, the presence of instances with missing values can lead to the degradation of accuracy and performance of the classification model. By dealing these missing values suitably the performance of the model can be improved. Case deletion is a simple and commonly used missing value handling techniques used to delete the instances with missing values. In the recent development, many meta-heuristic techniques have been inspired by the nature systems, especially computation-based algorithms named the evolutionary computing. The nature-inspired meta-heuristics techniques deals with complex computational problems and mainly used to find the best optimal solution for nonlinear problems. The CNN classified result parameter optimized using cuckoo search algorithm (CSO). Because of its non-greedy nature, this algorithm can achieve the global maxima without getting struck into local ones. Figure 2 explains the general block diagram for MRI brain cancer prediction using machine learning approach.

After acquiring a MRI of human head, neurologists inspect the series of 2-D slices to get information about the condition of tissues. Inspecting the slices manually will consume more time. Further, more and additional information can be obtained if the images are further processed using appropriate algorithms. An MRI slice contains not only brain tissues but also non-brain regions, scalp and skull. To study the brain tissues present in a MRI slice, the brain portion has to be segmented. Segmented brain portions can be easily processed and can be used for other application like compression, registration, etc. To segment and extract brain part from MRI of human head, several semi-automatic and automatic approaches have been developed. None of these approaches are capable of handling all image kinds in all orientations. Each has its own set of advantages and disadvantages. Brain extraction from MRI is still a difficult task. We offer four new strategies for extracting brain part from MRI of human head in this thesis [5].

The goal of this study is to provide a fully automated computer-assisted tumour detection approach for a real research gap in the field of medical MRI image processing, and to fill that gap by solving societal issues that arise in medical image analysis, reducing the need for manual contact. Efficient analysis of medical image is performed using the various bio-inspired techniques. This research apparently introduces several combinations of optimization and clustering techniques so as to ease the diagnosis; and detection of heterogeneous tumour effected region. This paper's reminder is organised as follows. Section 2: Prediction of MRI image datasets and related research Sect. 3 discusses the hybrid convolutional neural network algorithm with Cuckoo search optimization algorithm (HCNNCSO), and Sect. 4 compares the experimental results of the proposed and current systems. Finally, part five contains the work's concluding thoughts and future scope.

2 Literature Review

The problem of extracting brain portion from MRI of human head scans is a challenging task. In the beginning the segmentation process was done manually by an expert. Though segmentation by a human is still taken as gold standard, it consumed more time. Further, the segmentation results vary operator to operator. This process is also known as skull stripping process. As the computing facilities and several image processing algorithms were developed, researchers developed semi-automatic methods, in which human intervention is needed to run the algorithms. The refinements and improvements in semi-automatic methods lead to fully automatic methods. All segmentation methods either process in two-dimensional (2-D) or in three-dimensional (3-D) in MRI [25].

Hybrid approaches for brain segmentation are created by integrating two or more techniques. The researchers demonstrated a method for segmenting brain tissue from MRI images that combines three current techniques: expectation/maximization segmentation, binary mathematical morphology and active contour models. Eskildsen [6] used anisotropic filters, snake contouring approach and a priori information to create a fully automatic MRI brain segmentation system for axial PD/T2-W images. To handle coronal T1-W datasets, this approach was refined. This approach was designed for normal people, but it was unable to remove brains with aberrant anatomic features. A complex contouring algorithm is necessary to attain the desired results. Hassan et al. [9] used atlas-based active contour segmentation with a level set approach for skull stripping in T1-W, T1-contrast, T2-W, T2-flair and CT images. The hybrid watershed algorithm combines the watershed method and the deformable surface model (HWA). Using this technique, which is based on WM connectivity, the image is split into brain and non-brain components. The brain's boundary is then located in the image using a deformable surface model [29]. Researchers combined the results of various skull stripping techniques, such as BSE, BET, 3d intracranial and MRI watershed approaches, to extract the brain region from a T1-W image. Hwang and Sungon suggested a method for extracting brain from T1-W MR images (2019). It is a hybrid method that incorporates both the expectation maximisation (EM) algorithm and the active contour model. It also uses mathematical morphology and associated component analysis for final segmentation.

Using a combination of similarity and free-form transformation, Lee and Huh [18] suggested a method for 3-D segmentation of internal structures of the skull from MR images using a connectivity based method. For 3-D magnetic resonance images, an adaptive fuzzy segmentation technique was created. Zaho et al. [32] used a 3-D watershed algorithm to segment the picture into brain and non-brain components using an intensity based technique for T1-W images. However, it results in over-segmentation and noisy images, hence a skull stripping method based on graph cuts has been developed for T1-W images (GCUT). First, a threshold of intensity between the GM and the CSF is calculated. The graph cuts are then utilised to find a related sub mask and segment the image.

Iftikhar et al. [12] used SVM and ANN to combine disease dataset classifications. On the basis of accuracy and training time premise, an investigation was done between two systems. The Cleveland Heart Database and Stat log Database were used, which were downloaded from the UCI machine learning dataset vault. SVM and ANN were used to divide the data into two classes. In addition, it examined both dataset performances and used maximum a posteriori probabilistic (MAP) technique for identifying the lungs tumour. The proposed technique can be able to segment the portions of gliomas in the MRI slices. The proposed technique produced Dice Overlap Index (DOI) for tumours of high-grade data as 0.73, 0.56 and 0.5, and they can be further increased.

Huang et al. [10] and Deepak and Ameer [4] had demonstrated to analyse the 3-D anatomical structure of gallbladder pairs. In the developed model, the dense shape registration is implemented to the shape of the liver according to its complexity in gallbladder shape. In terms of multi-scale concavity and curvatures, the gallbladder shape is implemented in feature corresponding to the semantic shape decomposition. Hence, the set of liver-gallbladder CT data is analysed by the developed 3-D anatomical model. The experimental result shows that the in the developed model the data retrieval process is fast and effective in analysing the gallbladder pairs. The model has to be improved to incorporate it with other structures for data retrieval. Cherif [3] objectivized to optimize the study of cancer disease prognosis by using multiple data mining methods. The authors have offered a method to enhance the projected classifier pattern by feature selection. An feature selection approaches assisted to enhance the precision of every through minimizing few low ranked attributes that aided in attaining precision of 87.8, 86.80 and 79.9% in case of SMO, naïve Bayes and C4.5 decision tree algorithms correspondingly. Soni et al. [28] intended a system to identify the rules effectively to predict patients risk level on the basis of a provided parameter regarding their health. The rules had been prioritized on the basis of user's prerequisite. The system performance was assessed based on precision classification and the consequences revealed this system had greater potential to level more precisely. However, many methods exist for this objective which may further delay the working in fully automatic mode and to deliver much accuracy. The literature survey of various segmentation techniques for MRI brain image has been done. From the survey made, the major issues in the analysis of MRI image have been listed below.

a. Enormous volumes of data have to be diagnosed within less/minimum time duration.
b. The final decisions of treatment/surgery have been taken by getting the suggestion from two or more experts.

c. Identification/Segmentation of various pathologies in medical image is done manually by radiologists, leading to human errors, huge expenses and time consumption.
d. In the field of medical image analysis, several automatic techniques have been proposed by the researchers, and even though they was reliable, they have not focused on MR images with varied tumour dimensions of disproportioned boundaries with interceded noise signals. Dealing with noise images through typified and unique segmentation algorithms is the core-objective of this research.

3 System Design

In this research, we presents a work is done to design and evaluate approaches to handle missing values, attribute noise and imbalanced class distribution in datasets to predict. In this section, a brief description in HCNNCSO in knowledge discovery is presented. During the whole course, food resource is equal to the most optimist solution. This solution is worked on CSO algorithm by incorporating each individual pixel image element or particle selected region in medical image segmentation. The complexity of CSO algorithm lies in the behaviour pattern of each pixel and the volume served by each pixel. Figure 3 explain the proposed system architecture flow diagram for implementation, the steps to be followed for implementing HCNNCSO algorithm are as follows,

The main aim of CSO is to find pixel best or updated position that is, the pixel with different intensities are separated or highlighted by the algorithm. The evaluation of the fitness function results in finding the best position for pixels in the image. In general, each pixel represents a position in the 3-D space and follows flown technique [2]. In the search space, the best position is calculated iteratively by comparing the neighbourhood pixels in the multi-dimensional 3-D matrix of the preprocessed MRI brain image. Here in this work the optimized CSO along with the CNN are proposed. The blooming prominence and advancements seen in machine learning in the latest generation have inspired researchers to have a comprehensive investigation. There is multiples of promising datamining research issues, on which data classification is analysed as a major contest to focus on.

CSO starts with a bunch of random pixels, or solutions, and then updates generations to find the best solution. The particle is modified in each cycle by using the two best values. The first is the most effective option (fitness) thus far. This fitness value is likewise saved, and it is referred to as best p. The best value obtained so far by each particle in the population is another best value tracked by the CSO. This best value is referred to as best g because it is a global best. The particle swarm iterates until a stopping requirement is reached.

The differentiated intensity values are separated by the implementation of the objective function and updated with the help of CSO algorithm. On implementing CSO algorithm for the preprocessed MRI brain images, detection and prediction of

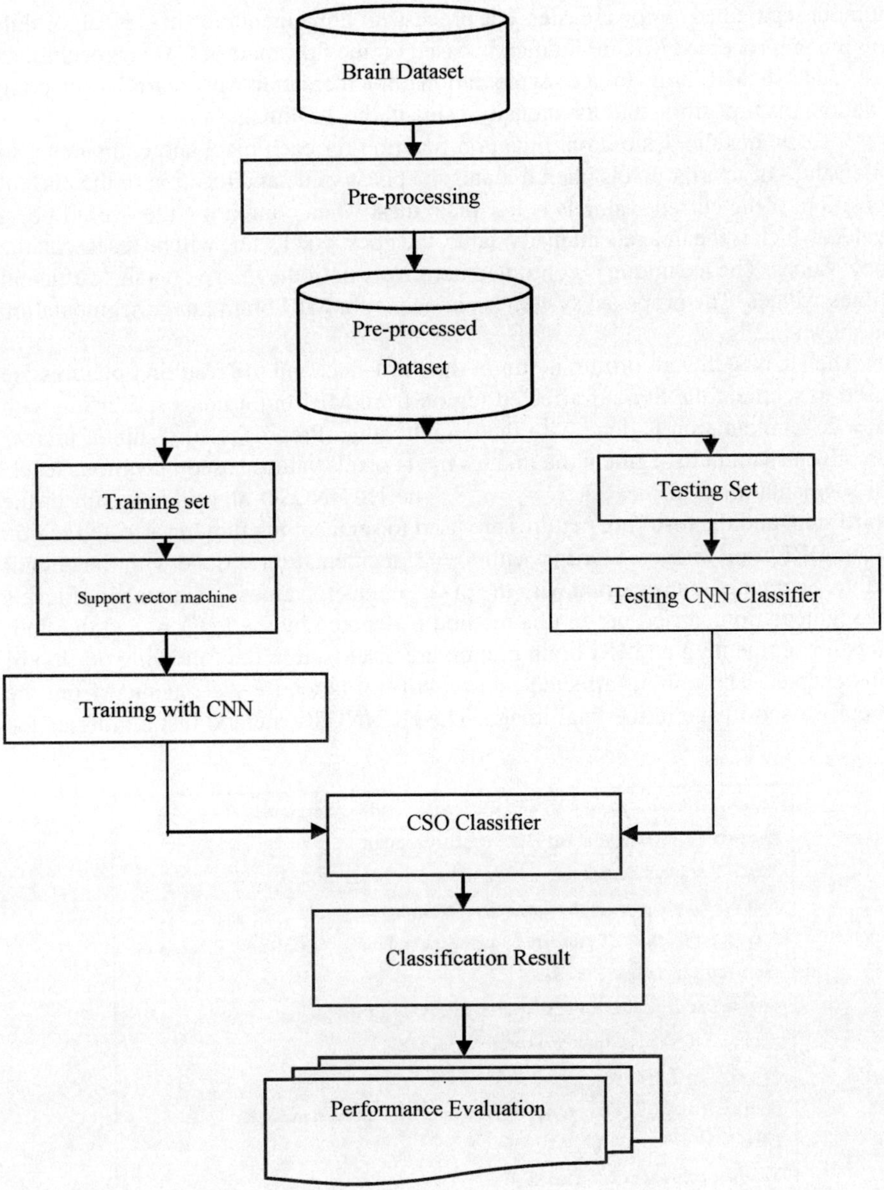

Fig. 3 System framework of HCNNCSO

tumour segmented region is easier. The process for implementation of CSO algorithm on the preprocessed MRI brain data is based on the flowchart of CSO algorithm, as given above. MRI brain image segmentation implementation work starts by selecting random pixel position and its intensity value in the 3-D image space.

It calculates the desired optimization function for each pixel and compares it to the values of nearby pixels, then updates the pbest value and location in the current position. If the current value is better than gbest when comparing the overall pbest value, which is the image's intensity value, the gbest pixel value will be reset with the new values. The technique is continued iteratively until the image contains sufficient fitness values. The proposed system pseudocode for MRI brain image segmentation is shown in Fig. 4.

The HCNNCSO algorithm is run in the third step, and the resulting pictures are used to segment the tumour-affected region from MRI brain images. Starting with $n = 2$, segmentation is done with the default value. Preprocessed resultant images are given as input to segment the images by its pixels values based on various levels of segmentation with $n = 2, 3, 4, \ldots, 8$. The HCNNCSO algorithm is run in the third step, and the resulting pictures are used to segment the tumour-affected region from MRI brain images. Starting with $n = 2$, segmentation is done with the default value. Segmentation is carried with the pixels intensity values. The number of levels of segmentation carried out in this method is denoted by the letter n, and the findings for segmenting an MRI brain picture are discussed in the following portion of this chapter. The tumour-affected region will not be entirely segmented from the preprocessed image in the final image. The HCNNCSO method just highlights the

Input: Medical Data set (Ds) with the features and the class label

Output: Feature Sub-Set For HCNNCSO Algorithm

Step1. Read middle slice

Step2. Find midpoint of the slice in the MR Image.

Step 3: Apply the SET property to define the regions having intensity values more than 180 adn assign to the Set.

Step 4: Draw the contour of the pixels satisfying the Set property.

Step 5: Is there more than one connected region?

IF yes,

then perform LCC (using eqn.(5.4)) and take LCC as brain mask(B)

Else

Take the contour as brain mask (B)

Step 6: Segment brain portion using B

Step 7: Repeat step 1 to step 6 for all slices from middle slice to top slices and middle slice to bottom slices.

Fig. 4 Pseudocode hybrids convolutional neural network with Cuckoo search optimization for algorithm contrast pixels in the generated image, i.e. the tumour afflicted region, and divides the image into black and white. The resultant image reveals the tumour-affected region of the MRI brain images approximately. The elapsed time and space complexity of HCNNCSO algorithm are also calculated for finding the efficiency of the algorithm. Figure 5 shows the segmented image as a 5 × 5 pixel rate and de-convolution techniques applied after produced binary values are displayed.

$$I_{\text{ero}} = I_{\text{bin}} \Theta s_6 \tag{1}$$

Equation 1 explained to the MRI eradicated image, the de-conventional digital images are again converted into the S6 binary valued mage it consist of total values of the edge-based values are detected and calculated to the overall detected values in that functions.

If a searching procedure in the CSO is executed, the ranger or the scrounger will have chances of discovering a location that is better and the current producer of other

$$\begin{pmatrix} 176 & 179 & 178 & 172 & 168 \\ 177 & 178 & 176 & 174 & 173 \\ 173 & 175 & 176 & 174 & 173 \\ 166 & 164 & 164 & 163 & 164 \\ 175 & 169 & 165 & 162 & 159 \end{pmatrix} \quad \begin{pmatrix} 177 & 180 & 179 & 170 & 166 \\ 178 & 178 & 175 & 174 & 174 \\ 173 & 177 & 178 & 176 & 174 \\ 163 & 161 & 162 & 161 & 164 \\ 175 & 168 & 163 & 161 & 157 \end{pmatrix}$$

$$\begin{pmatrix} 0 & 0 & 1 & 1 & 1 & 1 & 1 & 1 & 1 & 0 & 0 \\ 0 & 1 & 1 & 1 & 1 & 1 & 1 & 1 & 1 & 1 & 0 \\ 1 & 1 & 1 & 1 & 1 & 1 & 1 & 1 & 1 & 1 & 1 \\ 1 & 1 & 1 & 1 & 1 & 1 & 1 & 1 & 1 & 1 & 1 \\ 1 & 1 & 1 & 1 & 1 & 1 & 1 & 1 & 1 & 1 & 1 \\ 1 & 1 & 1 & 1 & 1 & 1 & 1 & 1 & 1 & 1 & 1 \\ 1 & 1 & 1 & 1 & 1 & 1 & 1 & 1 & 1 & 1 & 1 \\ 1 & 1 & 1 & 1 & 1 & 1 & 1 & 1 & 1 & 1 & 1 \\ 1 & 1 & 1 & 1 & 1 & 1 & 1 & 1 & 1 & 1 & 1 \\ 0 & 1 & 1 & 1 & 1 & 1 & 1 & 1 & 1 & 1 & 0 \\ 0 & 0 & 1 & 1 & 1 & 1 & 1 & 1 & 1 & 0 & 0 \end{pmatrix}$$

Fig. 5 **a** Image block of size 5 × 5 pixel, **b** De-convoluted values of the block, **c** structuring element S6 members will fail to discover a better location [1]. The factor of constriction is that one other variable that ensures convergence and over fitting being the problem acquires more specifications while training. The ranger or the scrounger having a better location in the next session and the producer and the other members in the previous search session carries out the activity of scrounging [31]. This fitness function is designated to ith individual is a least-squared error function as per Equation.

The error in the training set can be driven to a small value by means of minimizing the error function but as its side effect, the problems of over fitting may sometimes occur and result in a generalization error which may be large. The following equation are generated final calculation values for the entire HCNNCSO-based segmentation generations.

$$I_{\text{rough}}(x, y) = \begin{cases} 1 \text{ if } I(x, y)\, \varepsilon\, \text{LCA} \\ 0 \text{ otherwise} \end{cases}$$

$$I_B(x, y) = \begin{cases} I(x, y) \text{ if } I_{BM}(x, y) = 1 \\ 0 \quad\quad \text{otherwise} \end{cases}$$

As a result, the preceding halting tactics are recommended for increasing the performance of the CSO. During the training phase, the rate of error validation was monitored. If a validation error occurs for a specific set of iterations, the training will be completed.

4 Result and Discussion

The primary goal of this study is to identify the brain tumour-affected region using MRI brain imaging. In the process of finding tumour, CSO algorithm is implemented on images to detect tumour-affected regions. The inputs are taken from real patients' MRI brain 3-D images. Image categorisation and selection are specified and figures are displayed in the previous data selection, the experiments are carried out in the research work with MATLAB (R2008a) software. The system used in this study has an intelcore2Duo processor with 8 GB RAM and runs on the Windows 7 operating system. If no prior knowledge is provided, result validation is done before and after feature selection using learning methods in the dataset. The categorization performance of the reducts is used to make all of the comparisons. The absolute count of positive records for the infected person category and the perfectly recognised count of true positive records are used to calculate accuracy. As a result, the PSO algorithm is used with preprocessed MRI brain scans to identify the tumour afflicted region in the pictures, and accurate findings are obtained. These images will help physicians and radiologists provide better service to the medical community, and they can be used for clinical diagnosis and treatment planning. The suggested HCNNCSO output results achieved in each step implementation are detailed in Table 1.

Table 2 explain the model based on proposed HCNNCSO yields the better false positive, true positive, *F*-score for MRI image dataset of (28.15, 72.78 and 78.68), at most nearer random forest algorithm provides the only 41.25 of false positive, 68.35 of true positive and 76.54 of *F*-score values. Support vector machine provides the only 46.38 of false positive, 63.25 of true positive and 73.25 of *F*-score values. Logistic regression provides the only 51.25 of false positive, 48.65 of true positive and 71.71 of *F*-score values.

Table 1 Proposed HCNNCSO segmentation result

1	Input image	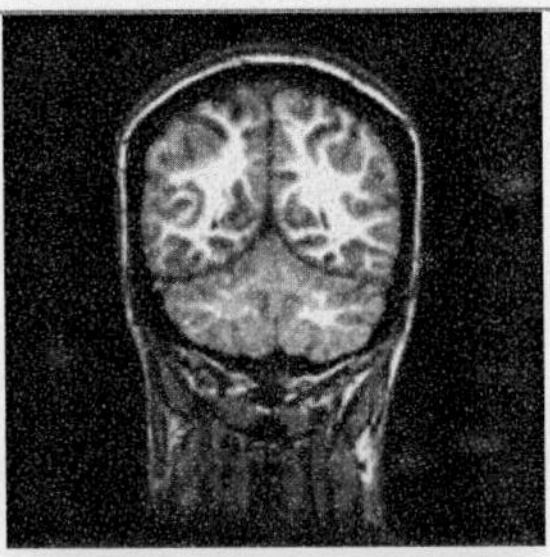
2	Find the a, b elements in SET A in all rows	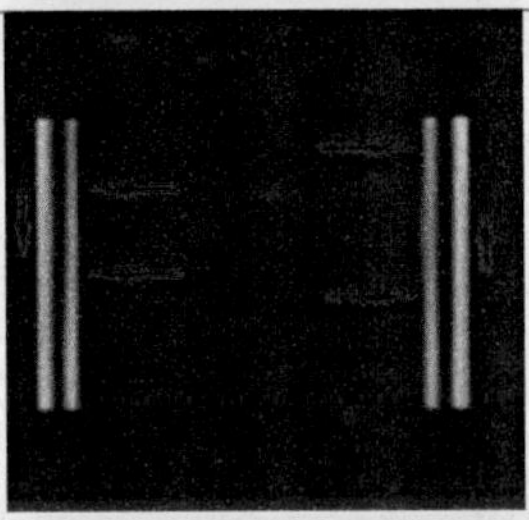
3	Point spread function	
4	De-convolution	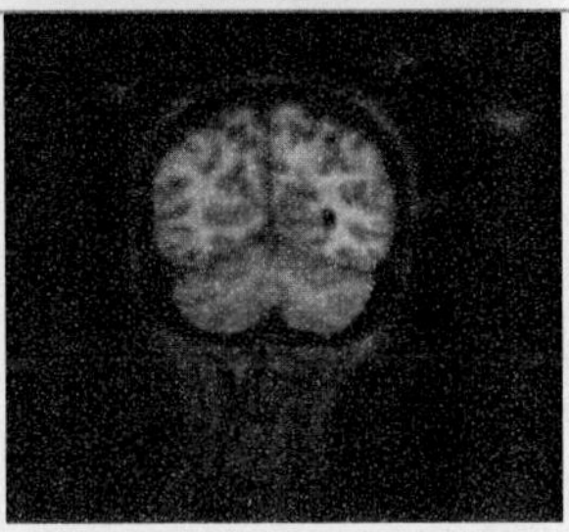

(continued)

Table 1 (continued)

5	Binary image	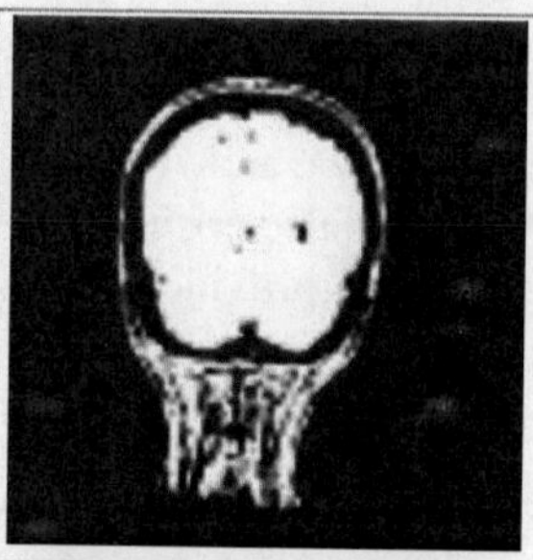
6	Eroded image	
7	Largest connected area	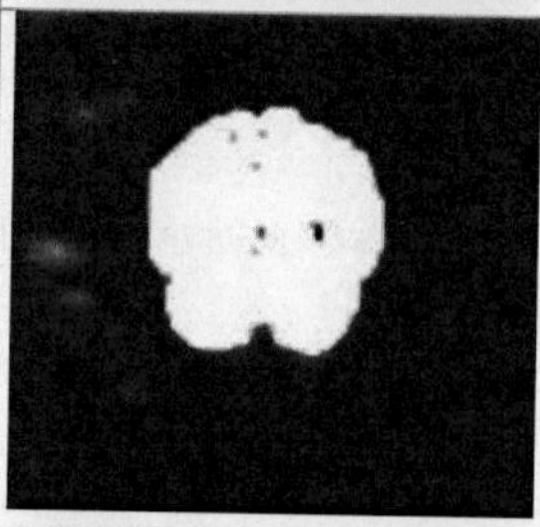
8	Dilated image	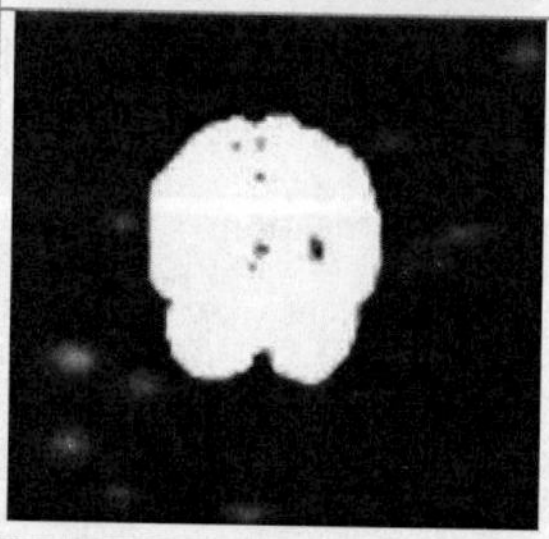

(continued)

While the proposed hybrid artificial neural network algorithm with chicken swarm optimization (HANNCSO) yields the quality matrix values for MRI dataset explained in Fig. 6. The false positive, true positive and F-score exposed to the comparatively hybrid convolutional neural network with Cuckoo search optimization is better than random forest, support vector machine and logistic regression.

Table 3 explains the model based on proposed HCNNCSO yields the better accuracy, sensitivity and precision for MRI image dataset of (94.18, 95.32 and 96.37),

Table 1 (continued)

9	Dilated image	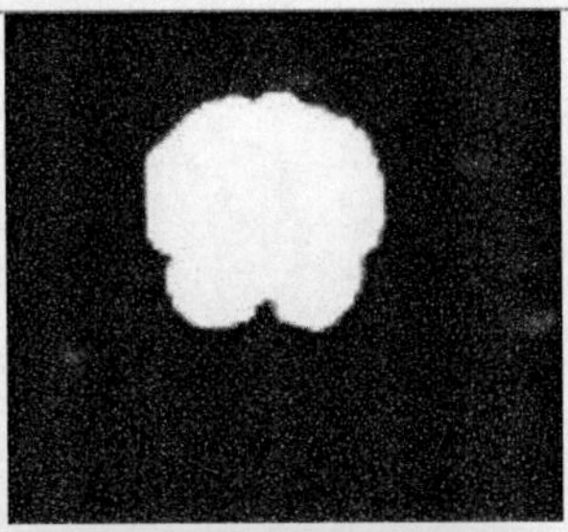
10	Constructing the contour for the Connected component	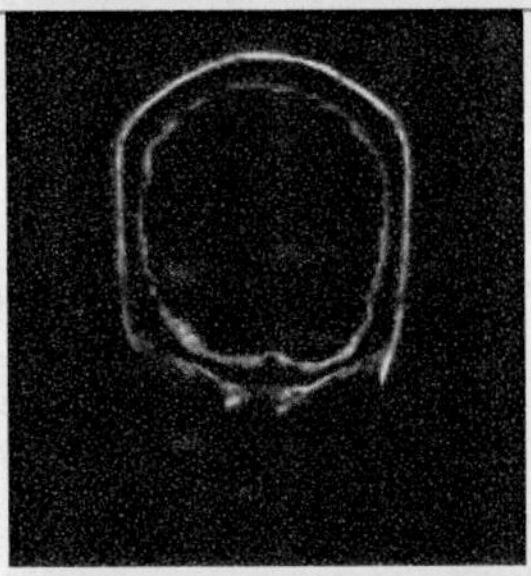
11	Brain mask	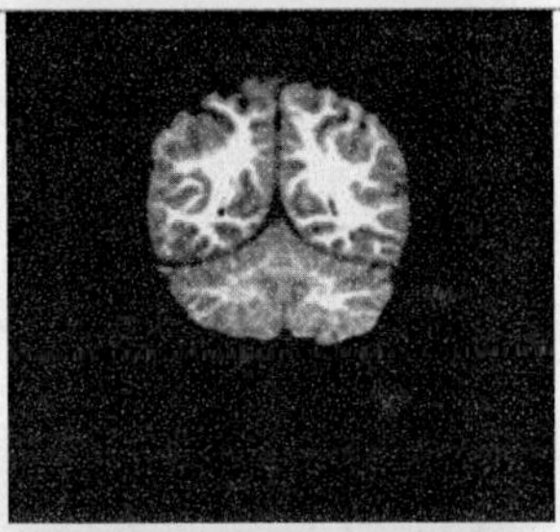
12	Extracting the brain portion	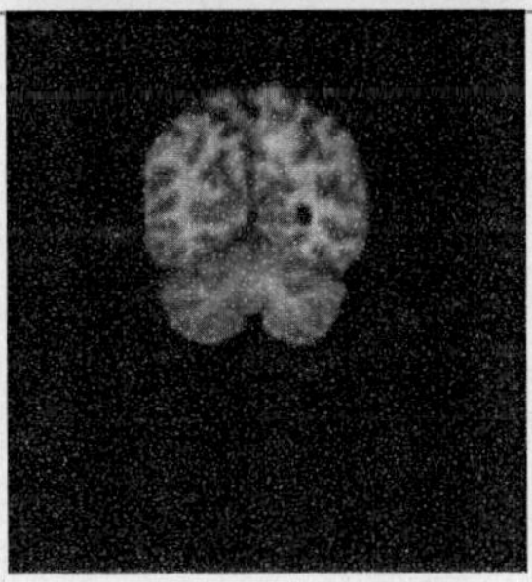

Table 2 Performance comparison brain MRI dataset

Methods	False positive	True positive	F-score
HCNNCSO	28.15	72.78	78.68
Random forest	41.25	68.35	76.54
Support vector machine	46.38	63.25	73.25
Logistic regression	51.25	48.65	71.71

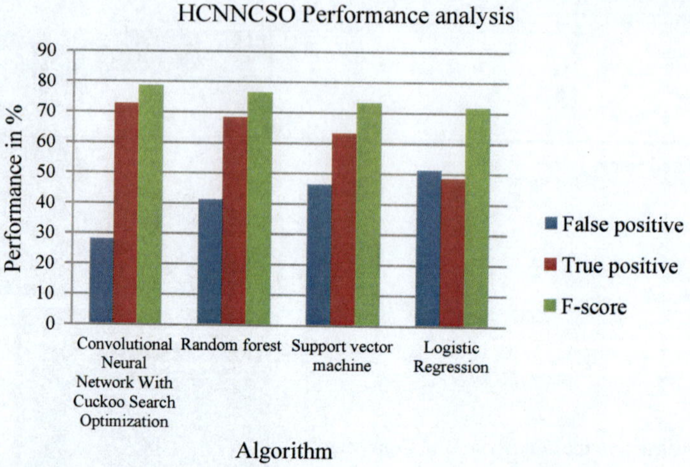

Fig. 6 Performance comparison brain MRI dataset

Table 3 HCNNCSO performance analysis

Approach	Accuracy	Sensitivity	Precision
HCNNCSO	94.18	95.32	96.37
Random forest	79.89	81.25	83.65
Support vector machine	78.65	79.25	81.65
Logistic regression	63.12	69.65	80.47

at most nearer random forest algorithm provides the only 79.89 of accuracy, 81.25 of sensitivity and 78.65 of precision values. Support vector machine provides the only 79.89 of accuracy, 79.25 of sensitivity and 81.65 of precision values. Logistic regression provides the only 63.12 of accuracy, 69.65 of sensitivity and 80.47 of precision values.

While the proposed hybrid artificial neural network algorithm with chicken swarm optimization (HANNCSO) yields the quality matrix values for MRI dataset explained in Fig. 7. The accuracy, sensitivity and precision exposed to the comparatively hybrid convolutional neural network with Cuckoo search optimization is better than random forest, support vector machine and logistic regression.

Table 4 explains the model based on proposed HCNNCSO yields the better accuracy, in minimum time duration for MRI image dataset of (14.56 s), at most nearer support vector machine algorithm 19.35 s to achieve desired performance. Random forest algorithm 21.56 s to achieve desired performance. Logistic regression algorithm 33.45 s to achieve desired performance.

While the proposed hybrid artificial neural network algorithm with chicken swarm optimization (HANNCSO) yields the quality matrix values for MRI dataset explained in Fig. 8. The running time exposed to the comparatively hybrid convolutional neural

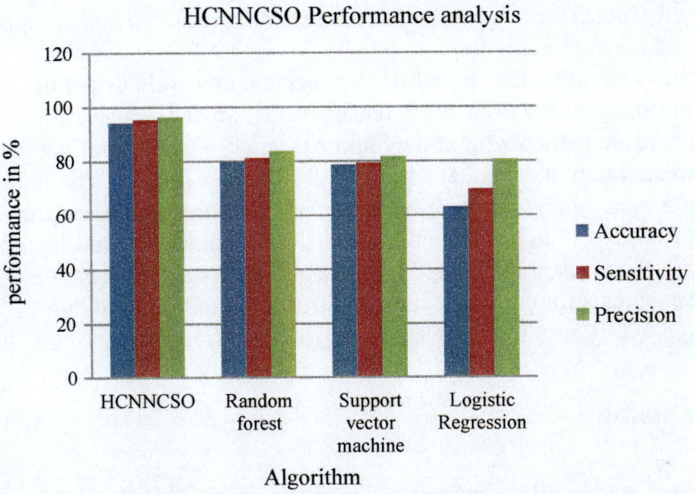

Fig. 7 HCNNCSO performance analyses

Table 4 HCNNCSO time duration analysis

Approach	Running time (s)
HCNNCSO	14.56
Random forest	21.56
Support vector machine	19.35
Logistic regression	33.45

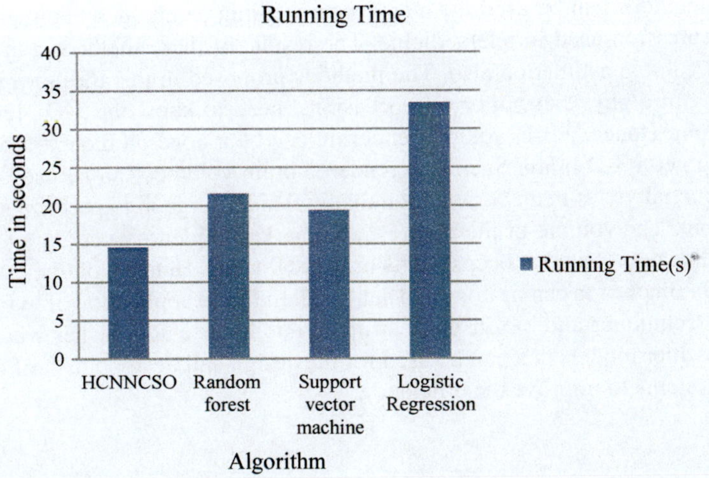

Fig. 8 HCNNCSO time duration analysis

network with Cuckoo search optimization is better than random forest, support vector machine and logistic regression.

The results are discussed based on the segmentation values and the algorithms segment the images based on the intensity values of the images. PSO and FCM algorithms are implemented and they approximately identify and detect the brain tumour-affected region separately from the MRI brain images. The final resultant images time, space and white pixel variation are measured and noted for the further comparison of algorithm results in the succeeding chapters. The results produced by the MRI brain images are validated by medical experts and assists in medical field to diagnose patients for early detection and treatment of brain tumour.

5 Conclusion

A big data patient record classification has been performed in this work for medical dataset by using HCNNCSO method, hybrid optimization-based method and max-margin classifier. The proposed method outperformed the widely used HCNNCSO method, but it also outperformed random forest, support vector machine and logistic regression. The proposed strategy, on the other hand, is more consistent in terms of performance from one volume to the next. The transform constructed from HCNNCSO can be further extended 5×5 size or higher to detect edges of the boundary, more accurately. The proposed methods can be used to study the brain injuries and brain deformities. By making use of these methods, the doctors can diagnose the disease quickly and can deliver fast health services to the public. All the four methods can be used for many preprocessing techniques like registration, ROI compression used in telemedicine. The results of these methods can be used for brain volume estimation also. The methods proposed in this thesis work for 2-D slices. But a physician, in certain occasions, need to know the 3-D view of the whole brain. Hence, a 3-D volume rendering can be a done on the extracted brain portion to get a 3-D brain. Such 3-D rendered brain volumes can be used to make cross-sectional views in any orientation that will aid a surgeon to perform surgeries in the brain. The volume of the brain is a biomarker to identify certain age related problem like dementia that occur adults of over 60 years. Hence, volume estimation from the brain portion can be done and analysed. In future application of hybrid optimization techniques and bio-inspired artificial intelligence approaches would yield better classifier models that can be used for the design and development of decision support systems to improve the efficiency.

References

1. Akanskha, E., Sahoo, A., Gulati, K., Sharma, N.: Hybrid classifier based on binary neural network and fuzzy ant colony optimization algorithm. In: 2021 5th International Conference

on Trends in Electronics and Informatics (ICOEI), pp. 1613–1619. IEEE (2021)
2. Chandra, S., Bhat, R., Singh, H.: A PSO based method for detection of brain tumors from MRI. In: 2009 World Congress on Nature & Biologically Inspired Computing (NaBIC), pp. 666–671. IEEE (2009)
3. Cherif, W.: Optimization of K-NN algorithm by clustering and reliability coefficients: application to breast-cancer diagnosis. Proc. Comput. Sci. **127**, 293–299 (2018)
4. Deepak, S., Ameer, P. M.: Brain tumor classification using deep CNN features via transfer learning. Comput. Biol. Med. **111** 103345 (2019)
5. El-Dahshan, E.S.A., Mohsen, H.M., Revett, K., Salem, A.B.M.: Computer-aided diagnosis of human brain tumor through MRI: a survey and a new algorithm. Expert Syst. Appl. **41**(11), 5526–5545 (2014)
6. Eskildsen, S.F., Coupé, P., Fonov, V., Manjón, J.V., Leung, K.K., Guizard, N., et al.: BEaST: brain extraction based on nonlocal segmentation technique. NeuroImage **59**(3), 2362–2373 (2012)
7. Gordillo, N., Montseny, E., Sobrevilla, P.: State of the art survey on MRI brain tumor segmentation. Magn. Reson. Imaging **31**(8), 1426–1438 (2013)
8. Greenspan, H., Van Ginneken, B., Summers, R.M.: Guest editorial deep learning in medical imaging: overview and future promise of an exciting new technique. IEEE Trans. Med. Imaging **35**(5), 1153–1159 (2016)
9. Hasan, A.M., Meziane, F., Aspin, R., Jalab, H.A.: Segmentation of brain tumors in MRI images using three-dimensional active contour without edge. Symmetry **8**(11), 132 (2016)
10. Huang, W., Xiong, W., Zhou, J., Zhang, J., Yang, T., Liu, J., et al.: 3D shape analysis for liver-gallbladder anatomical structure retrieval. In: International MICCAI Workshop on Computational and Clinical Challenges in Abdominal Imaging, pp. 178–187. Springer, Berlin, Heidelberg (2012)
11. Hwang, H., Rehman, H.Z.U., Lee, S.: 3D U-Net for skull stripping in brain MRI. Appl. Sci. **9**(3), 569 (2019)
12. Iftikhar, S., Fatima, K., Rehman, A., Almazyad, A.S., Saba, T.: An evolution based hybrid approach for heart diseases classification and associated risk factors identification. Biomed. Res. **28**(8), 3451–3455 (2017)
13. Islam, M.R., Rishad, N.: Effects of filter on the classification of brain mri image using convolutional neural network. In: 2018 4th International Conference on Electrical Engineering and Information & Communication Technology (iCEEiCT), pp. 489–494. IEEE (2018)
14. Jannin, P., Krupinski, E., Warfield, S.K.: Validation in medical image processing. IEEE Trans. Med. Imaging **25**(11), 1405–1409 (2006)
15. Kalimuthu, S., Naït-Abdesselam, F., Jaishankar, B.: Multimedia data protection using hybridized crystal payload algorithm with chicken swarm optimization. In: Multidisciplinary Approach to Modern Digital Steganography, pp. 235–257. IGI Global (2021)
16. Karegowda, A.G., Jayaram, M.A., Manjunath, A.S.: Feature subset selection problem using wrapper approach in supervised learning. Int. J. Comput. Appl. **1**(7), 13–17 (2010)
17. Kasban, H., El-Bendary, M.A.M., Salama, D.H.: A comparative study of medical imaging techniques. Int. J. Inf. Sci. Intell. Syst. **4**(2), 37–58 (2015)
18. Lee, C., Huh, S.: Unsupervised segmentation of 3D brain MR images. In: Applications of Digital Image Processing XXI, vol. 3460, pp. 687–694. International Society for Optics and Photonics (1998)
19. Machhale, K., Nandpuru, H.B., Kapur, V., Kosta, L.: MRI brain cancer classification using hybrid classifier (SVM-KNN). In: 2015 International Conference on Industrial Instrumentation and Control (ICIC), pp. 60–65. IEEE (2015)
20. Mazurowski, M.A., Habas, P.A., Zurada, J.M., Lo, J.Y., Baker, J.A., Tourassi, G.D.: Training neural network classifiers for medical decision making: the effects of imbalanced datasets on classification performance. Neural Netw. **21**(2–3), 427–436 (2008)
21. Melin, P., Olivas, F., Castillo, O., Valdez, F., Soria, J., Valdez, M.: Optimal design of fuzzy classification systems using PSO with dynamic parameter adaptation through fuzzy logic. Expert Syst. Appl. **40**(8), 3196–3206 (2013)

22. Naik, J., Patel, S.: Tumor detection and classification using decision tree in brain MRI. Int. J. Comput. Sci. Netw. Secur. (ijcsns) **14**(6), 87 (2014)
23. Rajini, N.H., Bhavani, R.: Classification of MRI brain images using k-nearest neighbor and artificial neural network. In 2011 International Conference on Recent Trends in Information Technology (ICRTIT), pp. 563–568. IEEE (2011)
24. Rousseau, F., Glenn, O.A., Iordanova, B., Rodriguez-Carranza, C., Vigneron, D.B., Barkovich, J.A., Studholme, C.: Registration-based approach for reconstruction of high-resolution in utero fetal MR brain images. Acad. Radiol. **13**(9), 1072–1081 (2006)
25. Rundo, L., Militello, C., Tangherloni, A., Russo, G., Vitabile, S., Gilardi, M.C., Mauri, G.: NeXt for neuro-radiosurgery: a fully automatic approach for necrosis extraction in brain tumor MRI using an unsupervised machine learning technique. Int. J. Imaging Syst. Technol. **28**(1), 21–37 (2018)
26. Sharma, M., Purohit, G.N., Mukherjee, S.: Information retrieves from brain MRI images for tumor detection using hybrid technique K-means and artificial neural network (KMANN). In: Networking Communication and Data Knowledge Engineering, pp. 145–157. Springer, Singapore (2018)
27. Sivanantham, K.: Sentiment analysis on social media for emotional prediction during COVID-19 pandemic using efficient machine learning approach. Comput. Intell. Healthcare Inf., 215–233 (2021)
28. Soni, J., Ansari, U., Sharma, D., Soni, S.: Predictive data mining for medical diagnosis: an overview of heart disease prediction. Int. J. Comput. Appl. **17**(8), 43–48 (2011)
29. Ségonne, F., Dale, A.M., Busa, E., Glessner, M., Salat, D., Hahn, H.K., Fischl, B.: A hybrid approach to the skull stripping problem in MRI. Neuroimage **22**(3), 1060–1075 (2004)
30. Tarhini, G.M., Shbib, R.: Detection of brain tumor in mri images using watershed and threshold-based segmentation. Int. J. Signal Process. Syst. **8**(1), 19–25 (2020)
31. Vieira, G., Bockheim, J., Guglielmin, M., Balks, M., Abramov, A.A., Boelhouwers, J., et al.: Thermal state of permafrost and active-layer monitoring in the antarctic: Advances during the international polar year 2007–2009. Permafrost Periglac. Process. **21**(2), 182–197 (2010)
32. Zhao, Y., Guo, S., Luo, M., Liu, Y., Bilello, M., Li, C.: An energy minimization method for MS lesion segmentation from T1-w and FLAIR images. Magn. Reson. Imaging **39**, 1–6 (2017)
33. Zhou, X., Wang, S., Xu, W., Ji, G., Phillips, P., Sun, P., & Zhang, Y.: Detection of pathological brain in MRI scanning based on wavelet-entropy and naive Bayes classifier. In International Conference on Bioinformatics and Biomedical Engineering, pp. 201–209. Springer, Cham (2015)

Role of Deep Learning for Smart Health Care

Moiz Khan Sherwani, Abdul Aziz, and Francesco Calimeri

Abstract Deep learning has become one of the trendiest fields in recent years. Nonetheless, it is an appealing and hard undertaking to do tasks related to computer vision, natural language processing, speech recognition, bioinformatics, and smart health care. Acquiring information and significant knowledge from complex, high-dimensional, and heterogeneous biomedical data sources remains a critical test in changing health care. Different kinds of data have been arising in present-day biomedical exploration, including electronic health records, imaging, sensor data, and information, which are perplexing, heterogeneous, inadequately explained and for the most part unstructured. Conventional data mining and machine learning approaches regularly need to initially perform feature designing to acquire robust and more vigorous features from those data and afterward build prediction or clustering models on top of them. There are several severe difficulties on the two stages in the case of complex data and lacking adequate domain information. Based on machine learning, medical IoT devices, detection and diagnosis through imaging, automated surgeries and other several techniques have been in use in real-world and some are still under development. A few potential issues like security, QoS improvement, and arrangement demonstrate the vital piece of deep learning. We reveal working principle, deep learning for health care including images and text, smart devices and privacy issues in health care data.

Keywords Smart health care · Deep learning · Internet of medical things · Privacy issues

M. K. Sherwani (✉) · F. Calimeri
Department of Mathematics and Computer Science, University of Calabria, Rende, Italy
e-mail: sherwani@mat.unical.it

A. Aziz
Department of Software Engineering, Dalian University of Technology, Dalian, China

M. Lahby et al. (eds.), *Computational Intelligence Techniques for Green Smart Cities*, Green Energy and Technology, https://doi.org/10.1007/978-3-030-96429-0_8

1 Introduction

Health care is one of the long-term and ever-evolving disciplines that has profoundly influenced many aspects of life. Information technology advancements have entered in a new era of innovation. Since last decade global health care market is dependent on Artificial Intelligence (AI) based application as described in Fig. 1. Deep learning is a revolutionary and new generation of information technology in the field of smart health care. The enormous accessibility of biomedical data brings tremendous freedoms to health care research. Specifically, investigating the relationship among the various data samples in the datasets is a fundamental issue in developing reliable medical instruments dependent on data-driven methodologies and Machine Learning (ML). To this point, past works attempted to connect various data sources to fabricate joint information bases that could be utilized for predictive analysis and discovery [1–3].

Even though existing models exhibit extraordinary guarantees, predictive tools dependent on Machine Learning procedures [4–7] have not been generally applied in the health care environment. There are many difficulties in utilizing the biomedical data, attributable to their high dimensionality, heterogeneity, temporal dependency, sparsity, and inconsistency. These difficulties are additionally puzzled by different medical ontologies used to generalize the data (for example, Systematized Nomenclature of Medicine (SNOMED-CT) [8], Unified Medical Language System (UMLS) [9], International Classification of Disease-ninth variant (ICD-9) [10]), which frequently contain clashes and irregularity [11].

A typical methodology in a biomedical research is to have a domain expert to determine the aggregates to use in a specially appointed way. Nonetheless, the directed meaning of the element space scales ineffectively and messes up the chances to find novel examples. On the other hand, representation learning techniques permit naturally finding the representations required to predict from the raw data. Deep learn-

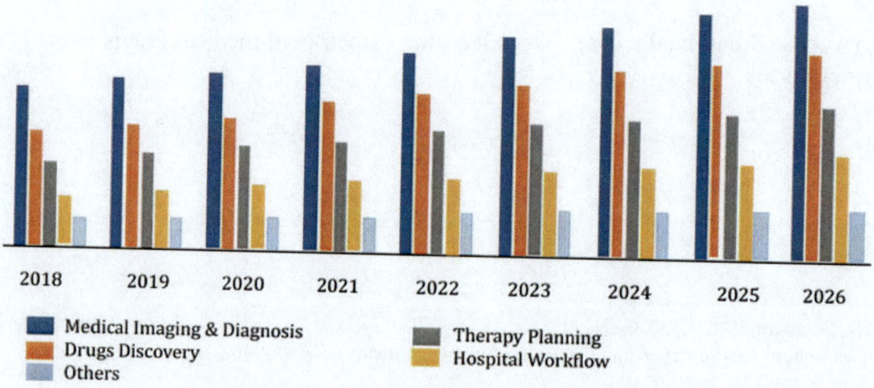

Fig. 1 Global health care market artificial intelligence market by application

ing (DL) strategies are representation learning calculations with numerous degrees of representation, created by creating specific yet nonlinear modules that change the representation at one level (beginning with the raw contribution to) a representation at a higher level, somewhat more conceptual level. Deep Learning models exhibited extraordinary execution and potential in PC vision, speech recognition and natural language processing. Given its demonstrated performance in various areas and the quick advances of methodological upgrades, Deep Learning ideal models present stimulating new freedoms for biomedical informatics. Attempts to apply Deep Learning techniques to health care are in practice or in progress. For instance, Google DeepMind has reported designs to apply its skill to health care and Enlitic is utilizing Deep Learning insight to spot health issues on medical images (X-ray, Computer Tomography (CT), and Machine Resonance Imaging(MRI)).

However, Deep Learning approaches have not been widely assessed for a wide scope of medical issues that could profit from its abilities. There are numerous parts of Deep Learning that could be useful in health care, like its superior performance, end-to-end learning scheme with integrated feature learning, the capability of handling complex and multi-modality data. To speed up these efforts, the Deep Learning research field, all in all, should address few difficulties identifying with the attributes of health care data (i.e. sparse, noisy, heterogeneous, time-dependent) as a need for improved techniques and apparatuses that empower deep Learning to interface with smart health care data work processes.

2 Working Principle of Deep Learning for Health Care

In conflict with conventional Machine Learning calculations, Deep Learning is filled by massive data measures and requires magnificent quality machines with GPUs to run in a reasonable time. In routine Machine Learning procedures, most of the applied features should be distinguished by a domain expert to diminish the intricacy of the data and make designs more noticeable to learning calculations to work.

The most significant benefit of deep Learning calculations is that they attempt to take in significant level features from data without any assistance. This hypothetically takes out the requirement for domain mastery and no unnecessary feature extraction. In complex issues where a significant degree of robotization is required, there is an absence of domain understanding for feature designing deep Learning methods are soaring to never-seen levels of exactness. Several applications of artificial intelligence in the field of health care is described in Fig. 2

3 Deep Learning for Health Care in Practice

By handling many data from different sources like X-rays, Computed Tomography (CT), Machine Resonance Imaging (MRI), genomic data and electronic health

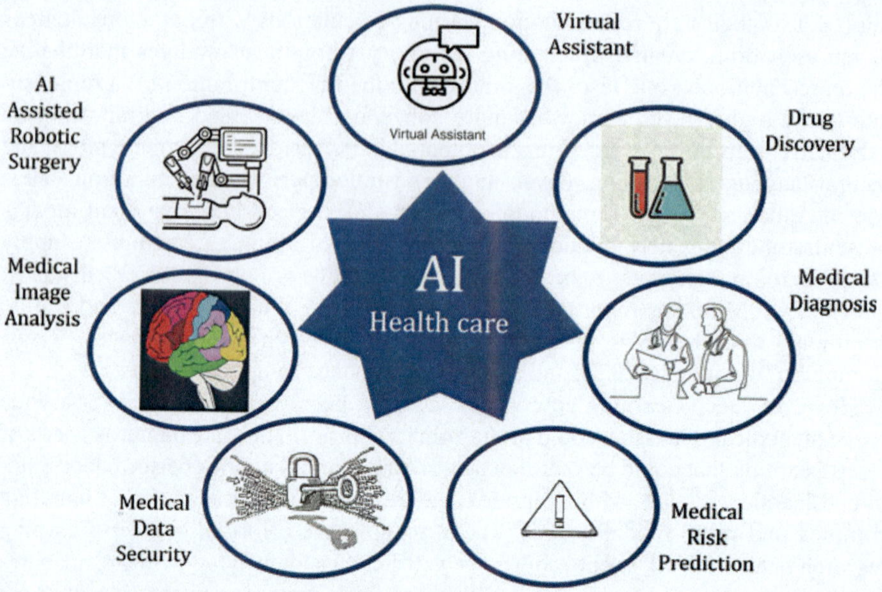

Fig. 2 Application of artificial intelligence in health care

records. Deep Learning can help doctors dissect data and distinguish numerous health conditions, attempting to address many required health care concerns like lessening the pace of misdiagnosis and foreseeing the result of methods.

Here there are some notable medical regions where Deep Learning is presently flaunting:

Medical Image Analysis: The reception of Convolutional Neural Networks (CNNs) essentially works on the early identification of disease, arriving at extremely high correctness in issues like breast cancer recognition [32], brain tumor [19] and so on medical images. In this field, deep Learning calculations are drawing significant performance. A detailed overview is provided in Sect. 3.1.

Predictive Models for Electronic Health Records: Predictive data model displaying with electronic health records (EHR) is expected to drive customized medication and improve medical service quality. Developing predictive models regularly requires the extraction of curated indicator factors from standardized EHR information, a concentrated work cycle that disposes of most of the data in every patient's record. EHR records are dependent on the Fast Health care Interoperability Resources (FHIR) design. Profound learning techniques utilizing this model are prepared to do precisely anticipating various medical occasions from numerous focuses without site-explicit data harmonization [33]. Detailed overview is provided in Sect. 3.2

Wearable devices: With fewer specialists, the effectiveness of medication will turn out to be progressively essential. Wearable devices are another wilderness for medical services, giving memorable longitudinal information and constant physiological 24-hour observing; and early recognition of persistent conditions and intense infection beginning dependent on customized physiological baselines. Wearable devices can permit specialists, medical attendants and other medical staff to screen patients–regardless of whether they are in the clinic or at home. With patients' physiology distantly estimated continuously, medical clinics can let loose beds, specialists can keep an eye on patients' activities with more prominent effectiveness.

With wearables, specialists and patients can acquire a consultation with intelligent devices. Several devices can provide informative data for cardiovascular, digestive, endocrine, immune, muscular, nervous, renal, reproductive, respiratory and skeletal systems and recently COVID-related issues. A detailed overview is provided in Sect. 3.3.

3.1 Deep Learning in Medical Imaging

Current medical images are digital in nature. To adequately use them in health care, there are few difficulties that should be considered. Medical imaging portrays an assortment of procedures to make visual portrayals of inside parts of the human body with the end goal of diagnosis, analysis, and medical intervention. This is useful in staying away from or decreasing the requirement for the more seasoned medical norm of experimental medical procedure. Since opening any part of the human body through careful means expanding the danger of diseases, strokes, and different intricacies, medical imaging is presently the favored method for starting determination in the medical setting.

The current medical norm of evaluating medical images is the utilization of doctors, pathologists, or radiologists who look at the pictures and decide the underlying driver of medical sicknesses. This medical standard is inclined to human mistake and is likewise exorbitant and costly, frequently requiring years or many years of involvement to accomplish a degree of understanding which can reliably evaluate these images. Taking into account that the showing of practical machine learning capacities in the advanced age, medical images were one of the main areas to be studied during the adoption of machine learning procedures in health care [16].

The precision of the diagnosis is critical in the medical field as ill-advised analysis could prompt serious outcomes and results. If a medical procedure is performed where none was required or a misdiagnosis prompts ill-advised doses of endorsed prescription, the chance of a deadly result increment. In the domain of image processing, most strategies depend essentially on deep learning (DL) and explicitly in Artificial Neural Networks (ANNs). Current strategies use upgrades to ANNs as Convolutional Neural Networks (CNNs) to support execution when classifying images.

Most of the current distributions are utilizing some type of CNNs with regards to object detection in medical images [17]. Graphical Processing Unit (GPU) speed increase has made the structure of deep CNNs more effective, but significant difficulties in making a capable model still exists. The most significant issue is the requirement for a lot of annotated medical image data. The expense to aggregate and

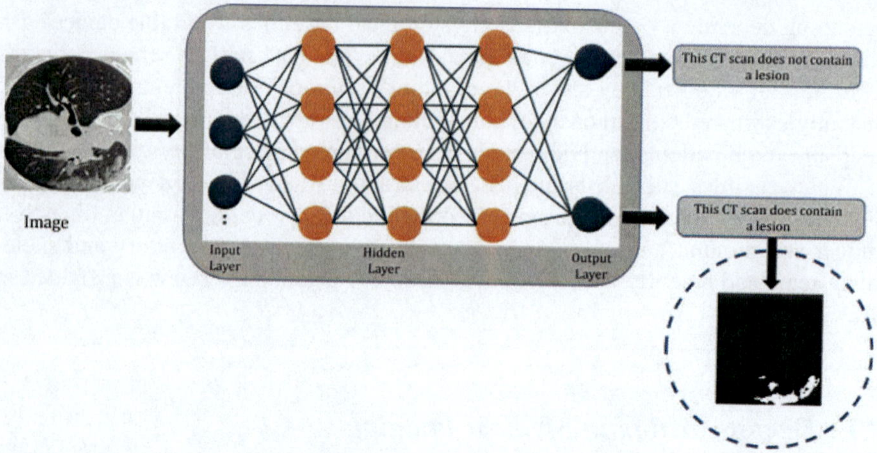

Fig. 3 Medical image analysis using deep learning

make such databases is frequently restrictive since it requires trained experts' time to annotate the images. Furthermore, concerns including patient privacy regularly hinder the capacity to make such databases open-source. Many investigations use around 100–1000 examples in preparing CNNs. This restricted example size expands the danger of overfitting and decreases the precision of the forecasts [18].

The most widely recognized utilization of current machine learning advancements in medication is for Computer Automated Detection (CAD) explicitly in the detection of lesions like those regularly found in mammograms, brain scans, and other body scans [19]. These techniques use CNNs to show the likelihood that a candidate lesion is indeed a lesion, regularly using a few 2D slices of 3D rotational scans of one or the other CT or MRI pictures. Architecture of medical image analysis is given in Fig. 3

Ultrasound pictures are additionally utilized to prepare an assortment of strategies, such as a randomized pivot of the pictures or focusing competitor sores in the picture's focal point. Particularly in mammography, CAD methods have arrived at a level where they are utilized as a "second assessment" for most radiologists, enormously working on the exactness of screenings without multiplying the expense of utilizing a human as the "second assessment". Computer-aided design is additionally presently parted into identification and analysis. An injury can be ordered as either harmless or dangerous, based on a doctor's information and appraisal. In any case, the genuine prediction is a significant initial phase in treating a patient.

Computer-aided detection helps in the acknowledgment of possible lesions from a medical image. For instance, detection and segmentation of glioblastoma is a troublesome task because of these cancers' intrusive and boundless nature. In contrast to other brain tumors, they are not easily localized. Deep learning has supported this by computerizing the appraisal of glioblastoma MRIs. These strategies are principally used to work on precision conclusions and early findings in the medical setting. Once more, these assignments have reliably been performed by machine learning, particularly in brain-related applications, because of the critical idea of evaluating mental health. Also, the finding of Alzheimer through medical imaging is a potential application for deep learning, which is showing some effective results [20].

3.2 Machine Learning for Textual Data in Health Care

Electronic Health Records (EHR), the new norm in numerous clinics, require a complex advanced framework. Unification of health data in an arranged way is a significant objective as it should build the proficiency of medical clinics to develop patient health results further. Notwithstanding, a considerable issue is the authentic existing actual documentation. Moving these current records into an electronic structure is troublesome and would be highly dreary and costly if individuals were recruited to physically information such data into an electronic framework. One utilization of machine learning, which might help with this issue, is Natural Language Processing (NLP). By checking these reports quickly and coordinating the subsequent pictures into a database, these frameworks endeavor to separate intelligible data from free text and consolidates picture preparation to recognize catchphrases and terms. Manually written doctor notes contain patient grievances, the doctors' perceptions, and patient family ancestry. This medical data can be clarified. However, the doctor's ineffectively phrased or mistaken composition can make it hard to dole out this data to suitable classes precisely. Structures and records with structure make for a lot simpler language handling; however, there is the danger of missing data.

Making a framework for further developed medical choice with the help of old patient records is practical. Any such framework is organized to help with the medical dynamic for individual patients dependent on a database of modernized information. Such a framework could be imagined as two-fold: 1. separating facts about the patient from their medical record, either through typed or written doctor notes or audio NLP, 2. Associated sickness states dependent on removed data from past known cases or through literature search employing NLP [12]. Reconciliation of a few specific NLP frameworks is needed for any evident and functional execution of such framework. Similarly, the gathering of the current scientific research into repositories is a troublesome task. Logical distributions have consistently been scattered across various diaries, and the current data blast has worsened the issue.

Examples of Natural Language Processing in Health Care Research: There are many invigorating potential outcomes where NLP could be utilized to develop medication and medical research further. We will talk about a couple of intriguing discoveries with comparative methodologies yet various objectives. This does not shape or form a thorough rundown yet features many potential machine learning applications.

In 2015, a research article distributed a paper detailing 100% precision of anticipating the beginning of psychosis utilizing recorded discourse of medically high-hazard youth. In light of the records, a machine learning calculation was prepared to anticipate whether a patient would foster psychosis. This was finished utilizing what is known as Latent Semantic Analysis to decide the intelligibility of discourse utilizing NLP. The example size for this review was fairly little be that as it may ($n = 34$) [13].

This technique broke down pathology reports and determination codes to distinguish patients from cancer using supervised machine learning. The accuracy was 0.872 with a precision of 0.843 and sensitivity of 0.848 [15]. The essential objective of this review was to mechanize the most common way of revealing disease patients to the National Program of Cancer Registries in the United States.

This review had the option to effectively recognize cirrhosis patients from electronic health records, ICD-9 code mixes, and radiological sweeps with a 95.71% sensitivity and 93.88% specificity [14]. This demonstrates that such a framework could accurately distinguish cirrhosis patients based on existing medical data in many medical clinics.

These instances of NLP use in health care feature a wide variety of utilizations inside medication. Language is essential for conveying complex data; specialists' notes and annotations on medical records hold meaningful experiences in population and individual patient health. The inconsistency and fluctuation of language and extraction of more elevated level data into relevant subcategories make investigation troublesome.

3.3 *Smart Internet of Medical Thing (IoMT) Devices*

The Internet of Medical Things (IoMT) is an associated foundation of medical devices, programming applications, and health frameworks and administrations.

IoT advances are helping numerous industries, and it is a flood of sensor-based devices – including wearables and independent devices for distant patient checking. Several applications of Internet of Medical Things devices is given in Fig. 4 The ascent of IoMT is driven by "an increment in the quantity of associated medical devices that can create, gather, analyze or send health data or pictures and interface with health care supplier organizations, communicating data to either a cloud storage or internal servers." The availability between medical devices and sensors is smoothing out the medical work process. Hence, the executives prompt a general improvement in health care, both inside health care centers and remote areas.

3.3.1 The Potential of IoMT in Health Care

The capacities of IoMT are more precise solutions, more minor mistakes and lower expenses of care. Recent developments in smartphone applications the innovation permits patients to send their health data to specialists to all the more likely watch disease and follow and track constant sicknesses. This kind of innovation is not just assisting with working on the patient experience by disposing of the requirement for in-person medical visits but also assisting with decreasing expenses. The Internet of Medical Things (IoMT) is a mixture of medical devices and applications

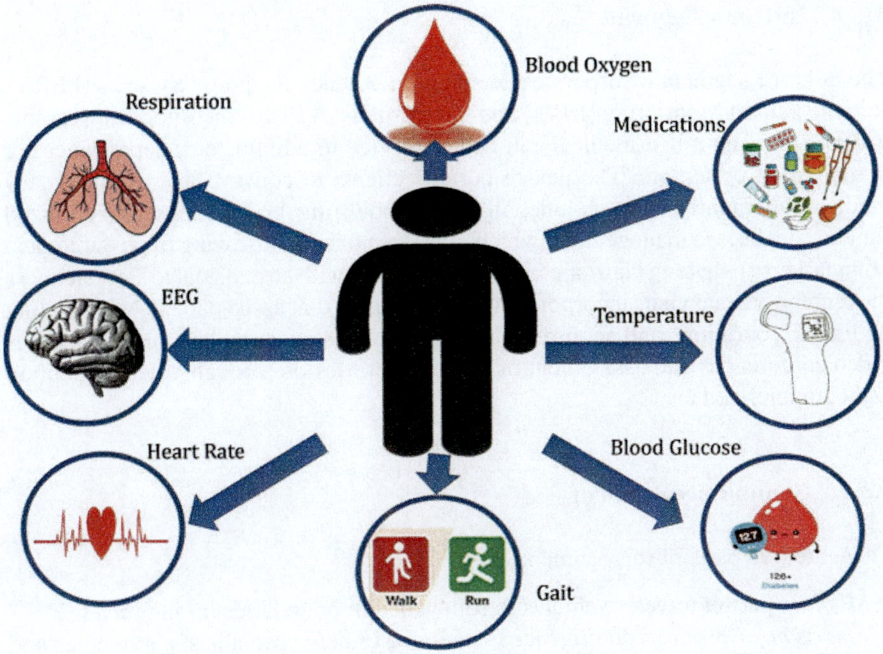

Fig. 4 Present smart internet of medical thing devices

that can interface with health care data innovation frameworks utilizing organizing advances. It can lessen superfluous emergency clinic visits and the weight on health care frameworks by interfacing patients with their doctors and permitting the exchange of medical data over a solid organization.

3.3.2 On-Body Segment

The on-body segment can be comprehensively separated into customer health wearables and medical and medical-grade wearables. Buyer health wearables incorporate purchaser-grade devices for individual well-being or wellness, like activity trackers, bands, wristbands, sports watches, and smart garments. Health specialists do not control most of these devices yet might be embraced by specialists for explicit health applications dependent on casual medical validation and customer studies. Organizations in this space incorporate Misfit (Fossil gathering), Fitbit, Withings, and Samsung Medical.

Medical-grade wearables incorporate cool devices and supporting stages that are for the most part certified/endorsed for use by at least one administrative or health specialist, as the U.S. Food and Drug Administration. The more significant part of these devices is utilized related to expert advice or a doctor's prescription.

3.3.3 In-Home Segment

The in-home segment incorporates Personal Emergency Response Systems (PERS), Remote Patient Monitoring (RPM) and virtual visits. A PERS incorporates wearable gadget units and a live medical call center service to administer independence for home-bound or seniors. The device permits clients to convey and get emergency medical care rapidly. RPM includes all home monitoring devices and sensors utilized for chronic disease management, which includes nonstop observing of physiological boundaries to help long haul care in a patient's home with an end goal to slow sickness movement. Virtual visits incorporate virtual counsels that assist patients with dealing with their conditions and acquire remedies or suggested care plans. It incorporates video conferences and assessment of symptoms or lesion through video perception and computerized tests.

3.3.4 Community Segment

There are five parts of this segment:

- *Mobility* permit traveler vehicles to follow health boundaries during travel.
- *Emergency Response Intelligence* is intended to help specialists, paramedics and medical clinic crisis division care suppliers.
- *Kiosks* are physical stalls, frequently with PC touchscreen shows, that can administer items or offer types of assistance like network to care providers.
- *Point of-care devices* are medical devices utilized by a supplier outside of the home or customary health care settings, for example, at a medical camp.
- *Logistics* includes the vehicle and conveyance of health care labor and products including drugs, medical and surgical supplies, medical devices and instruments and different items required via care providers.

3.3.5 In-Clinic Segment

This segment incorporates IoMT devices that are utilized for managerial or medical capacities. Point-of-care devices here vary from those local segments in one key perspective: rather than the care supplier utilizing a gadget, the supplier can be found distantly while a gadget is utilized by qualified staff. An example includes

- CloudMD, which is a cloud-based research platform for clinicians to survey patients any time of care;
- ThinkLabs' advanced stethoscope; and
- Tytocare's exhaustive telehealth patient assessment gadget for the heart, lungs, ears, skin, throat, and midsection can also measure temperature.

3.3.6 In-Hospital Segment

This segment is partitioned into IoMT-empowered devices and a bigger gathering of arrangements in a few administration regions:

- *Asset Management* screens and tracks high-esteem capital hardware and versatile resources, for example, for example, wheelchairs, all through the hospital.
- *Work force management* estimates staff productivity and efficiency.
- *Patients Flow Management* develops office tasks by forestalling bottlenecks and improving patient experience—for instance, observing patient arrival times from an operating space to present care in a wardroom.
- *Inventory Management* streamlines requesting, stockpiling, and utilization of emergency clinic supplies, consumables, and drugs and medical devices to diminish stock expenses and further develop staff proficiency.
- *Environment* (e.g., temperature and humidity) and energy checking supervises power use and guarantees ideal conditions in persistent regions and extra spaces.

3.4 Privacy Preserving Health Care Data Analytics Techniques

An individual's ability to decide what data can be shared and use access control is privacy. Because of the large amount of data generated by patients, deep learning in health care raises significant security and privacy problems. Harmonized data transmission protocols such as Digital Imaging and Communication in Medicine (DICOM) and electronic data storage are the industry standard in medical imaging, partially addressing the first issue, but privacy restrictions are as stringent. To safeguard patient privacy while supporting scientific studies on massive datasets aimed at improving patient care, technology solutions that simultaneously fulfill data security and utilization requirements are required [21]. Data governance will be the first step in controlling and managing health care data as the health care industry advances toward a value-based business model based on health care analytics [24]. Because of the amount of the data and the rate at which it is produced, privacy-preserving data sharing and privacy-preserving multidimensional big data analytics over big health care data face considerable problems.

Moreover, Big data analytics approaches are thought to be the process of extracting knowledge from databases (which can be centralized or distributed) [22]. The healthcare system has evolved into one of the most essential elements of people's life, resulting in a surge in medical big data. In order to improve and to reap the benefits of big data, information must be shared, analyzed, and processed without causing any harm and violation. However, the data transfer and storage pose several privacy and security concerns [24].

Medical data security and privacy issues could have severe consequences because of a pause in therapy, even endangering the patient's life. As a result, health care

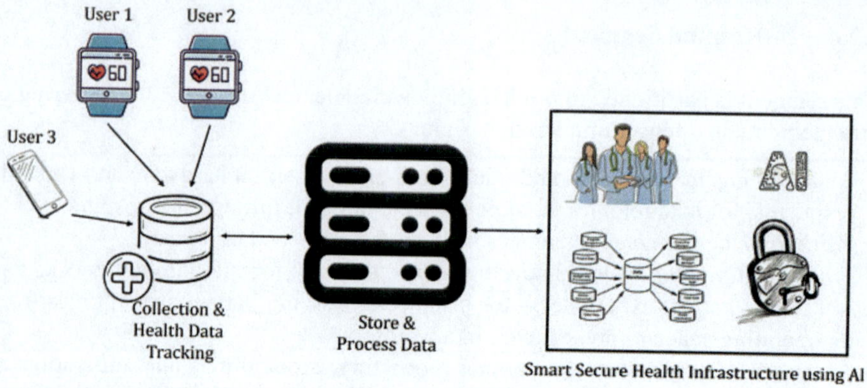

Fig. 5 Smart secure health care system based on artificial intelligence

providers must be fully equipped with sufficient infrastructure to systematically create and analyze big data in order to deliver relevant solutions for improving public health [24]. Big data management, analysis, and practical interpretation can change the game by allowing modern health care systems to explore new possibilities. While the health care business uses big data, security and privacy concerns are becoming more prominent as new threats and vulnerabilities emerge. Following are the techniques for big health care data analytics that protect privacy. Smart health care system architecture is defined in Fig. 5.

3.4.1 Privacy-Preserving Similarity Search

In this technique, the encryption acts like feature-rich multimedia data, allowing for fast file and index updates while maintaining privacy. Instead of using a keyword, we used a high-dimensional feature vector as the search criteria, and we constructed solutions using fuzzy Bloom filters that use locality-sensitive hashing to encode an index linking file identifiers and feature vectors. All-pairs locality-sensitive hashing and a unique encrypted index building are the other two strategies based on this privacy-preserving similarity search [25].

- k-anonymity: This is a more advanced variant of the standard deidentification approach. The smaller the likelihood of reidentification with this approach, the greater the value of k. However, owing to k-anonymization, it may cause data distortions and, therefore more information loss. Excessive anonymization might also make revealed data less helpful to receivers by making some analyses unfeasible or resulting in biased and erroneous conclusions. If the quasi-identifiers containing data are used to link with other publicly accessible data to identify persons in k-anonymization, the sensitive feature (such as illness) will be disclosed as one of the identifiers [25].

- Global versus Local Differential Privacy: Global differential privacy (GDP) and local differential privacy (LDP) are the two main approaches to differential privacy. A trustworthy curator in the GDP setting applies well-calibrated random noise to the actual data returned for a given query. The GDP model is sometimes referred to as the trustworthy curator model. Two of the most commonly utilized noise-generating methods in GDP are the Laplace and Gaussian mechanisms. LDP, on the other hand, does away with the necessity for a trusted curator by randomizing the data before it is accessed by the curator [23]. As a result, LDP is also known as the untrustworthy curator model. A responsible party may also use LDP to generate all records in a database simultaneously randomly. Because noise is often used to ensure individual record privacy, LDP algorithms may create excessively noisy data. LDP is a robust and rigorous concept of privacy that allows for plausible deniability. LDP is considered a state-of-the-art technique for privacy preservation due to the above features. Furthermore, the T closeness measure improves L diversity since an equivalence class is regarded to have 'T closeness' if the gap between the distributions of sensitive attributes in the class is less than a threshold and all equivalence classes have T closeness. Concerning sensitive attributes, T closeness may be computed on any attribute [25].
- Randomization technique: Randomization is adding noise to data, which is usually accomplished using a probability distribution. Randomization is used in surveys, sentiment analysis, and other applications. Randomization does not require any prior knowledge of the data's other records. It can be used during the data gathering and preparation stages [23]. Randomization has no overhead in terms of anonymization. Randomization on massive datasets, however, is not practicable due to temporal complexity and data usefulness.

3.4.2 Security and Privacy in IoMT

The Internet of Things (IoT) is a rapidly developing technology that allows infrastructures, computerized equipment, physical objects, apps, and humans to connect, communicate, collect, and share data via networking. As a result, the Internet of Medical Things (IoMT) applies the Internet of Things in the medical and health industries. The constant advances in IoT, such as the development of microprocessors, biosensor architecture, and emerging 5G technologies, are expected to result in a substantial increase in the efficiency and standards of treatment with the adoption of the IoMT [27–29].

Medical devices and biosensors are in charge of recording the body's vital signs and transmitting massive amounts of raw biological data in real-time, such as heart rate, brain activity, body temperature, and blood glucose level. Personal servers, which are either equipment near the patient's body, such as mobile phones, medical programs, and laptops, or devices remote from the body, such as gateways and routers, collect and process the raw data [28]. In addition, personal servers often feature a computing analysis facility linked to a local archiving database to store the patient's original records. The raw data is collected and processed by personal servers, either equipment near the patient's body, such as mobile phones, medical programs, and laptops, or devices remote from the body, such as gateways and routers. Furthermore, personal servers frequently include a computing analysis facility connected to a local archiving database for storing the patient's original records [27].

Traditional and zero-day attacks can compromise IoMT devices. This is mostly due to the absence of well-established security standards and procedures in device manufacturing and the nature of the devices and IoT networks. Due to their small size, the devices' computational resources and batteries are insufficient to handle cryptography and existing strong security mechanisms. Furthermore, as the number of internet-connected devices grows, so does the amount of data produced. It is commonly known that not only are IoMT devices vulnerable to cyber-attacks, but so is their data. The most pressing challenges in IoMT infrastructure right now are privacy, and data disclosure [29]. The security and privacy standards for the Internet of Things are distinct from traditional networks, known as the CIA-triad (confidentiality, integrity, and availability). Other criteria for the IoMT system include privacy and non-repudiation. Deep Learning uses multiple layers of artificial neural networks to simulate the working principle of the human brain when processing data for object detection, speech recognition, language translation, and decision-making. In contrast to standard ML approaches that require an additional feature selection method, deep learning can conduct feature selection/extraction based on its learning process without the requirement for another method [27].

3.5 Open Issues

This section summarizes several future directions, discussions, and open issues for deep learning in smart health care. Deep learning in smart health poses significant hurdles, but it is fundamental to smart health transformation. Before deep learning, previous methodologies did not consider end users' Quality of Service (QoS) to access the available data and data response. To perform well, deep learning algorithms need a large amount of data, and the development of deep learning-based tools and techniques to support real-time query processing and data analytics is critical for smart health care. Deep learning's main stumbling block is this. With enormous data comes a large number of factors that must be adjusted into the deep learning algorithm. The vast majority of the time, this massive quantity of data is not available, and it is frequently insufficient when it is.

Another difficulty is the problem of overfitting. This is common in neural networks, where the error that occurs during dataset training differs significantly from the error that occurs when a new dataset is introduced. This is an issue since training a model aims to guarantee that it works well with datasets other than the one with which it was developed. Deep learning frequently needs large-scale resource deployments to perform correctly. That is, the more powerful your computational resources are, the more probable the deep learning algorithm will provide a more effective outcome. To train models successfully, you will need a lot of storage space in addition to computer resources. In addition, training a dataset with a deep learning algorithm takes longer than with traditional machine learning approaches.

3.6 The Need and Requirements for Deep Learning in Health Care Systems

Recent health care trends emphasize accessing information at any time and from any location, stimulating health care data movement. Hence, deep learning gives the health care sector the capacity to examine data at breakneck rates while maintaining high accuracy. It is neither machine learning nor artificial intelligence; instead, it is a sophisticated hybrid of the two that sifts through data at breakneck speed thanks to a layered mathematical design. The advantages of deep learning in health care are numerous—quick, efficient, and accurate, to name a few—but they do not end there. More advantages may be found in the neural networks, which are created by several layers of artificial intelligence and machine learning, as well as their capacity to learn [29–31]. Therefore, it is obvious that the name of this game is when it comes to deep learning is learning. Even though deep learning has various advantages, it also poses unique privacy and security risks to health data concerning the usage of that health data in clouds [28].

Using deep learning, various techniques have been created to protect and enhance the privacy of health care systems. For example, in a spoofing identity attack, the attacker impersonates a legitimate user, whereas data tampering includes harmful changes and modifications to the information. To solve this problem, it is important to define the significance of deep learning in this context [29]. The exposing of information to organizations with no permission to access it is known as information disclosure. When discussing the idea and necessity for privacy in health care systems, several topics such as data integrity, confidentiality, accountability, authenticity, and anonymity are considered essential elements.

Furthermore, patients and practitioners are anticipated to be seamlessly connected across health care systems using digital health care systems that utilize electronic health records (EHRs) and technology like deep learning models/frameworks and IoT. As the deep learning models employ mathematical models that are intended to function similarly to the human brain. Hence, these systems are also becoming more and more linked to various types of medical wearable technology for real-time health care monitoring through the Internet [27].

3.6.1 Representation and Transformation of Medical Data

Everyone in today's modern and technological era is playing with data, and the majority of that data is textual, coming from a variety of sources such as web pages, emails, technical and corporate documents, books, digital libraries, and consumer complaints, letters, and patents, among others. However, medical data is not adequately structured for acquiring any information because medical terminologies richness in-text frequently contains rich context information. It is also tough to assess the similarity of the textual data's context [27–30]. Humans can quickly process and interpret the data, and when there is a simultaneous change, such as in intensity and quantity, it may be easier in understanding and adjustments made in response to the changes; temperature and light are two examples. In deep learning, similar processes and situations necessitate a great deal of encoding and careful mathematical expressions in deep learning and transformation. Meaningful representation for effective feature representation is necessary to explain data with multimodal property, a cross-domain feature learning technique based on stacked denoising. So, to address some aforementioned real-world scenarios, a deep learning architecture is a thing that is capable of simultaneously integrating several types of data and representations [31].

3.6.2 Complexity of Deep Learning in Multitasking

Multiple physiological signs may now be collected simultaneously and constantly, thanks to the growth in the usage of wearable devices in recent years. For various objectives, multiple deep learning techniques may be required to classify and analyze these signals. Future research should focus on a single generic deep learning technique that can handle a variety of classifications. This strategy will save time and effort that would otherwise be required to develop a unique solution for each category [28]. Designing deep learning algorithms that incorporate numerous physiological data for categorization is another potential consideration that is a crucial task. Multiple signals from wearable devices allow for the creation of a cohesive model by combining various signals. Constructive application of these signals will undoubtedly improve accuracy and will also serve numerous purposes, allowing the system to work even if one of the signals is unavailable [30, 31].

3.7 *Summary*

This chapter gives a high-level overview of deep learning techniques used in smart health care. There is a need to integrating deep learning with smart health. To allow AI-based current technology developments in smart health care, deep learning techniques are regularly used for smart health. In addition, the chapter discusses deep learning's problems and potential, notably in the health care field with related open issues and challenges. Deep learning is proven to be a proper and developing method in analyzing and using smart health data. Despite its drawbacks, its application in smart health is primarily recognized. At the start of this chapter, we briefly addressed the rise of smart cities and their connections to smart health. In the introduction of this chapter, we also discussed the connection between smart health, machine learning and deep learning. We also emphasized using deep learning techniques in smart health, ranging from cancer detection to health status forecasts. In the field of smart health care, deep learning plays a critical role in various applications, including bioinformatics, medical imaging, and predictive analysis applications. In the final portion of this chapter, the problems, challenges and open issues of deep learning were explored.

To conclude this, we outline technological difficulties impeding the widespread adoption of smart and connected health care systems in this chapter. We also explore how deep learning applications and recent frameworks may speed up all health care stakeholders' implementation, deployment, and adoption of linked health care. On the other hand, deep learning in smart health care is still in its infancy, and it has already shown substantial outcomes. Leading organizations and medical authorities have acknowledged the benefits it provides, and the solutions' popularity has reached a fever pitch.

References

1. Xu, R., Li, L., Wang, Q.: dRiskKB: a large-scale disease-disease risk relationship knowledge base constructed from biomedical text. BMC Bioinf. **15**, 105 (2014)
2. Chen, Y., Li, L., Zhang, G.-Q., et al.: Phenome-driven disease genetics prediction toward drug discovery. Bioinformatics **31**, i276-83 (2015)
3. Wang, B., Mezlini, A.M., Demir, F., et al.: Similarity network fusion for aggregating data types on a genomic scale. Nat. Methods **11**, 333–337 (2014)
4. Tatonetti, N.P., Ye, P.P., Daneshjou, R., et al.: Data-driven prediction of drug effects and interactions. Sci. Transl. Med. **4**, 125ra31 (2012)
5. Li, L., Cheng, W.-Y., Glicksberg, B.S., et al.: Identification of type 2 diabetes subgroups through topological analysis of patient similarity. Sci. Transl. Med. **7**, 311ra174 (2015)
6. Libbrecht, M.W., Noble, W.S.: Machine learning applications in genetics and genomics. Nat. Rev. Genet. **16**, 321–32 (2015)
7. Wang, F., Zhang, P., Wang, X., et al.: Clinical risk prediction by exploring high-order feature correlations. AMIA Annual Symp. **2014**, 1170–9 (2014)
8. SNOMED CT. https://www.nlm.nih.gov/healthit/snomedct/index.html
9. Unified Medical Language System (UMLS). https://www.nlm.nih.gov/research/umls/
10. ICD-9 Code. https://www.cms.gov/medicare-coverage-database/staticpages/icd-9-code-lookup.aspx
11. Mohan, A., Blough, D.M., Kurc, T., et al.: Detection of conflicts and inconsistencies in taxonomy-based authorization policies. In: 2011 IEEE International Conference on Bioinformatics and Biomedicine, Atlanta, GA, USA, pp. 590–594 (2011)
12. Demner-Fushman, D., Chapman, W.W., McDonald, C.J.: What can natural language processing do for clinical decision support? J. Biomed. Inf. **42**(5), 760–772 (2009)
13. Bedi, G., Carrillo, F., Cecchi, G.A., Slezak, D.F., Sigman, M., Mota, N.B., et al.: Automated analysis of free speech predicts psychosis onset in high-risk youths. NPJ Schizophrenia **1**(1), 15030 (2015)
14. Chang, E.K., Christine, Y.Y., Clarke, R., Hackbarth, A., Sanders, T., Esrailian, E., et al.: Defining a patient population with cirrhosis: an automated algorithm with natural language processing. J. Clin. Gastroenterol. **50**(10), 889–894 (2016)
15. Osborne, J.D., Wyatt, M., Westfall, A.O., Willig, J., Bethard, S., Gordon, G.: Efficient identification of nationally mandated reportable cancer cases using natural language processing and machine learning. J. Am. Med. Inf. Assoc. **23**(6), 1077–1084 (2016)

16. Cai, L., Gao, J., Zhao, D.: A review of the application of deep learning in medical image classification and segmentation. Annals Trans. Med. **8**, 713–713 (2020). https://doi.org/10.21037/atm.2020.02.44
17. Greenspan, H., Van Ginneken, B., Summers, R.M.: Guest editorial deep learning in medical imaging: overview and future promise of an exciting new technique. IEEE Trans. Med. Imag. **35**(5), 1153–1159 (2016)
18. Giger, M.L.: Machine learning in medical imaging. J. Am. College Radiol. **15**(3), 512–520 (2018)
19. Sherwani, M.K., Zaffino, P., Bruno, P., Spadea, M.F., Calimeri, F.: Evaluating the impact of training loss on MR to synthetic CT conversion. In: Nicosia, G., et al. (eds.) Machine Learning,Optimization, and Data Science. LOD 2020. Lecture Notes in Computer Science, vol. 12565. Springer, Cham (2020)
20. Islam, J., Zhang, Y.: Brain MRI analysis for Alzheimer's disease diagnosis using an ensemble system of deep convolutional neural networks. Brain Inf. **5**(2) (2018)
21. Gan, W., et al.: Privacy preserving utility mining: a survey. In: 2018 IEEE International Conference on Big Data (Big Data). IEEE (2018)
22. Chamikara, M.A.P., et al.: An efficient and scalable privacy preserving algorithm for big data and data streams. Comput. Sec. **87**, 101570 (2019)
23. Canbay, Y., Vural, Y., Sagiroglu, S.: Privacy preserving big data publishing. In: International Congress on Big Data. Deep Learning and Fighting Cyber Terrorism (IBIGDELFT). IEEE (2018)
24. Dash, Sabyasachi, et al.: Big data in healthcare: management, analysis and future prospects. J. Big Data 6.1 (2019)
25. Tran, H.-Y., Jiankun, Hu.: Privacy-preserving big data analytics a comprehensive survey. J. Parallel Distrib. Comput. **134**, 207–218 (2019)
26. Dai, W., et al.: Privacy preserving federated big data analysis. In: Guide to Big Data Applications, pp. 49–82. Springer, Cham (2018)
27. Hameed, S.S., et al.: A systematic review of security and privacy issues in the internet of medical things; the role of machine learning approaches. PeerJ Comput. Sci. (2021)
28. Tobore, I., Li, et al.: Deep learning intervention for health care challenges: some biomedical domain considerations. JMIR mHealth and uHealth (2019)
29. Obinikpo, A.A., et al.: Big sensed data meets deep learning for smarter health care in smart cities. J. Sens. Actuator Netw. (2019)
30. Holzinger, A., et al.: From smart health to smart hospitals. Springer book Smart health (2020)
31. Holzinger, A., et al.: Smart health: open problems and future challenges. Springer book series (2015)
32. Shen, L., Margolies, L.R., Rothstein, J.H., Fluder, E., McBride, R., Sieh, W.: Deep learning to improve breast cancer detection on screening mammography. Sci. Rep. **9**(1), 12495 (2019). https://doi.org/10.1038/s41598-019-48995-4. PMID: 31467326; PMCID: PMC6715802
33. Rajkomar, A., Oren, E., Chen, K., et al.: Scalable and accurate deep learning with electronic health records. npj Digital Med. **1**, 18 (2018). https://doi.org/10.1038/s41746-018-0029-1

The Solution of Computer Vision for Combating Covid-19

Ngoc Chi Le, Hue Vu, Long Van Nguyen, Duc Hoang Trinh, Dam Ngoc Nguyen, Phuong Huy Nguyen, and Tung Nguyen

Abstract Recently, the Covid-19 pandemic has become very complicated and seriously affecting the economy as well as society in every countries in the world. In this chapter, we explore the solution of Computer Vision for handling the Covid-19 pandemic situation. The given scenarios based on deep learning techniques are used to monitor the traffic of people and vehicles through the checkpoints to control the in-out movement in significant areas. In addition, we also need to pay attention to complying with the regulations on wearing masks and ensuring a safe social distance in public places. From there, the proposed system will effectively support organizations to deal with the Covid-19 pandemic.

1 Object Detection and Object Counting

1.1 Introduction

In June 2021, Afkaar Ar et al. proposed the idea of applying technology to identify and count people (vehicles) to support Covid-19 control in the article: *A computer vision-based object detection and counting for Covid-19 protocol compliance: a case study of Jakarta* [1].

The system detects and counts the number of people (vehicles) based on CCTV cameras installed at important roads. The author's research aims to analyze public surveillance activities by using existing CCTV cameras with integrated Artificial Intelligence to optimize regulatory compliance and monitoring systems Covid-19. Camera systems can estimate the density of people as well as vehicles in a certain area.

N. C. Le · H. Vu · L. V. Nguyen · D. H. Trinh (✉) · D. N. Nguyen · P. H. Nguyen · T. Nguyen
Hanoi University of Science and Technology, Hanoi, Vietnam
e-mail: trinhhoangduc.n@gmail.com

M. Lahby et al. (eds.), *Computational Intelligence Techniques for Green Smart Cities*, Green Energy and Technology, https://doi.org/10.1007/978-3-030-96429-0_9

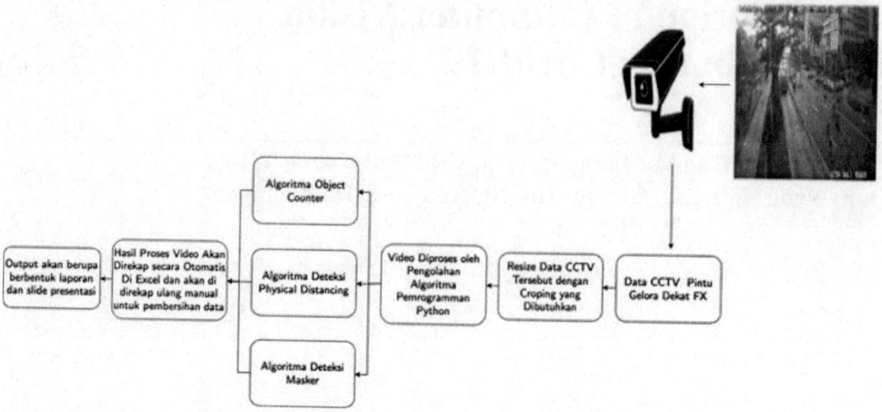

Fig. 1 System overview [1]

1.2 Method

System overview: The analysis procedure will be performed by using predefined cameras to facilitate data detection and extraction. Image data collected by the camera will be transmitted to a computer system running a Python program for processing. Artificial intelligence programs that recognize and count the number of people (vehicles) will be implemented here. These artificial intelligence modules will be discussed in more detail in the following sections. Finally, the results from the programs will be transferred to the Tableau system to generate analysis and synthesis charts for easy monitoring (Fig. 1).

The MobileNet architecture is defined in Table 1. All layers are followed by a batchnorm [2] and ReLU nonlinearity with the exception of the final fully connected layer which has no nonlinearity and feeds into a softmax layer for classification.
Object Detection: The author uses the proposed MobileNet-SSD object recognition model in 2015. The operation process of the MobileNet-SSD model is divided into two phases: Perform Feature extraction through the MobileNet model (*Phase one*), Apply convolutional filter for object recognition (*Phase two*).
Phase one: Feature extraction through the MobileNet model.

In MobileNet, the concept of Depth Separable Convolutions was introduced for the first time. This is a convolution technique that helps the model reduce the number of weights significantly, there by speeding up the training process and providing the ability to run the model on low-end devices.
Phase two: Apply convolutional filter for object recognition.

After extracting feature maps based on MobileNet model, SSD model [4] uses one kernel per cell to make predictions. Each filter output is a vector of $C + 4$ values, where C values correspond to the score of each class and 4 values represent the bounding box for the object.

Table 1 MobileNet body architecture [3]

Type/Stride	Filter shape	Input size
Conv/s2	$3 \times 3 \times 3 \times 32$	$224 \times 224 \times 3$
Conv dw/s1	$3 \times 3 \times 32$ dw	$112 \times 112 \times 32$
Conv/s1	$1 \times 1 \times 32 \times 64$	$112 \times 112 \times 32$
Conv/s1	$1 \times 1 \times 64 \times 128$	$56 \times 56 \times 64$
Conv dw/s1	$3 \times 3 \times 128$ dw	$56 \times 56 \times 128$
Conv/s1	$1 \times 1 \times 128 \times 128$	$56 \times 56 \times 128$
Conv dw/s2	$3 \times 3 \times 128$ dw	$56 \times 56 \times 128$
Conv/s1	$1 \times 1 \times 128 \times 256$	$28 \times 28 \times 128$
Conv dw/s1	$3 \times 3 \times 256$ dw	$28 \times 28 \times 256$
Conv/s1	$1 \times 1 \times 256 \times 256$	$28 \times 28 \times 256$
Conv dw/s2	$3 \times 3 \times 256$ dw	$28 \times 28 \times 256$
Conv/s1	$1 \times 1 \times 256 \times 512$	$14 \times 14 \times 256$
5 × Conv dw/s1 Conv/s1	$3 \times 3 \times 512$ dw $1 \times 1 \times 512 \times 512$	$14 \times 14 \times 512$ $14 \times 14 \times 512$
Conv/s1	$1 \times 1 \times 512 \times 1024$	$7 \times 7 \times 512$
Conv dw/s2	$3 \times 3 \times 1024$ dw	$7 \times 7 \times 1024$
Conv/s1	$1 \times 1 \times 1024 \times 1024$	$7 \times 7 \times 1024$
Avg Pool/s1	Pool 7×7	$7 \times 7 \times 1024$
FC/s1	1024×1000	$1 \times 1 \times 1024$
Softmax/s1	Classifier	$1 \times 1 \times 1000$

Object Counting: The object counting algorithm is performed after knowing the objects on the processing frame. The processing is shown:

- The recognition program will select bounding boxes with 40% confidence threshold.
- Assign each of these objects an ID and try to maintain this ID with subsequent frames.
- Select a boundary line, when the object crosses this boundary line is counted in the quantity.

The ID assignment algorithm used is called *Centroid Tracking Algorithm* [5]. How the *Centroid Tracking Algorithm* works:

- Bounding processing is performed at each frame.
- Calculate the distance between each pair of bounding box centers in the current frame and the bounding box centers assigned the ID in the previous frame.
- Assign the object ID in the new frame with the ID of the object closest to it by Euclidean distance.
- Generate additional IDs for unassigned objects in the old frame.

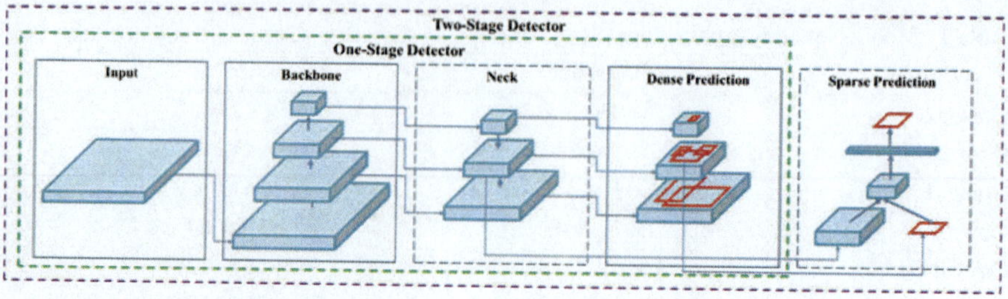

Fig. 2 Overview of object detectors [6]

1.3 Proposal to Improvement

Object detection—Yolo v4: Yolo—You Only Look Once is a family of one stage objects detection models, they are also the best object detection models at the moment. Up to now, there are at least five versions of Yolo are announced (include official and unoffical). Although all of them called Yolo, the versions of this model have very significant improvements after each version. After three main versions of first author Joseph Redmon, from Yolo v1 to Yolo v3, the Yolo v4 was published by Alexey Bochkovskiy and the Yolo v5 is being in progress of development (Fig. 2).

The main goal of the Yolo v4 is becoming a fast operating speed of an object detector in production systems and optimization for parallel computations. At the time of publication, Yolo v4 claims to have state-of-the-art precision while maintains a high processing speed. It achieves an accuracy of 43.5% AP (65.7% AP_{50}) for the MS COCO dataset (see Fig. 3) with a real-time inference speed of approximately 65 FPS on NVIDIA Tesla V100 GPU [6].

Before going to main architecture of Yolo v4 , there is a natural problem: conventional object detectors that have better performance usually require more inference cost. **Bag of Freebies** (BoF) is a common name for a group of methods that only change the training strategy or only increase the training cost [6]. The method often used is data augmentation, which increase the variability of the input images. On the other hand, some BoF are dedicated to solving the problem that the semantic distribution in the dataset may have bias. Finally, the last BoF is the objective function of Bounding Box regression. The traditional object detector usually uses Mean Square Error, while there are some recently proposed methods such as IoU Loss and its improvements [7]. As regards **Bag of specials** (BoS), they contains different plugins and the post-processing modules that only increase the inference cost by a small amount but can significantly improve the accuracy of the object detector. In general, these methods or can be considered as an add-on for any object detectors present right now to make them more accurate on benchmark datasets.

The overall **architecture of Yolo v4** was designed from a GPU training perspective which is unheard of. The model architecture optimized for efficient use of parallel computational power of conventional GPU(s) which is not the case with other object detectors. Therefore, the objective of model is to find the optimal balance among [6].

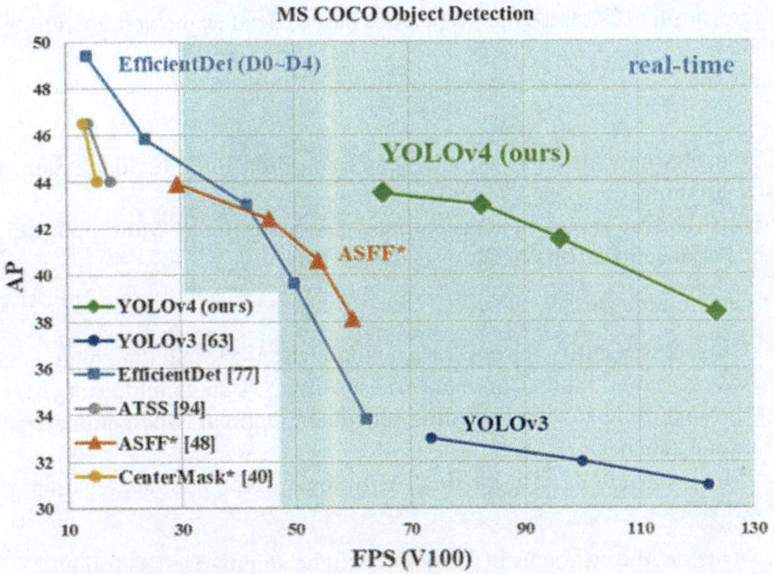

Fig. 3 Comparison of the Yolo v4 and other state-of-the-art object detectors. Yolo v4 runs twice faster than EfficientDet with comparable performance [6]

- Input resolution
- Parameter number
- Convolutional layer number
- The number of layer outputs.

Yolo v4 **Backbone** here refers to the feature-extraction architecture. In Yolo v4, backbones can be feature-extractor of any image classification networks such as VGG16, ResNet, EfficientNet, ResNeXt, or Darknet53. After numerous analysis of different parameters on standard benchmarks [6], the final backbone was chosen is CSPDarknet53 . Yolo v4 utilizes the CSP connections [8] above with the Darknet-53 , which has been introduced in Yolo v3 as an improvement[9], below as the backbone in feature extraction. In experiemnts, the CSPDarknet53 model has higher accuracy in object detection compared with ResNet based designs even they have a better classification performance, and the classification accuracy of CSPDarknet53 can be improved with Mish activation [10] (the author used it as BoS for detector [6]).

In terms of **Neck**, which is used in backbones for the extraction of wealthy semantic features that are further used for precise predictions. In Yolo v4, the author use Spatial Pyramid Pooling (SPP) and Modified Path Aggregation Networks (PAN) [6] as a rise of the receptive field and an alternative choice of Feature Pyramid Network (FPN, which was used in the previous version), respectively.

As regards **Head**, Yolo v3 Head is taken into consideration for loss predictions. Finally, the author choose CSPDarknet53 backbone, SPP additional module, PANet path-aggregation neck, and Yolo v3 (anchor based) head as the architecture of Yolo v4 [6].

Finally, there are some selecttions of BoF and BoS. [6] For backbone, Yolo v4 uses:

- BoF: CutMix and Mosaic data augmentation, DropBlock regularization, Class label smoothing
- BoS: Mish activation, Cross-stage partial connections (CSP), Multiinput weighted residual connections (MiWRC).

and for detector, it uses:

- BoF: CIoU-loss, CmBN, DropBlock regularization, Mosaic data augmentation, Self-Adversarial Training, Eliminate grid sensitivity, Using multiple anchors for a single ground truth, Cosine annealing scheduler, Optimal hyperparameters, Random training shapes
- BoS: Mish activation, SPP-block, SAM-block, PAN path-aggregation block, DIoU-NMS.

In conclusion, the attributes of Yolo v4 could be summarized as follows:

- An efficient and powerful object detection model. In experiemnts, the object detector can be fast trained on a 1080Ti or 2080Ti GPU [6].
- Some state-of-the-art Bag of Freebies and Bag of Specials are used during training.

Yolo family has become very famous for its real-time object detection problem. However, since the Yolo v3 version, the first author of Yolo, Joseph Redmon, has no longer studied and developed this architecture. He claimed that he had stopped researching computer vision due to concerns about the technology being misused. After the success of Yolo v4 and its variants, there is a version of Yolo v5 is developing by Ultralytics LLC. This version is currently quite promising according to the figures provided by the development company. However, this version of Yolo v5 has not yet been officially accepted and there is a lot of controversy surrounding the effectiveness of this model under development.

Object Counting—DeepSort: *The Centroid Tracking Algorithm* [5] has the disadvantage of being inaccurate in the case that this object is covered by another object, or in a certain frame where the object recognition model does not work well. In these cases, the result is that the ID is changed when the object is re-recognized. In 2017, Nicolai Wojke, Alex Bewley, Dietrich Paulus introduced the DeepSort model [11] to help effectively handle the Object Tracking problem. The DeepSort model is a combination of the Sort [12] model and the deep learning network. In Sort algorithm, we use a standard Kalman filter and take the bounding coordinates as (u, v, γ, h), where (u, v) is the bounding box center position, h is height of bounding box and γ is ration between width and height. $A_{\max}$ is max number to have left the scene and are deleted from the track set. **The Mahalanobis distance**, the way to solve the association between the predicted Kalman states and newly arrived measurements is to build an assignment problem that can be solved using the Hungarian algorithm:

$$d^{(1)}(i, j) = \left(\boldsymbol{d}_j - \boldsymbol{y}_i\right)^{\mathrm{T}} \mathbf{S}_i^{-1} \left(\boldsymbol{d}_j - \boldsymbol{y}_i\right), \tag{1}$$

where $\boldsymbol{d}_j$ is the jth bounding box, $\boldsymbol{y}_i$ is the ith track.

Further, using this metric it is possible to exclude unlikely associations by thresholding the Mahalanobis distance at a 95% confidence interval computed from the inverse χ distribution. The author denote this decision with an indicator:

$$b_{i,j}^{(1)} = \mathbb{1}[d^{(1)}(i, j) \leq t^{(1)}], \tag{2}$$

where $t^{(1)} = 9.4877$.

Each detection d_j, the author compute an appearance descriptor by a pretrained-model with architecture in Table 2. Then, the author defines the consine distance $d^{(2)}(i, j)$ to limit the dependence of Mahalanobis:

$$d^{(2)}(i, j) = \min\{1 - \boldsymbol{r}_j{}^{\mathrm{T}} \boldsymbol{r}_k^{(i)} \mid \boldsymbol{r}_k^{(i)} \in \mathcal{R}_i\}, \tag{3}$$

where:

- $\boldsymbol{r}_j$ is an appearance descriptor of the jth dection.
- $\mathcal{R}_i = \{r_k^{(*i)}\}_{k=1}^{100}$ is set of appearance descriptors for each track k.

Again, we have a binary variable to indicate if an association is admissible according to this metric:

$$b_{i,j}^{(2)} = \mathbb{1}[d^{(2)}(i, j) \leq t^{(2)}] \tag{4}$$

Table 2 Deep learning model in DeepSort [11]

Name	Patch size/stride	Output size
Conv 1	$3 \times 3/1$	$32 \times 128 \times 64$
Conv 2	$3 \times 3/1$	$32 \times 128 \times 64$
Max Pool 3	$3 \times 3/2$	$32 \times 64 \times 32$
Residual 4	$3 \times 3/1$	$32 \times 64 \times 32$
Residual 5	$3 \times 3/1$	$32 \times 64 \times 32$
Residual 6	$3 \times 3/2$	$64 \times 32 \times 16$
Residual 7	$3 \times 3/1$	$64 \times 32 \times 16$
Residual 8	$3 \times 3/2$	$128 \times 16 \times 8$
Residual 9	$3 \times 3/1$	$128 \times 16 \times 8$
Dense 10		128
Batch and ℓ_2 normalization		128

This Deep Learning model is trained on a dataset containing more than 1,100,000 images of 1261 people

Table 3 Performance comparison table of DeepSort and other models [11]

Model	MOTA	MOTP	Speed (Hz)
SORT [13]	59.8	79.6	60
EAMTT [14]	52.5	78.8	12
POI [15]	66.1	79.5	10
DeepSORT	61.4	79.1	40

After that, the author combine metrics using a weight sum:

$$c_{i,j} = \lambda d^{(1)}(i, j) + (1 - \lambda)d^{(2)}(i, j) \tag{5}$$

and has an association admissible if it is within the gating region of both metrics:

$$b_{i,j} = b^{(1)}_{i,j} + b^{(2)}_{i,j} \tag{6}$$

The results of our evaluation are shown in Table 3.

Sort is a relatively simple Object-Tracking model. The idea of Sort is to be able to predict where the object will be in the next frame. Then use the predicted position to compare with the bounding box in the new frame through the IoU (Intersection on Union) value to calculate the match. DeepSort is an upgraded version of Sort when using more deep learning to calculate the match point. Before calculating the match point, the bounding boxes predicted from Sort will be fed into a deep learning model to extract features into vectors. Tracking details are shown in Algorithm 1:

Algorithm 1: Matching Cascade [11]

Input: Track indices $T = \{(1, ..., N\}$, Detection indices $D = \{1, ..., M\}$, Maximum age A_{max}

1: Compute cost matrix $C = [c_{i,j}]$, where: $c_{i,j} = \lambda d^{(1)}(i, j) + (1 - \lambda)d^{(2)}(i, j)$

2: Compute gate matrix $B = [b_{i,j}]$, where: $b_{i,j} = \prod_{m=1}^{2} b^{(m)}_{i,j}$

3: Initialize set of matches $M \leftarrow \emptyset$

4: Initialize set of unmatched detections $U \leftarrow D$

5: **for** $n \in \{1, ..., A_{max}\}$ **do**

6: Select tracks by age $T_n \leftarrow \{i \in T | a_i = n\}$

7: $[x_{i,j}] \leftarrow min\_cost\_matching(C, T_n, U)$

8: $M \leftarrow M \cup \{(i, j) | b_{i,j} \cdot x_{i,j} > 0\}$

9: $U \leftarrow U \setminus \{j | \sum_i b_{i,j} \cdot x_{i,j} > 0\}$

10: **end for**

11: **return** M, U

2 Mask Face Recognition

2.1 Introduction

Face recognition is an important problem in the field of machine learning and deep learning. Along with the development of the big data of age and the increasingly high-performance hardware configuration, facial recognition systems are increasingly achieving improved accuracy even when subject to limitations about the outside, the image quality such as blur, wearing a mask, ... There are two research directions to train DCNNs model in face recognition. The first is training a multi-classes to recognize many different identities from the training dataset. For example, using softmax classifier and the second is learning directly from vector embedding, such as Triplet loss. Based on large-scale data, both methods can achieve high accuracy, but at the same time there are some limitations. For using softmax classifier, the linear matrix size W increases linearly with the number of identities and the model can classify well closed-set but it is not good enough when applied to open-set problems. For using Triplet loss, we may encounter the problem of combinatorial explosion of data pairs when using large-scale datasets and applying semi-hard example mining is a difficult problem to train the model effectively.

2.2 Related Work

To minimize the disadvantages of the two above-mentioned models, some studies have observed the last fully connected layer of the training DCNN classification model with softmax loss and found that there is a similarity with the centers of the classes. The researchers [16] [17] have proposed angular margin penalty to improve the association between classes and within each class. Sphererface [16] introduces an important idea of angular magin, uses some approximations and proposes a hybrid loss function to stabilize the training process. Cosface [18] adds a penatly called as the cosine margin penalty to the target logit, which provides better performance and is also easier to deploy. Arcface [19] uses arc-cosine function to calculate angle between current feature vector and target weights, then applies additive angular magin to target angle and uses cosine function to get new target logit value. The above methods are collectively known as using angular margin in order to achieve improved accuracy of the classification model and at the same time can be easily applied on large-scale training datasets.

2.3 System Overview

The basic system using facial recognition usually consists of three main modules. The first is the face detector, the second is the feature extractor and the last is the embedding matching model. For face detector, we use Retinaface to achieve high accuracy in real time. The Feature Extractor we use here is a classification model based on angular magin, specifically R100 trained on the Glint360k (Insightface) dataset. In the match embedding part, we use the KNN model.

2.4 Proposal to Improvement

Face recognition wearing a mask is an important issue in contactless identification. We have researched that using the DCNNs model applying angular magin technique, it is still possible to achieve an accuracy of over 90% when removing the lower half of the face and replacing it with a black background (Table 4). This analysis helps us propose a simple idea to improve the face mask recognition. Assuming that faces are capable of wearing different types of masks, it is possible to increase the distance between the faces of the same person but with a difference in the lower half of the face (mask). We conducted the average (or weight) vector of original face and the face after removing the lower part of the match. Or if the database of faces to match is not too large, we create many different face variations in the database, each face has a different weight depending on the level of augmentation. This weight will multiply directly by the Euclidean distance when matching the face embedding vector to the top-k result in the database. For example, randomly blackening $\frac{1}{8}, \frac{2}{8}, \frac{3}{8}, \frac{4}{8}$ of the lower half of the face creates four different variants with increasing weights 1.05, 1.1, 1.2, 1.25 (with a weighted original face of 1.0). This weight makes it easy for us to adjust between precision and recall rates depending on the needs of the system.

Table 4 Verification accuracy when removing the lower half of the face by blackening using Arcface R100 with dataset MS1Mv2 (Insightface)

	Full face	All half face
LFW	0.99750	0.99517
AGE30_DB	0.97800	0.95966
CFP_FP	0.97786	0.95070
CRP_FF	0.99743	0.99328
VGG2_FP	0.95360	0.93839
CALFW	0.95850	0.95416
Average	0.97715	0.96522

3 Control Wearing Masks and Social Distance

3.1 Detect and Track People Movement

At each frame, object detection and tracking model is applied to detect people and assign each person an ID to track them throughout their moving process in the camera. Assigning each person an ID and a high accuracy model makes it so that there are not too many IDs of the same person, then the accuracy of reporting statistics will not be too biased. Simultaneously, because the face of the person carrying that ID can not always be seen, assigning an ID to each person makes it possible for a moment when that person looks in the direction of the camera to be able to capture the face, from that conduct analysis and get facial information such as: wearing a mask, age, …

To ensure accuracy, the analysis and detection of people only applies in the pre-zoned area (Figs. 4 and 5).

Tracking Combined with Detection Most multi-object trackers do this independently of object detection, and that does not take advantage of the already analyzed image data for object detection, but must continue. Performing analysis in a neural network without any input from tracking. Thus, with the detection and tracking model, TraDes [20] (Tracking for Detection and Segmentation), exploit trace clues to aid end-to-end detection. TraDes inferred object tracking is offset by a volume cost, it is used with previous audience features to improve detection and segmentation of the current audience. The efficiency and superiority of TraDes is demonstrated on

Fig. 4 Results from the first frame

Fig. 5 Results from the 50th frame

four datasets, including MOT[1] (2D tracking), nuScenes[2] (3D tracking), MOTS[3] and Youtube-VIS[4] (track version segmentation).

- MOT: Dataset for 2D multi-object tracking problems
- nuScenes: Dataset for 3D multi-object tracking problems
- MOTS and Youtube-VIS: Two datasets for object segmentation in images or videos.

The special feature of the model is that it is possible to transmit features from previous frames to improve object detection and tracking.

3.2 *Estimate Distance*

Calculating the normal distance between objects in the camera using pixel will lead to certain deviations. Because by using a single camera, projecting a $3D$ world scene onto a $2D$ perspective image plane results in unrealistic pixel spacing between objects. This is called the perspective effect, because of the uniform distribution of distances throughout the image. For example, parallel lines intersect at the horizon and those farther away from the camera appear to be much shorter than those closer to the camera coordinate center.

[1] https://motchallenge.net/
[2] https://www.org/
[3] https://motchallenge.net/
[4] https://youtube-vos.org/dataset/vis

In $3D$ space, the center or reference point of each bounding box is associated with three parameters (x, y, z), while in the image received from the camera, the original $3D$ space is reduced to two dimensions of (x, y) and the depth parameter (z) are not available. In such a low-dimensional space, it would be misleading to use the Euclidean distance criterion directly to estimate distances between people. To apply the bird eye view mode transition process (bird eye) corrected, it is necessary to calibrate the camera first by setting $z = 0$ to remove the perspective effect. To do this, it is necessary to know the camera's position, its height, its angle of view, as well as the optical specifications (i.e. the camera's intrinsic parameters) [21].

By applying transform, the $2D$ pixels (u, v) will be mapped to the corresponding world coordinate points X_w, Y_w, Z_w:

$$[u, v, 1]^T = K \times R \times T[X_w, Y_w, Z_w],$$

where R, T, K is rotation matrix, translation matrix and matrix of the camera's intrinsic parameters, respectively.

$$R = \begin{bmatrix} 1 & 0 & 0 & 0 \\ 0 & \cos\theta & -\sin\theta & 0 \\ 0 & \sin\theta & \cos\theta & 0 \\ 0 & 1 & 1 & 1 \end{bmatrix}, T = \begin{bmatrix} 1\,0\,0 & 0 \\ 0\,1\,0 & 0 \\ 0\,0\,1 & \frac{-h}{\sin\theta} \\ 0\,1\,1 & 1 \end{bmatrix}, K = \begin{bmatrix} f * ku & s & c_x\ 0 \\ 0 & f * kv & c_y\ 0 \\ 0 & 0 & 1\ 0 \end{bmatrix},$$

where h is the camera height, f is the focal length, and k_u and k_v are the measured calibration factor values in horizontal and vertical pixel units, (c_x, c_y) is the key point move that adjusts the optical axis of the image plane.

The camera creates an image with the projection of three-dimensional points in world coordinates falling on a retinal plane. Using uniform coordinates, the relationship between the three-dimensional points and the obtained projection points can be shown as follows:

$$\begin{bmatrix} u \\ v \\ 1 \end{bmatrix} = \begin{bmatrix} m_{11} m_{12} m_{13} m_{14} \\ m_{21} m_{22} m_{23} m_{24} \\ m_{31} m_{32} m_{33} m_{34} \end{bmatrix},$$

where $M \in R^{3\times4}$ is the transformation matrix, which maps world coordinate points to pixels based on camera position and reference system, provided by camera intrinsic matrix, rotation matrix R and translation matrix T.

And finally the transition from perspective space to inverse perspective space (BEV) can also be represented as the following scalar:

$$(u, v) = \left(\frac{m_{11}x_w + m_{12}y_w + m_{13}}{m_{31}x_w + m_{32}y_w + m_{33}}, \frac{m_{21}x_w + m_{22}y_w + m_{23}}{m_{31}x_w + m_{32}y_w + m_{33}} \right)$$

In fact, to get the camera settings as well as the affected external conditions makes the deployment extremely difficult. To be able to apply to any arbitrary camera with the idea of converting the view from 3D to 2D, the simplest conversion method can be applied, which is to select four points corresponding to a rectangle in the mode bird eye view. In addition, these points will form parallel lines in the real world. That assumes everyone is standing in a single plane, and that this top-down or bird eye view has the characteristic that the points are uniformly distributed horizontally and vertically (ratio to the orientation horizontal and vertical will be different). From this mapping, we can derive a transformation that can be applied to the entire perspective image, so the distance between two people can easily be calculated.

The perspective frame selection includes four points corresponding to coordinates $(0, 0)$, $(0, w)$, $(0, h)$, (w, h) and two points in the perspective frame corresponding to the distance two meters in the real world.

In which,

- w: length of image in bird eye view
- h: width of the bird eye view.

And the ratio between w and h should correspond to the ratio of the length and width of the real world perspective.

The process of finding transformation matrix: Let X be the four corresponding points on the camera, T is the point with degrees $(0, 0)$, $(0, w)$, $(0, h)$, (w, h). Then X and T are related and it is necessary to calculate a transformation matrix M such that:

$$T = M \times X \Rightarrow M = T \times X^{-1}$$

For each person detected in the frame, the distance calculation is done based on the bottom center of the box enclosing the person as the standing position of each person. Then, estimating (x, y) position in the bird eye view by applying a transformation to the bottom midpoint of each person's bounding box, resulting in their position in the bird eye view.

Calculate Distance and Give Warning: To estimate distances with low error, there are two steps:

- Select four corresponding points that form a rectangle in the real world
- Select two points that correspond to two meters in the real world to map to the distance on pixels (Figs. 6 and 7).

4 Conclusion

In this chapter, we discussed three main issues, which are the problem of detecting and counting people, and the problem of face recognition wearing a mask, and the problem of measuring social distance to help control the Covid-19 epidemic. Currently, the above solution have been deployed and can be put into the application directly to support the control of the Covid-19 epidemic. The above problems have achieved certain successes and are well evaluated. Specifically, the problem of detecting and counting people has achieved high accuracy as mentioned in section

Fig. 6 Example of settings for distance prediction

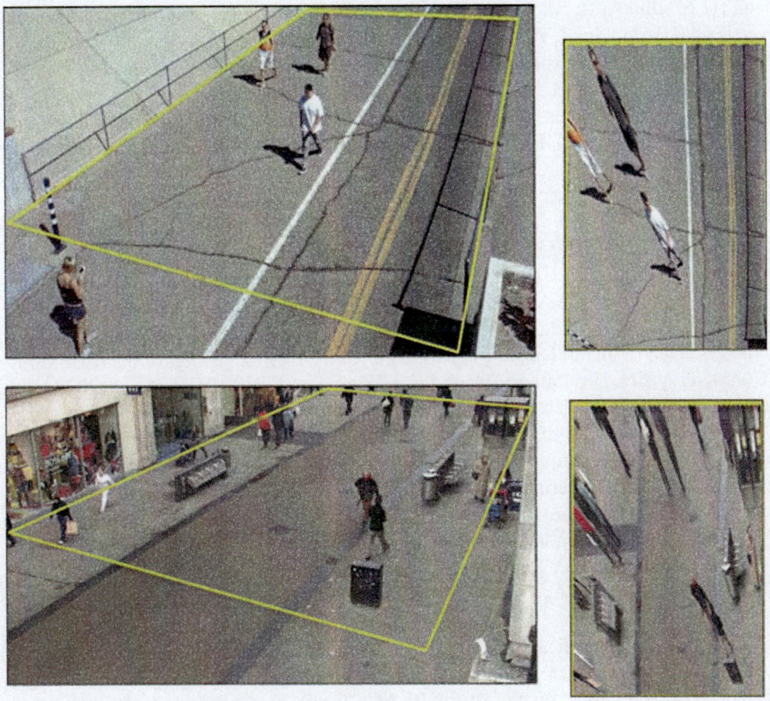

Fig. 7 Example for converting from camera view to bird eye view

12.1. With the problem of face wearing mask recognition, there are good ideas for improvement, the model has a balance between accuracy and performance. And with the problem of measuring social distance, there have been significant improvements in the results from the good calibration of the model's parameters. From the above results, these problems will be improved and more successful in the future.

References

1. Afkaar Ar, M.L., Sulthan Muzakki Adytia, S., Nu-graha, Y., Farizah Rizka, R., Ernesto, A., Kanggrawan, J.I.: A computer vision-based object detection and counting for COVID-19 protocol compliance: a case study of Jakarta. In: 2020 International Conference on ICT for Smart Society (ICISS), pp. 1–5 (2020). https://doi.org/10.1109/ICISS50791.2020.9307594
2. Ioffe, S., Szegedy, C.: Batch normalization: accelerating deep network training by reducing internal covariate shift, pp. 1, 3, 7 (2015). arXiv:1502.03167
3. Howard, A.G., Zhu, M., Chen, B., Kalenichenko, D.,Wang, W., Weyand, T., Andreetto, M., Adam, H.: Mo-bileNets: efficient convolutional neural networks for mobile vision applications. arXiv:1704.04861v1 [cs.CV] 17 Apr 2017 (2017)
4. Liu, W., Anguelov, D., Erhan, D., Szegedy, B., Ree, S., Cheng-Yang, F., Berg, A.C.: SSD: Single shot multi box detector (2015)
5. Nascimento, J.C., Abrantes, A.J., Marques, J.S.: An algorithm-for centroid-based tracking of moving objects. In: 1999 IEEE International Conference on Acoustics, Speech, and Signal Processing. Proceedings. ICASSP99 (Cat. No.99CH36258), pp. 3305–3308. Phoenix, AZ, USA (1999). https://doi.org/10.1109/ICASSP.1999.757548
6. Bochkovskiy, A., Wang, C.-Y., Liao, H.-Y.M.: YOLO v4: optimal speed and accuracy of object detection. arXiv:2004.10934 v1 [cs.CV] 23 Apr 2020 (2020)
7. Yu, J., Jiang, Y., Wang, Z., Cao, Z., Huang, T.: UnitBox: an advanced object detection network. In: Proceedings of the 24th ACM International Conference on Multimedia, pp. 516–520 (2016)
8. Wang, C.-Y., Liao, H.-Y.M., Wu, Y.-H., Chen, P.-Y., Hsieh, J.-W., Yeh, I-H.: CSPNet: a new backbone that can enhance learning capability of CNN. In: Proceedings of the IEEE Conference on Computer Vision and Pattern Recognition Workshop (CVPR Workshop) (2020)
9. Redmon, J., Farhadi, A.: YOLOv3: an incremental improvement. In: CVPR (2018)
10. Misra, D.: Mish: a self regularized non-monotonic activation function. In: BMVC (2020)
11. Wojke, N., Bewley, A., Paulus, D.: Simple online and real-time tracking with a deep association metric. In: 2017 IEEE International Conference on Image Processing (ICIP) (2017)
12. Bewley, A., Ge, Z., Ott, L., Ramos, F., Upcroft, B.: Simple online and realtime tracking. In: IEEE International Conference on Image Processing (2016)
13. Bewley, A., Zongyuan, G., Ramos, F., Upcroft, B.: Simple online and realtime tracking. In: ICIP, pp. 3464–3468 (2016)
14. Sanchez-Matilla, R., Poiesi, F., Cavallaro, A.: Online multi-target tracking with strong and weak detections. In: European Conference on Computer Vision. Springer, pp. 84–99 (2016)
15. Yu, F., Li, W., Li, Q., Liu, Y., Shi, X., Yan, J.: Poi: Multiple object tracking with high performance detection and appearance feature. In: ECCV, pp. 36–42. Springer (2016)
16. Liu, W., Wen, Y., Yu, Z., Li, M., Raj, B., Song, L.: Sphereface: deep hypersphere embedding for face recognition. In: CVPR (2017)
17. Liu, W., Wen, Y., Yu, Z., Yang, M.: Large-margin softmax loss for convolutional neural networks. In: ICML (2016)
18. Wang, H., Wang, Y., Zhou, Z., Ji, X., Liu, Z., Gong, D., Zhou, J., Liu, W.: Cosface: large margin cosine loss for deep face recognition. In: CVPR (2018)
19. Deng, J., Guo, J., Xue, N., Zafeiriou, A.: ArcFace: additive angular margin loss for deep face recognition. In: CVPR (2019)

20. Wu, J., Cao, J., Song L., Wang, Y., Yang, M., Yuan, J.: Track to detect and segment: an online multi-object tracker. In: CVPR (2021)
21. Ahmed, I., Ahmad, M., Rodrigues, J.J.P.C., Jeon, G., Dinf, S.: A deep learning-based social distance monitoring framework for COVID-19 (2021)

Machine Learning for Green Smart Health Toward Improving Cancer Data Feature Awareness

Md Rajib Hasan, Noor H. S. Alani, and Rashedul Hasan

Abstract Radiation therapy and chemotherapy may be considered life-saving treatments for cancer patients. Though this treatment is not always successful, it left very higher effects on our environment. Drugs used in cancer treatment stop the growth and division of cells; released into the environment, they can affect the ecosystem through altered fertility and increased genetic defects. For a safer greener world undoubtedly, we can rely on machine learning technology to diagnose cancer in early stages. Hence, we may call that choosing the right influencing feature may reduce morbidity and green technology for early cancer diagnosis. Cervical cancer is an excellent example for such a study, as well as impacting individuals, families, and the environment. Cervical cancer presents almost no symptoms at the early stages of development of this condition. Because multi-factors may be involved, this demands a lot of research and analysis to identify causative or linked features. Choosing the right influencing feature is a challenging field in data science due to the presence and complexity of multi-dimensional data. The researchers have applied and optimized an ensemble learning algorithm as it is the best model for multi-modal medical data when relatively high dimensionality is present. The main objective of this study was to minimize the dependency on data pre-processing techniques, while analyzing the data (filling/ignoring missing values with the statistical method). Also, increasing such awareness of feature selection will immensely impact the environment (e.g., chemotherapy-free, less radiation therapy). Main factors were studied and validated using root mean square error (RSME) and mean absolute error (MAE). The classification accuracy for features was obtained by tenfold cross-validation and test (where 66% is training data and 34% test data).

M. R. Hasan (✉)
BICC—Cyber Security & IT Solution, Auckland, New Zealand
e-mail: rajib@bicc.co.nz

N. H. S. Alani
Eastern Institute of Technology, Napier, New Zealand
e-mail: nalani@eit.ac.nz

R. Hasan
SUNY Upstate Medical University, New York, USA

M. Lahby et al. (eds.), *Computational Intelligence Techniques for Green Smart Cities*, Green Energy and Technology, https://doi.org/10.1007/978-3-030-96429-0_10

Keywords Feature selection · Ensemble learning · Chemotherapy-free · Green smart education · Machine learning · WEKA · MATLAB · Cervical cancer · Green smart health

1 Introduction

Radiation therapy and chemotherapy are common treatments for cervical cancer and are commonly used for gynecologic tumors remedy. The increase in radiative forcing or radiation therapy results in higher surface temperatures, increased water-holding capacity of the atmosphere, increased evaporation, and larger water vapor amounts which affect the greenhouse. Similarly, it is widely believed that radiation causes an impact on the environment and society [1]. To prevent such consequences, early detection of cervical cancer is seeing an observed decrease in cervical cancer incidence and mortality, and one of these approaches can be used to urge more screening guidelines and more analysis of medical data [2].

Medical data is our research interest as medical data classification is acknowledged as an area of increasing importance, yet it also poses many difficulties as highlighted by Krawczyk and Schaefer [3], Hasan et al. [4]. Multiple classifier systems are considered as one of the most promising in medical data classification [3, 5]. In this context, ensemble learning is a better learning method for improving the accuracy where multi-modal medical data with a high relative dimensionality is present [6–8]. Finding the accurate features in cervical cancer data means early detection and prevention of cervical cancer. Early detection of cervical cancer will offer more greener environment as it will have less radiation therapy required.

Furthermore, as we can observe in Fig. 1, the pattern of cervical cancer around the world is alarming. If we consider the southern part of Africa, there is a high indication that the spread of such an incident is not random or scattered around the

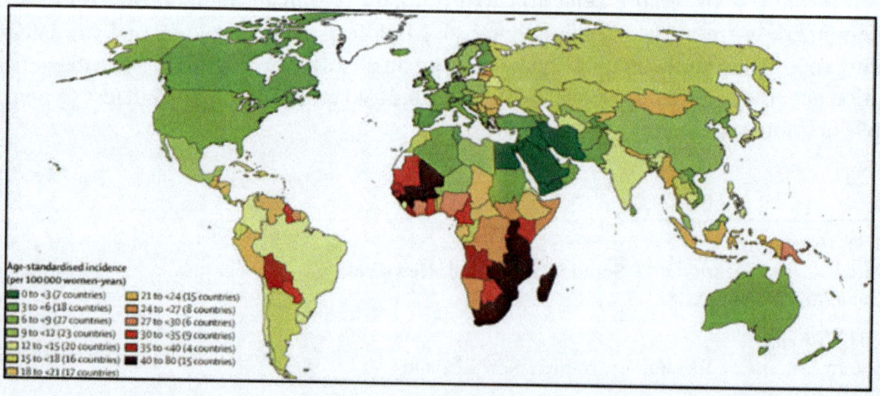

Fig. 1 Geographical distribution of world age-standardized incidence of cervical cancer by country estimated for 2018. *Source* [9]

region. When it is exposed in one region, it is prominent and that is because cervical cancer has no symptoms at early stages. Hence, a green smart health technique using data analytics methods would support the endeavors to understand the pattern, and from the analysis, we can establish a data product that can provide a thorough insight into factors, reasons, and relationships among all these factors.

The ensemble method has favorable properties that make them suitable for datasets with high dimensionality [7, 10–12], or missing values [13]. Data from medical studies (i.e., cervical cancer data from UCI will be used for this study, and this data poses the high dimensionality) typically suffer from one or more of the above conditions, due to the difficulty and cost of acquiring clinical data [7, 14, 15]. Ensemble methods are, therefore, suitable to be applied to medical datasets [5, 16].

This study employed an ensemble method to extract the influential and related features to the human papillomavirus (HPV) with the minimum dependency on data pre-processing techniques. Because medical professionals believe that ignoring or filling up, the missing value may change or bias the diagnosis outcome [17].

The novelty of this research is to minimize the dependency on data pre-processing techniques while analyzing the data (filling/ignoring missing values with the statistical method). This study identified that the feature STDs: HPV is not influenced by STDs: AIDS but influenced by STDs: HIV. To validate our findings, we have employed statistical methods: root mean square error (RSME) and mean absolute error (MAE); also, we have validated our findings from medical journals published PubMed to support our findings. This research bridges the knowledge of data mining, expert rue mining, and case-based reasoning to explain the relationship between the feature STDs: HPV, STDs: HIV, and STDs: AIDS. MATLAB and WEKA were employed as data analysis tools to develop and interpret the experimental results.

The rest of the chapter is structured as follows: Sect. 2 introduces common methods for feature extraction used in medical data, current context, and challenges, Sect. 3 discusses the proposed method, factors, and the analysis of the results, and Sect. 4 concludes.

2 The Findings from the Medical Literature

Human papillomavirus (HPV) and human immunodeficiency virus (HIV) are both infections that can be transmitted sexually, and it seems there is no medical link between the two conditions (see Fig. 2). The viruses cause different conditions, though people with HIV are more susceptible to HPV than others [18]. Thus, the behaviors that put someone at risk of contracting HIV can also raise the risk of getting HPV [19]. In addition to a high HIV burden, women have the burden of human papillomavirus (HPV) infection and related cancers worldwide [20].

In 2010, Rwanda showed that HPV infection increased the risk of HIV acquisition twofold. Collectively, these data have led to a new concept in which HPV and HIV infections may be bi-directional, each increasing the risk of the other [21]. Several medical trials have proved that HPV vaccination may prevent HIV infection [22].

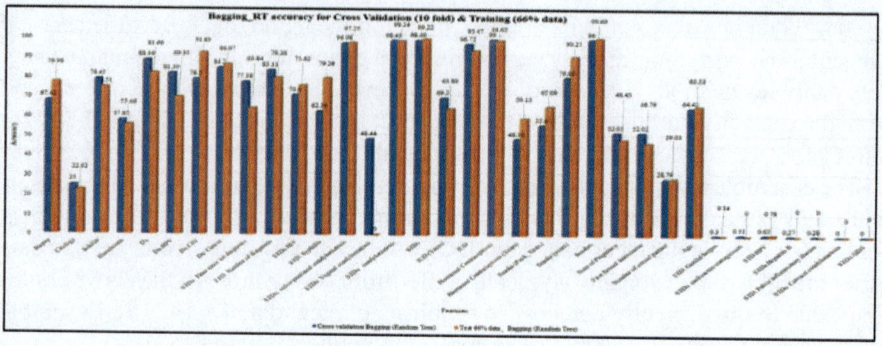

Fig. 2 Influence of all features studied in this research

HIV is an important enhancer of HPV carcinogenesis. There is also evidence that the unfavorable impact of HIV-related immunodeficiency has an impact on HPV. Clifford studied the effect of HIV infection on human papillomavirus and found that HPV detection was similar in 770 HIV positive (91.2%) from his sample data.

Researchers believe that it can take between 10 and 30 years from the time of an initial HPV infection until a tumor forms [23]. Persons with HIV can develop signs of infection anywhere from months to years after being infected. About half of the people with HIV develop AIDS within 10 years, but the time between infection with HIV and the onset of AIDS can vary greatly. After becoming infected with HIV, people could expect to get AIDS within about ten years, and then live only one to two years more [24]. From teaching perspectives, we include the HIV feature to provide a better understanding of the current relevancy of such a factor on cervical cancer analysis. Finally, Table 1 is provided to better summarize these approaches and findings (Table 2).

Table 1 Summary of facts about cervical cancer detection

Authors	Findings
[25]	Three machine learning models (decision tree, random forest, and XGBoost) are proposed to detect cervical cancer. The outcome shows a 93.33% accuracy rate
Fletcher et al.	The HPV and HIV viruses cause different conditions, though people with HIV are more susceptible to HPV than others
[26]	Studied the effect of HIV infection on human papillomavirus and found that HPV detection was similar in 770 HIV positive (91.2%) from his sample data
[23]	Researchers believe that it can take between 10 and 30 years from the time of an initial HPV infection until a tumor forms
[7, 27]	The ensemble method has favorable properties that make them suitable for datasets with high dimensionality or missing data

Table 2 Missing value dealing methods

Authors	Article title	Missing value dealing method
[28]	ILA4: overcoming missing values in machine learning datasets—An inductive learning approach	The inductive learning approach (ILA) is used to produce a set of classification rules by analyzing discrete training data with no missing values
[29]	SICE: an improved missing data imputation technique	An extension of multivariate imputation by chained equation (MICE) is proposed to deal with categorical and numerical data. The domain includes sixty hospitals around Bangladesh
[30]	Proper imputation of missing values in proteomics datasets for differential expression analysis	The study deals with a composition of missing values (random and not random). A public dataset was investigated to extract the composition of missing values. This is followed by a simulated dataset that emulates the missing value in the real-life dataset
[31]	Machine learning-based imputation soft computing approach for large missing scale and non-reference data imputation	The study focused immensely on pre-processing to split the dataset into small continuous segments. This is followed by employing multi-criteria decision-making (MCD) to choose which segment can represent the entire broken segments
[13]	A classifier ensemble approach for the missing feature problem	Multiple imputation methods based on random subspace
Khan et al.	Bootstrapping and multiple imputation ensemble approaches for missing data	Single imputation methods such as expectation maximization imputation, Gaussian random imputation, bagging single imputation, multiple imputations
Hassan et al.	Regression in the presence of missing data using ensemble methods	Generating missing values based on their probability density

3 Approach

The study design combines several phases including Phase 1: data analysis and knowledge acquisition, Phase 2: methods to build the ensemble model, Phase 3: primary data analysis and knowledge acquisition, and Phase 4: output/contribution as shown in Fig. 3. Throughout the study, MATLAB and WEKA have been used as data analysis tools.

Data has been obtained from UC Irvine (known as UCI) machine learning repository that is openly accessible from [32]. There are 858 instances and 35 attributes in this data. There are missing values in 23 features (total 35 features) in a different instance. STDs the time since the last diagnosis and first diagnosis features have 92%

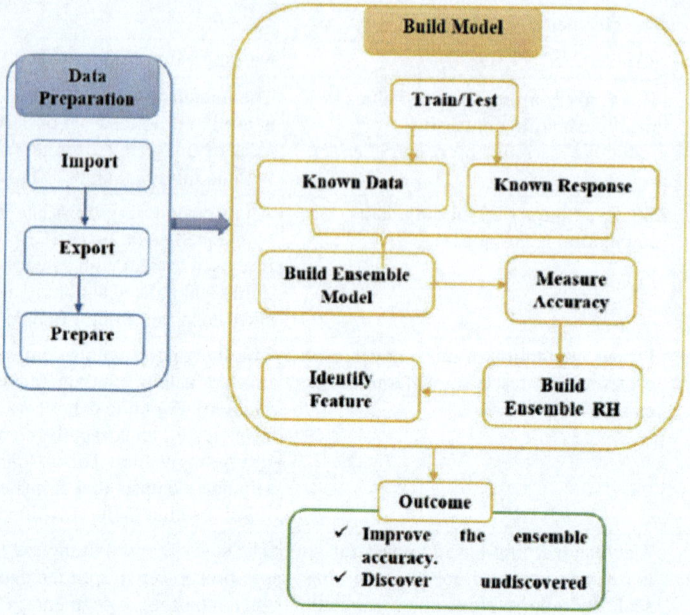

Fig. 3 Proposed approach

missing values, which is 787 out of 858 instances. Therefore, the number of missing values is very high in this cervical cancer data (Table 3).

In this research, we have fed all the 858 instances and 35 attributes/features into our preliminary analysis. Moreover, we have tried to select suitable features by feature selection techniques (ensemble modeling). The interesting and challenging part of this research is, we do not involve any data pre-processing techniques because it may change the result of the 45 medical diagnoses. Finally, the most challenging part of this research is "how to get better accuracy when data is suffering from outliers, missing values, and so on." From the data, the relationship between the attributes with cervical cancer will be identified.

In this research, we are employing ensemble methods. Bagging and boosting are the ensemble method that combine multiple classifiers, and it is the most robust machine learning method in medical settings [5]. Several studies dealt with a single classifier and only one class problem when all information was available, but in our case, the data is multivariate and suffers from missing values, outliers, and multi-classes.

The ensemble classifier is a combination of multiple classifiers (See Fig. 4) whose accuracy varies according to the accuracy of every single classifier, [5]. Since the ensemble method is itself a classifier, the combination of two or more ensemble classifiers can create a new ensemble method as follows. In summary, the brand-new ensemble method can be represented mathematically:

Table 3 Missing values in the feature (UCI cervical cancer dataset)

Feature name	Number of missing values	Missing value (%)
Number of sexual partners	26	3
First sexual intercourse	71	1
Number of pregnancies	56	7
Smokes	13	2
Smokes (year)	13	2
Hormonal contraceptives	108	13
Hormonal contraceptives (years)	108	13
IUD	117	14
IUD (years)	105	12
STDs (number)	105	12
STDs: condylomatosis	105	12
STDs: cervical condylomatosis	105	14
STDs: vaginal condylomatosis	105	12
STDs: vulvo-perineal condylomatosis	105	12
STDS: syphilis	105	12
STDs: pelvic inflammatory diseases	105	12
STDs: genital herpes	105	12
STDs: molluscum contagiosum	105	12
STDs: AIDS	105	12
STDs: HIV	105	12
STDs: Hepatitis B	105	12
STDs: time since first diagnosis	787	92
STDs: time since first diagnosis	787	92

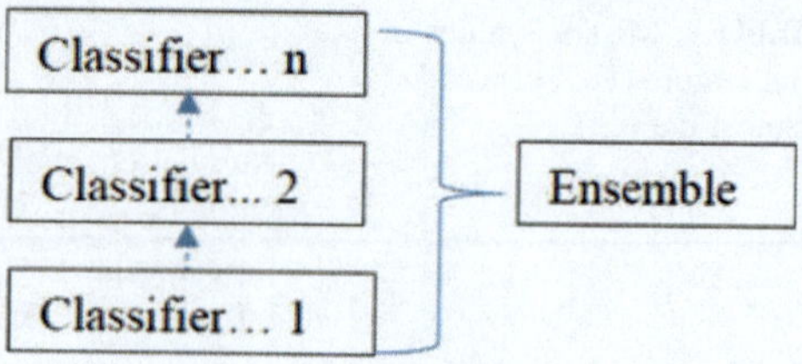

Fig. 4 Ensemble model

$$\text{Ensemble}_{\text{new}} = \frac{\text{Ensemble}_1 + \text{Ensemble}_2 + \text{Ensemble}_n}{N} \quad (1)$$

where

Ensemble_1 = first ensemble method (i.e., bagging)
Ensemble_2 = second ensemble method (i.e., boosting)
$\text{Ensemble}_n = n$ number of ensemble method where n is a positive real number
N = number of ensemble methods.

Alternatively, the formation of the new ensemble that can be written for this study is:

$$\text{Ensemble}_{\text{new}} = \frac{\text{Ensemble}_1 + \text{Decision tree}_1 + \text{Algorithm}_n}{N} \quad (2)$$

where

Ensemble_1 = first ensemble method (i.e., bagging or boosting)
Decision tree_1 = it could be a simple tree, complex tree, or any other tree
Algorithm_n = tt could be either n number of single decision tree or ensemble tree where n is a real positive number
N = number of algorithms.

More specifically, this study employed the following equation to develop Ensemble_RH which is:

$$\text{Ensemble}_{\text{RH}} = \frac{\text{Ensemble}_{\text{bagging}} + \text{Complex tree}}{N} \quad (3)$$

where

$\text{Ensemble}_{\text{bagging}}$ = ensemble bagging method
Complex tree = decision tree algorithm
N = number of decision tree algorithms.

The primary research challenge here is which decision tree or which ensemble method should be chosen. We considered the test results for the preliminary analysis and selected a complex tree and bagged tree to build the new ensemble model known as Ensemble_RH because both obtained near-perfect accuracy.

4 Experimental Results

To evaluate the performance of the proposed ensemble classifier method, several experiments with different classifiers were carried out. The results were then compared with the results of the proposed ensemble classifier method. Finally, the ensemble method was enhanced with Ensemble_RH and validated the outcome with statistical methods such as root squared error, mean absolute error.

4.1 Experimental Results

The preliminary results are divided into subsections: modeling the classification techniques, proposed modeling technique, and a summary of the preliminary results. In this section, simple tree, complex tree, linear SVM, boosted tree, and bagged tree have been employed to model classification. In the figure, the left side of the x-axis is the accuracy in terms of a correlation coefficient in percentage; the right side of the x-axis is the average error where the error can be from 0 to 1000. The highest coefficient accuracy with the least error is the optimal feature selection. Sixty-six percent of training and 34% testing regime have been employed in this study.

4.2 Modeling the Classification Techniques

Several classification techniques have been employed using experimental data. Figure 5 is a training analysis of decision tree classifiers, which depicts the medical data modeling problem for this study. We have chosen simple tree (95.6%), linear SVM (51.7%), boosted tree (97%), bagged tree (94.97%), and complex tree (97.9%). We noticed that linear SVM obtained the lowest accuracy of 51.17%, while other classifiers obtained an accuracy close to 95%. Surprisingly, two or more classifiers obtained nearly similar accuracy; complex tree and boosted tree obtained closely 97%. A similar pattern was also observed in a simple tree and bagged tree, which is closely 95%. Most of the literature identified that the different performance may be obtained due to multivariate, missing values, outliers, and multi-classes. Some literature suggests that employing an ensemble model may validate accuracy. However, most of the previous studies focused on only one classifier problem and either eliminated missing values or filled up missing values with the statistical method. However, medical experts do not agree with these methods of data handling. In this research,

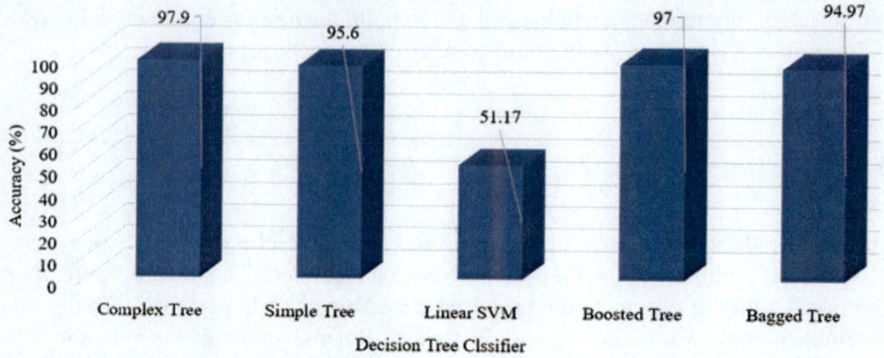

Fig. 5 Accuracy of different decision tree classifiers

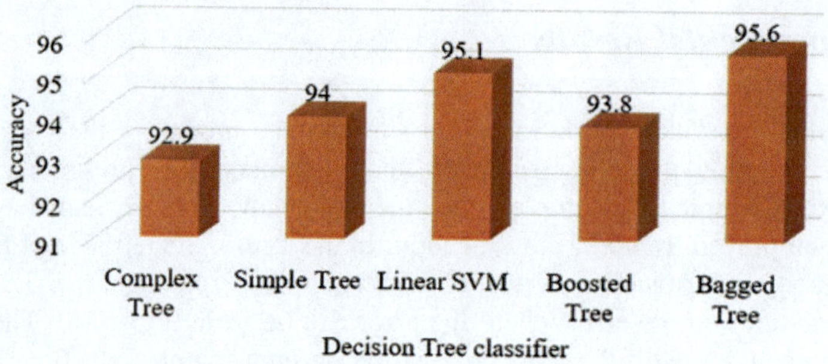

Fig. 6 Accuracy of different decision tree classifiers

we have tried to propose an ensemble method without involving data pre-processing techniques that are in line with the medical professionals' views.

Figure 6 explains the test results of the same data used above by decision tree classifiers. We have noticed a similar pattern of results like Fig. 5 but a huge change of performance of linear SVM, which is 95.1%.

Figure 7 is the performance between test and train analyses. The interesting point here is linear SVM improved accuracy by nearly 40% more than the test. The point may be noted that complex tree performance has been reduced by more than 5% in testing results, while it was the highest performer (97.9%) during test analysis.

4.3 *Missing Values*

Figure 8 depicts that the number of maximum missing values is 117, and the minimum is 0. This means we cannot ignore any feature, and this is very high-dimensional data.

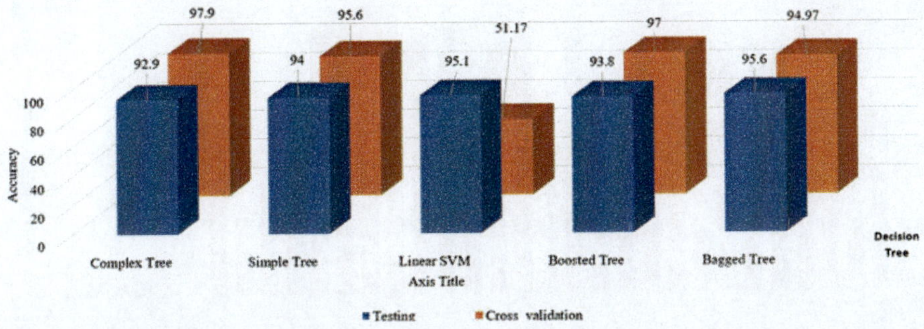

Fig. 7 Comparison between test and training analyses

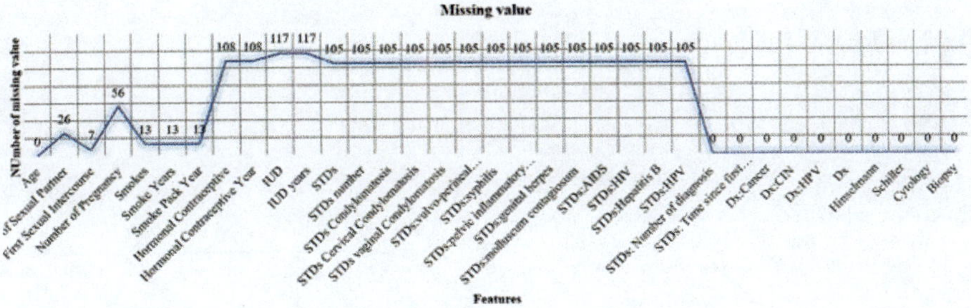

Fig. 8 Feature success rate and missing values

4.4 *Experimental Feature Selection*

This study will use the feature selection method of data modeling technique and identify the relationship in real-life problems by employing knowledge acquisition, symptom mining, and case-based reasoning. Figure 9 clarifies all the influential features that may be closely related to cervical cancer. The classification accuracy is for features obtained by test and train. During feature selection, "biopsy" is chosen as a predictor and other features as a predictor. Feature "biopsy" is chosen because in this study, we are proposing an intervention framework that will predict cervical cancer earlier than the cancerous stage.

4.5 *Relevancy of STD Features with Cervical Cancer*

The interesting point from Fig. 9 is it reveals that all features may not be equally influential in cervical cancer such as we can notice here that all STD features are not important. Hence, this study has a deeper look at the features of STD. Figure 10 shows that during the test analysis, all features in STDs are not very influential in cervical

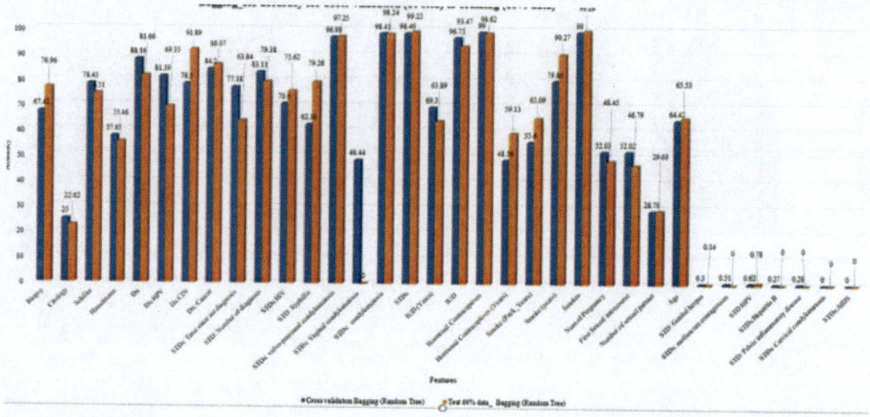

Fig. 9 Influence of all features

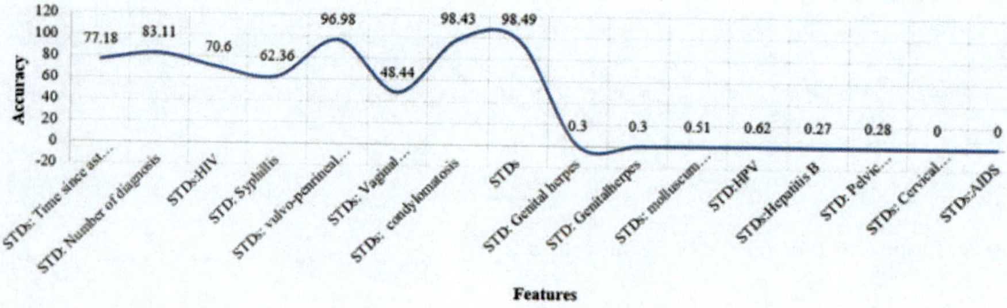

Fig. 10 Influence of STD features based on training analysis

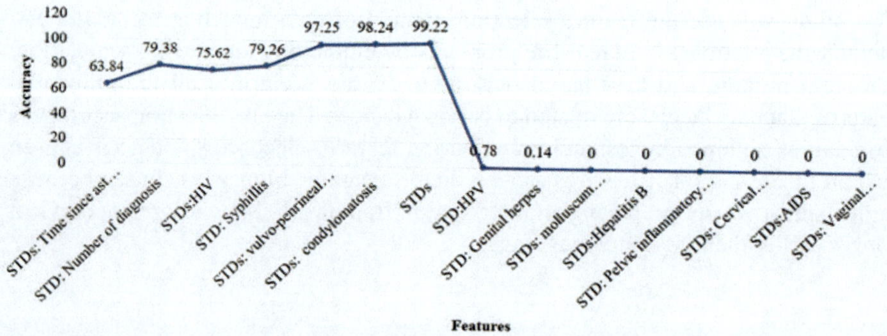

Fig. 11 Influence of STD features based on test analysis

cancer except STDs (98.49%), STDs: vulvo-perineal condylomatosis (96.98%), and STDs: condylomatosis (98.43).

In Fig. 10, it is clearly shown that STDs: vaginal condylomatosis influences 48.4%, while in the test analysis of the same feature extraction (Fig. 11), we have identified that STDs: vaginal condylomatosis has no relation (0%) with cervical cancer.

4.6 Relevancy of HIV and AIDS Features with Cervical Cancer

Figures 10 and 11 show that cervical cancer may be influenced by STDs: HIV but not by STDs: AIDS which seems unusual but in reality, it is not. Hence, data mining knowledge alone may not be sufficient to extract the relationship between features for an accurate intervention framework to identify the risk factor of having cervical cancer. For this reason, this research proposed that expert rule mining and case-based reasoning are important to identify and clarify the relationship between features. Once we have mined the rules of STs: HPV, STDs: HIV, and STDs: AIDS from an expert and utilized case-based reasoning, we have the answer for these unusual results from data mining. We found that HPV causes cervical cancer, but it requires a minimum of 10 years. If HPV is detected, early cervical cancer is preventable. Similarly, if anyone is infected with HIV, he or she may be diagnosed as an AIDS patient after 10 years. Hence, we see an HPV-infected person requires a minimum of 10 years to suffer from cervical cancer, while it is preventable if detected early, and an HIV-infected person requires 10 years to suffer from AIDS. On the other hand, the lifespan of AIDS patients is normally no more than two years. So, the AIDS patients normally do not survive for another 10 years as they may get cervical cancer after HPV infection. Hence, this research is a bridge between data mining and expert rule mining and case-based reasoning.

4.7 Relevancy of HPV Features with Cervical Cancer

Figure 12 employs several machine learning algorithms and shows the relevancy of the HPV feature to cervical cancer. The lowest relevancy rate is 63.07% by the M5P algorithm, and the highest is above 80% with several machine learning algorithms: decision stump, RepTree, ANN, SVM, and bagging.

4.8 Relevancy of Smoking Features with Cervical Cancer

Figure 13 reveals that smoking may be one of the relevant features that may influence cervical cancer, yet it is controversial (refer to the literature review section). It depicts the influence of having cervical cancer by smoking patterns such as what is the

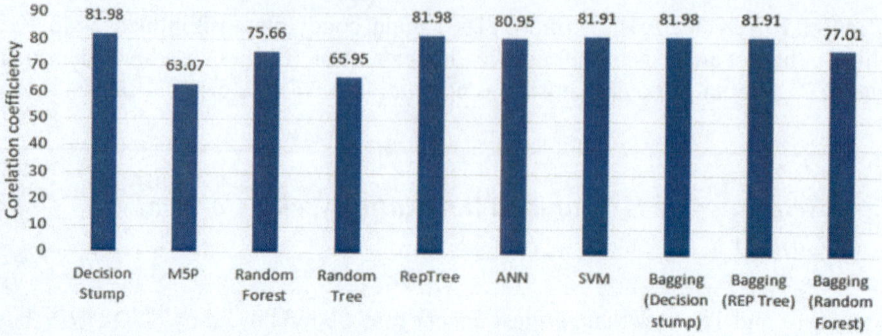

Fig. 12 Influence of STD features based on test analysis

Fig. 13 Influence of smoking

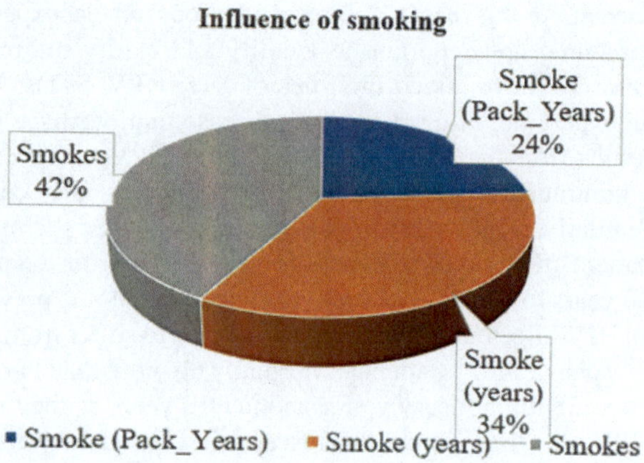

influence of a pack of cigarettes per day for a year, more than a packet per day for a year, and chain smoking. Our empirical study shows that if anyone continues smoking a packet of cigarettes per annum, he/she may have a chance of 24% to suffer from cervical cancer, smoking more than one packet in a year increases the chance of suffering from cervical cancer to 34%, and they may have an increased chance by 42% of having cervical cancer if they are a chain smoker.

4.9 Justification of the Feature Relevancy by the Average Error of Root Mean Squared Error and Mean Absolute Error)

The following qualitative analysis provides a comprehensive justification of the feature relevancy by the average error of RMSE and MAE (see Figs. 14, 15, 16, 17, 18, and 19).

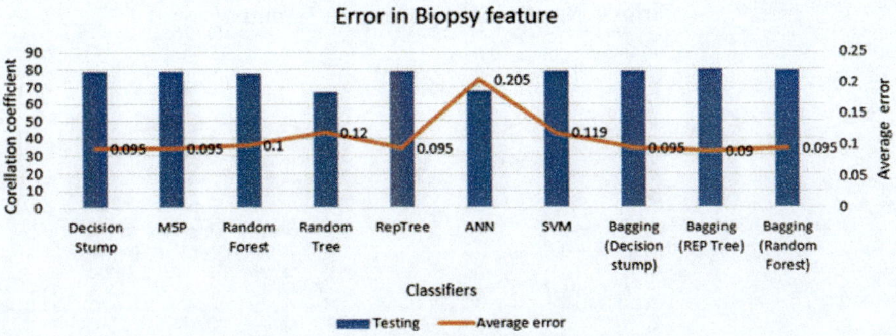

Fig. 14 Accuracy and average error in biopsy feature

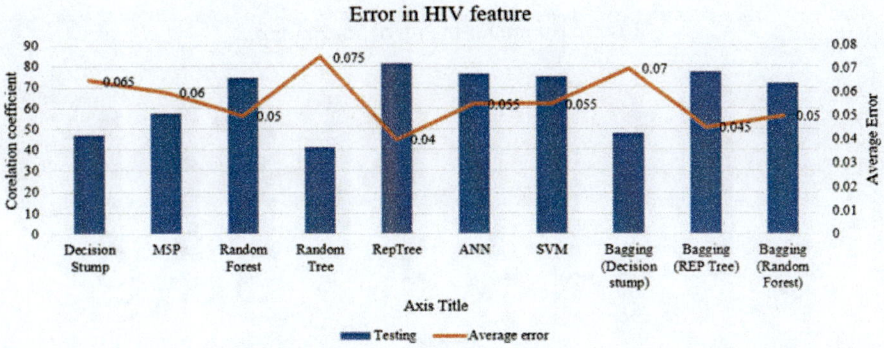

Fig. 15 Accuracy and average error in HIV feature

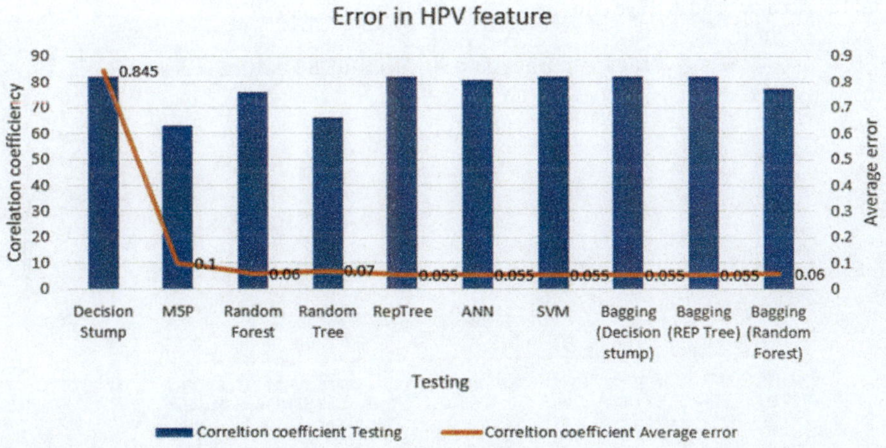

Fig. 16 Accuracy and average error in HPV feature

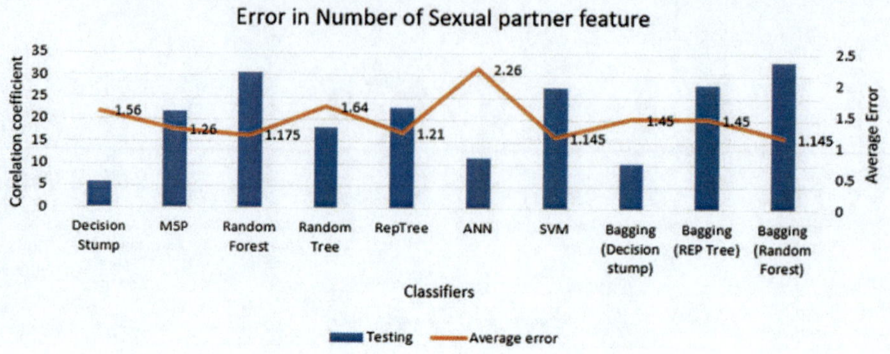

Fig. 17 Accuracy and average error in number of sexual partner feature

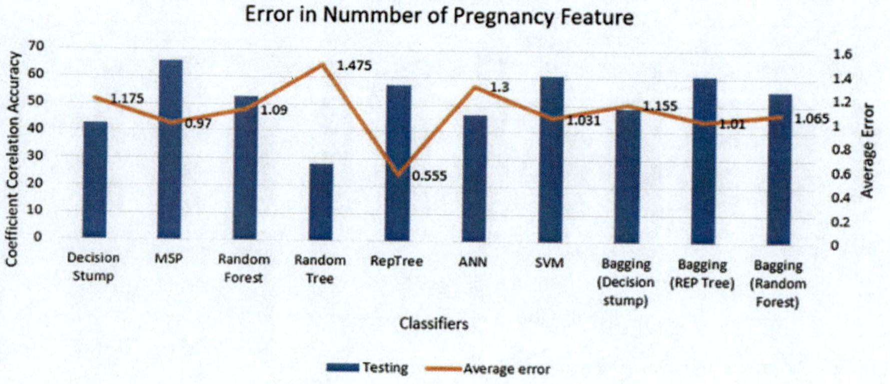

Fig. 18 Accuracy and average error in number of pregnancy feature

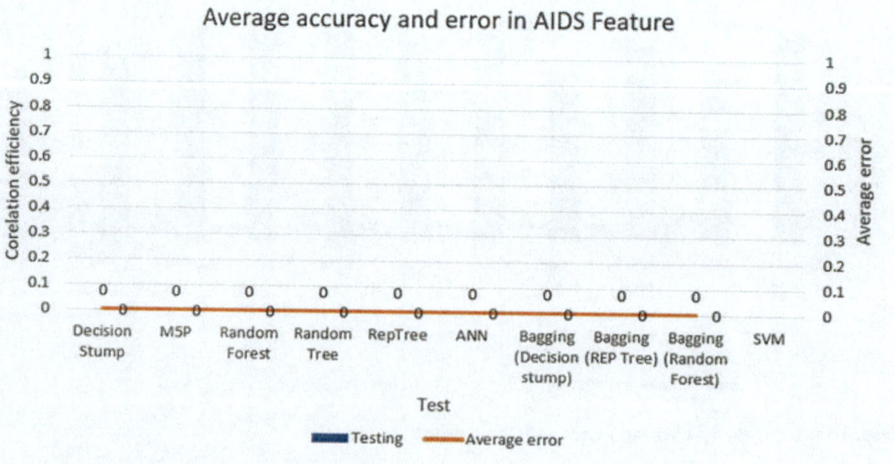

Fig. 19 Accuracy and average error in AIDS feature

Table 4 True positive (TP) and false positive (FP) error

Features	TP rate	FP rate
STD all	1	0.051
STDs: condylomatosis	1	0.068
STDs: vulvo-perineal condylomatosis	0.999	0.001
STD: HIV	0.999	0.889
STD: HPV	1	1
Smokes	1	0.618
Number of sexual partners	0.34	0.217
Number of pregnancy	0	0.011

4.10 Justification of the Feature Selection Accuracy by True Positive (TP) and False Positive (FP) Error

We have identified a few features that are related to cervical cancer and a few that are not. Table 4 shows the true positive (TP) and false positive (FP) rates. In the previous section, we claimed that not all STD features are related such as STD: HIV and STD: HPV and obtained high TP, but the rate of FP is also very high. The number of sexual partners and number of pregnancies obtained high accuracy with high error in a previous analysis. The TP and FP analyses also support that the identification from the previous analysis was correct. My analysis shows that smoking may be related to cervical cancer, but it obtained a bit high FP though TP is perfect. From the literature survey, the relevancy of smoking with cervical cancer is still debatable. Our findings from an ensemble perspective and statistical methods give a hint that medical researchers need to have a deep look into smoking features to identify their relevance with cervical cancer.

4.11 Classification Algorithm Performance for Cervical Cancer Data

Several classification techniques have been employed using experimental data. It is noticed that the linear SVM achieved the least accuracy of 50.8% (Figs. 20 and 21). In this stage, four decision tree classifiers have been applied: complex decision tree (97.9%), simple decision tree (95%), simple tree (95.6%), boosted tree (97%), bagged tree (max 95.7%). Figure 22 uncovers that the best performance is acquired by the bagged tree (min 95.5–max 95.7%) among the other classifiers (because the accuracy is free from bias). However, a consistent accuracy of the simple tree classifier (95.6%), complex tree classifier (97.9%), and boosted tree classifier (97%) reveals that the outcome is biased to get a positive result. This is a direct result of the data suffering from missing values, and it is multivariate data. Hence, among the classifiers, strange constant accuracy has been noticed, which is irregular.

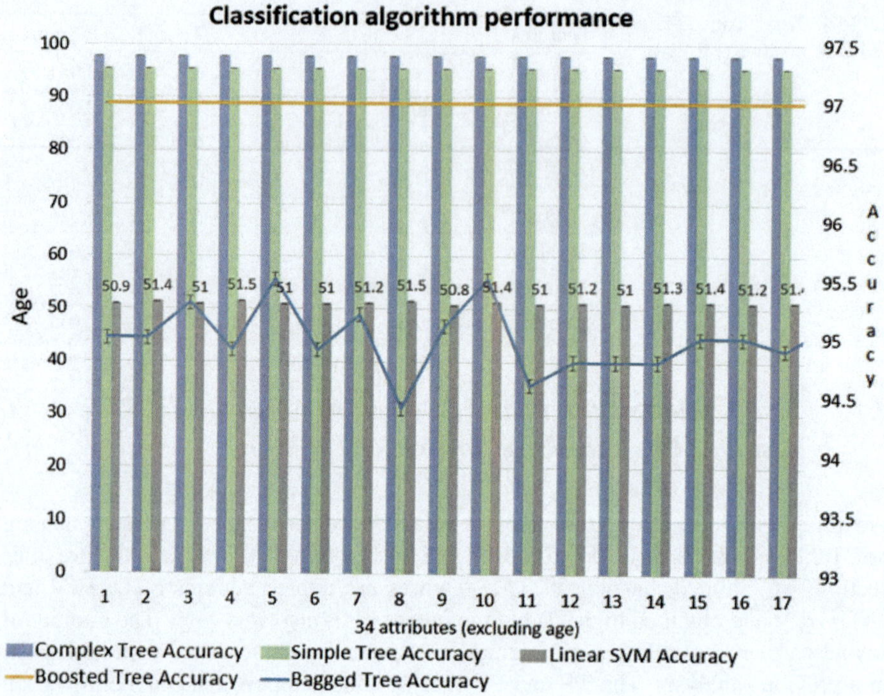

Fig. 20 Classification algorithm performance (First half)

From the analyses, it is noted that the ensemble bagged tree performs better with an accuracy of more than 95% (see Fig. 23).

5 Conclusion

Radiation therapy and chemotherapy may not be required when morbidity is reduced by early diagnosis. Less radiation therapy and chemotherapy may offer better health and be less viable for the environment. This study shows that identifying the features that are related to cervical cancer may change the cervical cancer treatment as it is early diagnosed and not required radiation therapy. Life critical data like cervical cancer data where missing values are present and missing value imputation is not a favorable option by the medical professionals (as it changes the diagnosis outcome), this research shows that choosing the right base classifiers in the ensemble, and the proposed new ensemble, Ensemble_RH, offers better accuracy without missing value imputation. Based on the literature review, it was found that ensemble is one of the best methods for early detection of cervical cancer. In this study, we have employed a decision tree classifier. We proposed a novel ensemble method called "Ensemble_RH," which offers expected 96.3% accuracy in the experimental analysis,

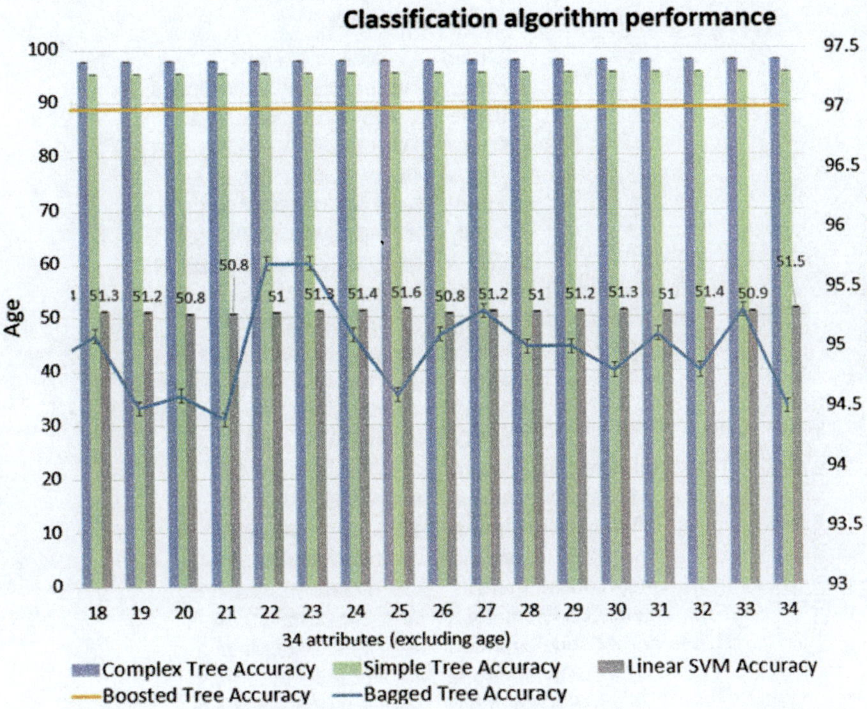

Fig. 21 Classification algorithm performance (Second half)

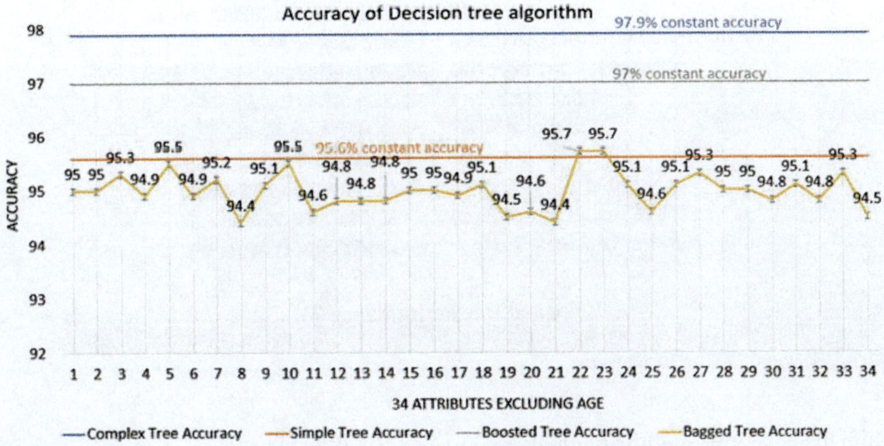

Fig. 22 Decision tree accuracy

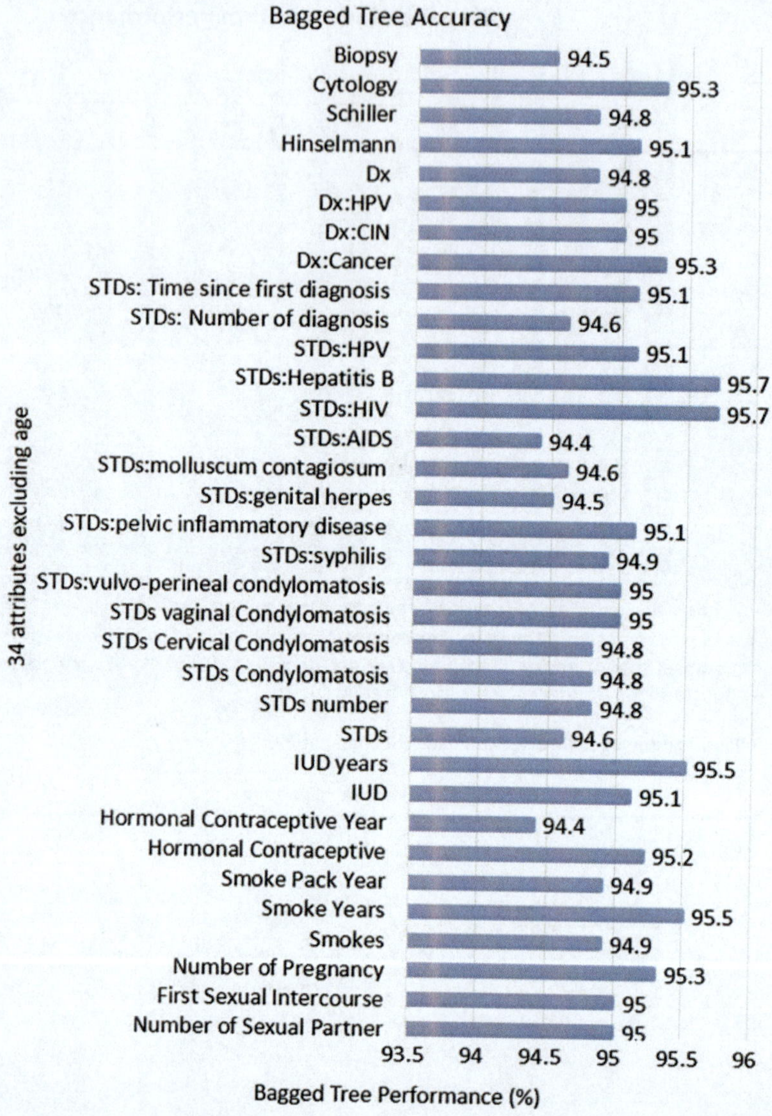

Fig. 23 Bagged tree performance

while the ensemble method bagged tree obtained 94.97%. The most challenging part was to evade data pre-processing as it is not a favorable and agreeable option by physicians. Our proposed ensemble model outperformed without employing data pre-processing techniques.

Acknowledgements We acknowledge Dr. Shariful Hasan from Ampang Puteri Hospital, Malaysia, for his guidance and consultancy during this research. Furthermore, we like to say thanks to Dr.

Safiuzzaman, in Chittagong medical college hospitals, who had given to test the system. We also would like to acknowledge Ian Purdon (EIT, New Zealand) for the fruitful discussions.

References

1. Coombs, N.J., et al.: Environmental and social benefits of the targeted intraoperative radiotherapy for breast cancer: data from UK TARGIT-A trial centres and two UK NHS hospitals offering TARGIT IORT. BMJ Open **6**(5), e010703 (2016).https://doi.org/10.1136/bmjopen-2015-010703
2. Saslow, D., et al.: American Cancer Society, American Society for Colposcopy and Cervical Pathology, and American Society for Clinical Pathology screening guidelines for the prevention and early detection of cervical cancer. CA Cancer J Clin **62**(3), 147–172 (2012)
3. Krawczyk, B., Schaefer, G.: Dealing with the difficult learning situation. Neural Netw. Appl. Electr. Eng. **1**(1), 12–15 (2012)
4. Hasan, M.R., Bakar, N.A.A., Siraj, F., Sainin, M.S., Hasan, M.S.: Single decision tree classifiers' accuracy on medical data. In: Proceedings of 5th International Conference on Computings and Informatics, ICOCI 2015, no. 188, pp. 671–676 (2015a)
5. Hasan, M.R., Siraj, F., Sainin, M.S.: Improving ensemble decision tree performance using Adaboost and Bagging. AIP Conf. Proc. **1691**, 1–7 (2015b)
6. Wu, G., Shen, D., Sabuncu, M.R.: Machine Learning, and Medical Imaging. Elsevier Inc. (2016)
7. Hasan, M.R., Golamhosseini, H., Sarkar, N.I., Safiuzzaman, S.M.: Intrinsic motivated cervical cancer screening intervention framework. Humanit. Technol. Conf., 506–509 (2017a)
8. Tay, W., Chui, C., Ong, S., Ng, A.C.: Expert systems with applications ensemble-based regression analysis of multimodal medical data for osteopenia diagnosis. Expert Syst. Appl. **40**(2), 811–819 (2013)
9. Arbyn, M., Weiderpass, E., Bruni, L., Sanjosé, S., Saraiya, M., Ferlay, J., Bray, F.: Estimates of incidence and mortality of cervical cancer in 2018: a worldwide analysis. Lancet Global Health **8**(2), e191–e203. ISSN: 2214-109X (2019)
10. Dittman, D.J., Khoshgoftaar, T.M., Napolitano, A.: Selecting the appropriate ensemble learning approach for balanced bioinformatics data. Int. Florida Artif. Intell. Res. Soc., 329 334 (2015)
11. Blagus, R., Lusa, L.: Boosting for high-dimensional two-class prediction. BMC Bioinform. **16**(1), 1–17 (2015)
12. Ojha, V.K., Jackowski, K., Abraham, A., Snášel, V.: Dimensionality reduction, and function approximation of poly (lactic-co-glycolic acid) micro-and nanoparticle dissolution rate. Int. J. Nanomed. **10**, 1119 (2015)
13. Nanni, L., Lumini, A., Brahnam, S.: A classifier ensemble approach for the missing feature problem. Artif. Intell. Med. **55**(1), 37–50 (2012)
14. Lee, C.H., Yoon, H.-J.: Medical big data: promise and challenges. Kidney Res. Clin. Pract. **36**(1), 3–11 (2017)
15. Kang, H.: The prevention and handling of the missing data. Korean J. Anesthesiol. **64**(5), 402–406 (2013)
16. Polikar, R., et al.: An ensemble-based data fusion approach for early diagnosis of Alzheimer's disease. Inf. Fusion **9**(1), 83–95 (2008)
17. Groenwold, R.H.H., Dekkers, O.M.: Missing data: the impact of what is not there. Eur. J. Endocrinol. **183**(4), E7–E9 (2020)
18. Fletcher, J., Murrell, D.: What is the link between HPV and HIV. Medical News Today, Sussex (2018)
19. Pietrangelo, N., Ernst, H.: HPV and HIV: What Are the Differences. Healthline media, San Francisco (2018)

20. Denny, L., Adewole, I., Anorlu, R.: Human papillomavirus prevalence and type distribution in invasive cervical cancer in sub-Saharan Africa. Int. J. Cancer J. Int. du cancer **1**(1), 1–7 (2013)
21. Vyankandondera, V., van de Wijgert.: HIV acquisition is associated with prior high-risk human papillomavirus infection among high-risk women in Rwanda. AIDS **24**(1), 2289–2292 (2010)
22. Schim van der Loeff, M., Nyitray, A., Giuliano, A.: HPV vaccination to prevent HIV infection: time for randomized controlled trials. Sex. Transm. Dis. **38**(7), 640–643 (2011)
23. McCredie, M.R.E., Sharples, K.J., Paul, C.: Natural history of cervical neoplasia and risk of invasive cancer in women with cervical intraepithelial neoplasia. A Retrosp. Cohort Study. Lancet Oncol. **9**(5), 425–434 (2008)
24. Peiperl, L., Coffey, S.: How long can people infected with HIV expect to live. US department of Veteras affair. [Online]. Available: https://www.hiv.va.gov/patient/faqs/life-expectancy-with-HIV.asp. (2017). Accessed 09 Feb 2019
25. Akter, L., Ferdib-Al-Islam, Islam, M.M., et al.: Prediction of cervical cancer from behavior risk using machine learning techniques. SN Comput. Sci. **2**, 177 (2021).https://doi.org/10.1007/s42979-021-00551-6
26. Clifford, G.M., De Vuyst, H., Tenet, V., Plummer, M., Tully, S., Franceschi, S.: Effect of HIV Infection on human papillomavirus types causing invasive cervical cancer in Africa. Epidemiol. Prev. **73**(3), 332–339 (2016)
27. Hasan, M.R., Gholamhosseini, H., Sarkar, N.I.: A new ensemble model for multivariate medical data. In: International Telecommunication Networks And Applications Conference, p. In press. (2017b)
28. Elhassan, A., Abu-Soud, S., Alghanim, F., Walid, A.S.: ILA4: overcoming missing values in machine learning datasets—an inductive learning approach. J. King Saud Univ. Comput. Inf. Sci. (2021). https://doi.org/10.1016/j.jksuci.2021.02.011
29. Khan, S.I., Hoque, A.S.M.L.: SICE: an improved missing data imputation technique. J. Big Data **7**, 37 (2020). https://doi.org/10.1186/s40537-020-00313-w
30. Liu, M., Dongre, A.: Proper imputation of missing values in proteomics datasets for differential expression analysis. Briefings Bioinform. **22**(3) (2021). https://doi.org/10.1093/bib/bbaa112
31. Alamoodi, A.H., Zaidan, B.B., Zaidan, A.A., Albahri, O.S., Chen, J., Chyad, M.A., Garfan, S., Aleesa, A.M.: Machine learning-based imputation soft computing approach for large missing scale and non-reference data imputation. Chaos Solitons Fractals **151** (2021)
32. Fernandes, K., Cardoso, J., Fernandes, J.: Transfer learning with partial observability applied to cervical cancer screening. In: Iberian Conference on Pattern Recognition and Image Analysis. Springer International Publishing (2017)
33. Moon, H., Ahn, H., Kodell, R.L., Baek, S., Lin, C.-J., Chen, J.J.: Ensemble methods for classification of patients for personalized medicine with high-dimensional data. Artif. Intell. Med. **41**(3), 197–207 (2007)
34. Deeks, S.G., Lewin, S.R., Ross, A.L.: International AIDS Society global scientific strategy: towards an HIV cure 2016. Nat. Med. **22**(1), 839–850 (2016)

Machine Learning and Green Environment

Solar Radiation Forecasting for Smart Building Applications

Gilles Notton, Ghjuvan Antone Faggianelli, Cyril Voyant, Sarah Ouedraogo, Guillaume Pigelet, and Jean-Laurent Duchaud

Abstract The development of solar energy and the concept of net zero or positive energy buildings are being increased in a sort of symbiotic relationship. To make buildings more energy efficient, smart and optimal energy management systems called predictive building control were developed both for heat and electricity utilization. In these energy management systems (EMSs), a weather forecasting platform is often incorporated and allows to anticipate the meteorological events influencing the electrical and thermal energy consumption and to react accordingly. The objectives of this chapter consist in showing how the introduction of a solar radiation forecasting tool into the EMS improves its performances and saves energy and money. A brief overview of solar radiation forecasting methods is shown and focuses on solar prediction at short time horizon using statistical and artificial intelligences technics. These solar radiation prediction models are applied, validated, and compared on the Mediterranean site of Ajaccio, France. The most reliable forecasting tool, ARMA model, is then incorporated into the EMS which manages electricity into a micro-grid composed of a photovoltaic/battery energy system and supplying a building and an electrical vehicle. The electricity cost benefit is then estimated and discussed. It appeared that the addition of PV forecasting based on an ARMA model into the EMS increases the gain of about 7% compared to a EMS without forecasting; this gain could reach 10% with a perfect forecasting reaches.

Keywords Energy management system · Solar energy · Solar radiation forecasting methods · Energy savings · Optimization

---

G. Notton (✉) · G. A. Faggianelli · C. Voyant · S. Ouedraogo · G. Pigelet · J.-L. Duchaud
Laboratory Sciences for Environmental, UMR CNRS 6134, University of Corsica, Route des Sanguinaires, 20000 Ajaccio, France
e-mail: notton_g@univ-corse.fr

C. Voyant
Castelluccio Hospital, Radiotherapy Unit, BP 85, 20177 Ajaccio, France

M. Lahby et al. (eds.), *Computational Intelligence Techniques for Green Smart Cities*, Green Energy and Technology, https://doi.org/10.1007/978-3-030-96429-0_11

1 Introduction

New solar architectural building concepts are increasing for several years with the objective to achieve a net zero or positive energy level for houses and possibly for a whole city both for a thermal and electrical energy point of view: these buildings produce themselves electricity and heat for their own needs, thanks to solar energy. This concept of zero energy building is already outdated, being replaced by positive energy building which produces more on-site energy from renewable sources than it consumes and sends the production in excess to the neighborhood. These buildings have for objective to achieve appropriate thermal comfort levels [1].

Due to the dynamic change in solar radiation, reliable energy generation forecasting is necessary for grid operation in the case of solar energy generation and also for passive solar architectural design for the optimal thermal performance of buildings [2].

The rapid developments in information and communication technology (ICT) make possible today a smarter management of energy, through the utilization of improved sensoring, monitoring, and automation capabilities with an improving energy efficiency for a better utilization of resources.

Smart homes or buildings interconnected in a smart grid are energy systems, generally comprising a local generation (often solar production), storage, and loads. Their inhabitants are usually considered as prosumer (both producing and consuming (buying) electricity and heat) with communication and control abilities aiming to increase their social benefits (e.g., reducing their bills and sometimes their environmental impact) [3]. The operation of these smart solar homes requires an optimal control system called predictive building control (PBC) or model predictive control (MPC) [3]. PBC and MPC both require a model of the system, real-time controllers, and solar irradiation forecasts. Good models for buildings and control scheme were well studied and developed in literature but solar forecasting requirements for such applications have not yet been well studied [4].

Demand response (DR) is well-known and consists of controlling only the building electricity consumption by adjusting and/or curtailing load consumption (decreasing their load during critical network congestion period with higher energy price or to shift it to off-peak periods) [5, 6]. DR only optimizes the load but not the renewable energy sources and energy storage systems utilization. Reinisch et al. [7] developed the ThinkHome approach realizing a smart system integrating devices, protocols, parameters, data to achieve higher energy efficiency and comfort.

To make a home smart, the building management system must control and manage both load, production, and storage. A predictive building control is a supervisory control technology which forecasts weather and energy use, generation, and storage data fully operational in real time which leads to significant improvements in building energy performance [8]; it improves the heating, ventilation, and air conditioning system performance and minimizes the building energy use, costs, and carbon footprint while achieving better comfort conditions in buildings. An overview of MPC for building control and modeling paradigms and a presentation of algorithms used for real-life implementation was realized by Drgona et al. [9]. MPCs applied to active energy storage systems, with optimal management of on-site renewable energy sources were shown by Serale et al. [10].

If the previous works were developed, rarely, the description and the choice of the forecasting method were presented and rarely, the benefit of its introduction into an EMS was estimated and compared with a strategy without an anticipation of the production and the consumption.

This chapter has for objective to estimate the benefit in term of energy and money savings obtained by the introduction of a solar production forecasting tools into a smart energy management algorithm; the other weather variables forecasting will not be discussed in this chapter.

Firstly, a short review of solar radiation forecasting methods available in the literature is realized according to the prediction horizon and time step data. Eleven models for the prediction of hourly solar irradiations at short time horizon (1–6 h) will be presented in more detail.

Then, in a second paragraph, these forecasting methods are compared in term of reliability level on hourly solar data measured in Ajaccio, Corsica, a Mediterranean site with a medium solar variability. It will appear that the more complex methodology is not always the more efficient and it must be kept in mind that the method must be integrated with a relative simplicity into the EMS.

At last, the solar forecasting tool, that showed its effectiveness, is integrated into a energy management system applied to a microgrid with photovoltaic (PV) production and a battery storage, providing electricity to a building and an electrical vehicle; the efficiencies gained by the introduction of solar energy forecasting (ARMA and perfect one) are thus shown in term of kilowatt-hour cost savings.

The benefit of the use of forecasting tools is clearly shown with a decreasing of 7% in the total energy cost, benefit that could increase for applications in location with higher solar variability and/or with a more complex or variable electricity consumption profile.

2 Solar Radiation Forecasting Overview

The intermittence of solar radiation and its randomness induces some difficulties for managing the powers produced and consumed in an electrical (or thermal) grid. We do not know what will be the future available solar production (the same problem occurs for future load consumption); it can be, for instance, more profitable to store solar energy for a future use, when the tariff of the electricity supplied by the electricity company will be higher, than to use it immediately. We must keep in mind that a perfect balance between production and consumption must be reached at every moment, and it is not easy when the production is random and intermittent. Already, it appears that a predicted and anticipated event is easier to manage both for technical and cost reasons. The electrical energy operator or the EMS needs to anticipate the future of the electrical production and consumption at various temporal horizons [11, 12].

A short overview of solar forecasting methods is firstly presented versus the time horizon of the prediction; the time horizon interesting for the energy management in a small smart grid being a medium time horizon (1–6 h), statistical, and artificial intelligence methods will be described more in detail.

2.1 Solar Energy Forecasting Methods Categories

The forecasting method is chosen according to the forecasting time horizon and the temporal resolution (See Fig. 1).

The existing methods can be classified in four different sets [11, 12]:

- Time series-based methods: They are based on statistical models solely applied to ground data measured in the past and at the instant of the prediction;
- Numerical weather forecast (NWP)-based methods: They are based on weather forecasts that employ a set of equations that describe the flow of fluids inside the atmosphere;
- Satellite imagery-based methods: They use images of the earth taken by satellite;
- Sky image-based methods: Based on observations of cloud cover from the ground with an in-situ camera in view to observe the clouds movement and predict its positions some minute later.

These four families of models are not in competition and each one shows good performances within the range of its time horizon; moreover, they are increasingly

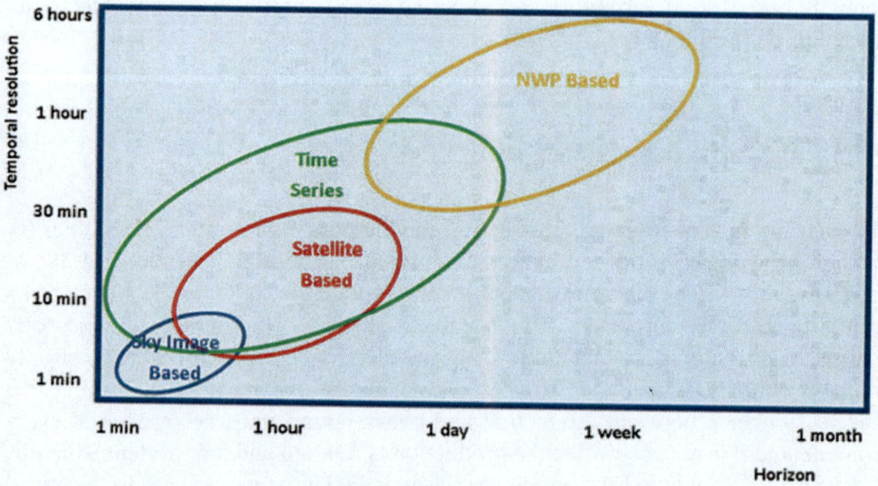

Fig. 1 Forecasting time horizon versus temporal resolution [12]

coupled together in hybrid methods generally more complex to implement but also more efficient [14].

For building applications, the forecasting time horizon lies generally between some minutes to some hours [13]. In this gap of prediction horizons, time series-based methods are particularly well adapted. It appears important to add some other information before describing in more detail the models:

- To forecast PV or solar thermal power directly from the previous power data is more complicated than to predict solar irradiance and to transform it in PV or thermal power via a physical model: in statistical and artificial intelligence models, a data history is required for training the forecasting model and it is easier to access to an historical set of solar radiation than of solar plant production (the solar plant has to be installed for a very long time if we want to use the data production for training the model). If there is any change in the solar plant peak power or if technical problems or maintenances occur, the historical data set is no more reliable nor homogenous [15] and then the training data set will be difficult to be used efficiently.
- It is simpler to forecast the production of a group of solar plants spread over a large area than that of a single solar plant; the aggregate effect, repartition of solar systems over a large territory, averaging, and smoothing the production.

2.2 Time Series-based Methods

These methods predict the future solar irradiance $G(t+h)$ from the p past observed data $G(t-p)$ [16]; thus, according to Eq. (1):

$$G(t+h) = f[G(t), G(t-1), G(t-2), \ldots, G(t-p)] \tag{1}$$

The various forecasting methods based on time series use

- Naïve models as persistence and smart persistence, easy to implement but not very efficient, and often used as a reference for comparison with more complex models;
- Statistical models as autoregressive models (AR), moving average (MA), autoregressive moving average (ARMA) or Markov chains models;
- Artificial intelligence models as artificial neural network (ANN), fuzzy logic (FL), and other hybrid methods.
- In the following part of this paragraph, the methods will be briefly presented and we suggest for more information to read references [17, 18].

Preliminary Step: Data Cleaning and Stationarity: Before developing and implementing a forecasting method, it is first necessary to control the data quality because some measuring errors may appear due to problems with the acquisition system.

Secondly, the data must be filtered by removing night-time, sunset, and sunrise data (these last ones are unreliable due to the bad cosine response of the pyranometer and the mask effect of the surrounding). All data with a zenith angle higher than 80–85° are deleted.

Artificial and statistical models are efficient for predicting time series with a stationary behavior but solar radiation has a seasonality and periodicity [19]. To remove these periodicities and to make the solar data stationary, they are converted in non-dimensional data k_t (clearness index) defined as the ratio of the solar radiation G on the ground to the clear sky solar radiation G_{CS} considering the climatic conditions of the meteorological site. The clear sky solar irradiance is the solar irradiance received on the ground when the sky is clear (without clouds). k_t is defined by Eq. (2):

$$k_t(t) = \frac{G(t)}{G_{cs}(t)} \tag{2}$$

Numerous models of clear sky solar radiations were developed in the literature with a more or less high complexity. Between these models, we find Solis model developed by Mueller et al. [20] and simplified by Ineichen [21] and the European Solar Radiation Atlas (ESRA) model [22].

The next step consists in choosing the number of input data for each model (i.e., the number p of Eq. (1)) using for example an auto-mutual information method [23]. The auto-mutual information determines the degree of statistical dependence between the inputs and the output of the model.

At last, the data are divided randomly into a training data set (80%) and a test data set (20%) and the process is redone k times (often k is taken equal to 10). Hence, the results become independent of the training data set which increases their robustness. This method is called k-fold sampling [24].

In what follows, the symbol $\left(\widehat{G}\right)$ means that the value is forecasted; without this symbol, it is measured. Only some information and references are given for each model because describing in detail each method would be long and fastidious.

Naïve Models. Two simple models are called Naïve predictor. The first one, the persistence model, considers that the solar irradiation predicted at a time $t + h$ is identical to that at time t. Its accuracy decreases greatly with forecasting horizon. It is considered that it cannot be used for a time horizon greater than 1. The persistence model can be written as

$$\hat{G}(t + h) = G(t) \tag{3}$$

To consider the fact that the sun position changes between t and $t + 1$, the persistence model is corrected with a clear sky ratio term k_t [25] and is then called scaled or "smart" persistence.

$$\hat{k}_t(t + h) = k_t(t) \rightarrow \hat{G}(t + h) = G(t) \times \frac{G_{CS}(t + h)}{G_{CS}(t)} \tag{4}$$

These two models are used as reference models: The performance of a complex predictor will be compared to the performances of the scaled persistence in view to judge if the complexity improves more or less the efficiency of the prediction.

ARMA Model. The autoregressive mobile average (ARMA) model is well-known to predict the future value of solar radiation [26]. It is divided into an autoregressive part (AR) and a moving average part (MA); p and q are, respectively, the order of the AR and the MA, and the ARMA model is noted as ARMA (p, q). It can be written according Eq. (5):

$$\hat{k}_t(t+h) = \varepsilon(t) + \sum_{i=0}^{p} \varphi_i \cdot k_t(t-i) + \sum_{i=0}^{q} \theta_i \cdot \varepsilon(t-i) \tag{5}$$

φ and θ are the ARMA parameters deduced by a least square method, and $\varepsilon(t)$ is the error related to a normal distribution.

Artificial Neural Network (ANN): Multilayer Perceptron (MLP): ANN is probably the more known machine learning method and is a non-linear predictor. The multilayer perceptron (MLP) with feed-forward back propagation has proven to be a powerful tool to forecast solar radiation [27, 28]. A hidden layer (often only one) receives input data and sends an output signal to the output layer. A neuron collects signals from prior neurons or input data in only one direction for a feed-forward MLP configuration (Fig. 2).

The predicted k_t at a time horizon $t+h$ is then:

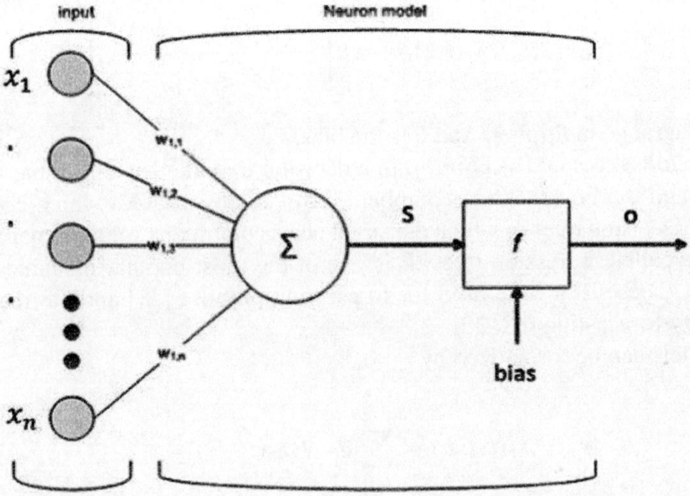

Fig. 2 Neuron model

$$\hat{k}_t(t+h) = \sum_{j=1}^{m} \omega_j.(g\left(\sum_{i=0}^{n-1} \omega_{i,j}.k_t(t-j) + b_j\right) \tag{6}$$

b_j is the biases of the hidden neuron j and $\omega_{i,j}$ the weights between the input i and the hidden node j, g is the transfer function, ω_j the weight between the output and the hidden neuron j.

Gaussian Process (GP). A Gaussian process is a Gaussian distribution with an infinity of variables. The functioning and implementation of GP are described by Rassmusen [29]. Each forecasted k_t is considered as the sum of a function $f(k_t(\tau))$ and an independent Gaussian noise $\mathcal{N}(\mathrm{O}, \sigma_{\mathrm{n}}^2)$ with a variance σ_n^2 and $k_t(\tau) = (k_t(t),$ $k_t(t-1), \cdots, k_t(t-n))$, it can be presented as

$$\widehat{k_t}(t+h) = f(k_t(\tau)) + \mathcal{N}(\mathrm{O}, \sigma_{\mathrm{n}}^2) \tag{7}$$

Support Vector Regression (SVR): SVR is a kernel-based model which was developed to solve regression problems [30] but also to predict solar radiation [18]. $\widehat{k_t}$ is predicted using

$$\widehat{k_t}(t+h) = \sum_{\tau=1}^{t-1} \alpha_\tau \cdot f_{rbf}(k_t(t+h), k_t(t-\tau)) + b \tag{8}$$

where f_{rbf} is the kernel radial basis function given by

$$f_{rbf}(k_t(t_p), k_t(t_q)) = exp\left[\frac{-(k_t(t_p) - k_t(t_q))^2}{2\sigma_f^2}\right] \tag{9}$$

α_i is the Lagrange multipliers and b is the bias.

Regression Trees (RTs) Family: In a decision tree, the leaves are the values of the target and the branches are combinations of input variables which conduce to this target. Decision trees in which the target takes continuous values (solar radiation values) are called regression trees. It is one of the most popular machine learning algorithms. These RTs were used for forecasting purpose [31] and particularly for solar energy forecasting [32, 33].

RT models can be formalized by

$$\widehat{k_t}(t+h) = \sum_{i=1}^{t-1} a_i \cdot I(k_t(t-i)) \tag{10}$$

with a_i constant factors, I a function equal to 1 if input is used and 0 if not; a regression model is developed for each node.

Several submodels are available:

- Pruned RT: This method allows to decrease the RT size by removing sections of the tree because a too large tree can over fit the data; a cross-validation is used: At each pair of leaf nodes with the same parent, the error with real data is calculated, and if the sum of squares is smaller than a threshold, the two nodes are deleted and the parent becomes a leaf [34].
- Boosted and bagged RT: In a bagged RT, several trees are built in parallel and connected together [35]; in a boosted one, a new tree is built to correct the error of the previous one [36].
- Random forest (RF): Random forest combines the output of multiple decision trees to reach a single result [37]. The data set is equally divided in samples but each regression tree grows differently, each node is split using the best among a subset of predictors randomly chosen at that node. RF is well-known as a very efficient machine learning model [38].

3 Application of Forecasting Methods to a Mediterranean Site and Comparison

3.1 Presentation of the Meteorological Station

The eleven previous forecasting methods were applied to hourly data of horizontal global solar irradiation measured in the laboratory Sciences for Environment of the University of Corsica Pasquale Paoli, located in Ajaccio (latitude: 41°55'N; longitude: 8°55'E at 200 m from the sea and at an altitude of 70 m (Fig. 3). The site has an insular Mediterranean climate. The global irradiance on horizontal surface is measured by a Kipp & Zonen (CM11) pyranometer (Fig. 3). The period of the solar data collect is between 01/01/1998 and 12/31/2000, i.e., 3 years of solar data. The variability of the solar radiation data set was estimated using the concept of mean absolute log return [39]; in the previous work [40] comparing the variability of various sites, the variability was calculated equal to 0.1961 for Ajaccio which corresponds to a weak variability.

Fig. 3 Situation of the meteorological station and solar measurement devices

3.2 Performances of Forecasting Models

As explained in paragraph 2.2.1., a preprocessing has been applied for the set of three years of hourly global solar irradiation. The Solis model was used to determine the hourly global irradiation by clear sky; the optimal number of input data (first minimum of the auto-mutual information criteria, see paragraph 2.2.1) is $n = 8$; consequently, for estimating the future solar irradiation, the 8 previous measured solar irradiations are used. The training set is about 80% of the data set and the testing set about 20% considering a k-fold sampling equal to 10.

The eleven models were applied to predict the hourly solar irradiation for time horizon between h + 1 and h + 6 h.

In Fig. 4, the relative root mean square error nRMSE for all the models is plotted and we see clearly than three models are not adapted: the persistence (as expected), the support vector regression, and the regression tree. A zoom was realized on the eight "best" models and is shown in Fig. 5. The smart persistence shows good results for h + 1 to h + 3 and then its nRMSE increases greatly. Clearly, Gaussian process and pruned RT are not very efficient. The five other models show performances very close to each other. For each time horizon, the best result was represented by a dot, and the leadership model is different for each horizon. However, regarding Table 1 in which all nRMSE values was introduced, it appears that ARMA model is always the first- or second-best model or very close (as for h + 3); it is then possible to conclude that this ARMA model is well adapted for the site of measures.

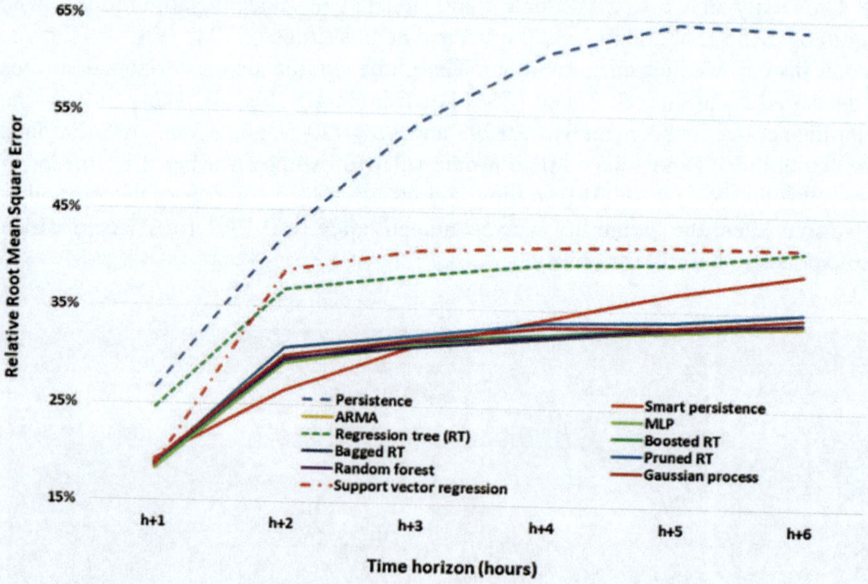

Fig. 4 nRMSE values for the eleven forecasting models

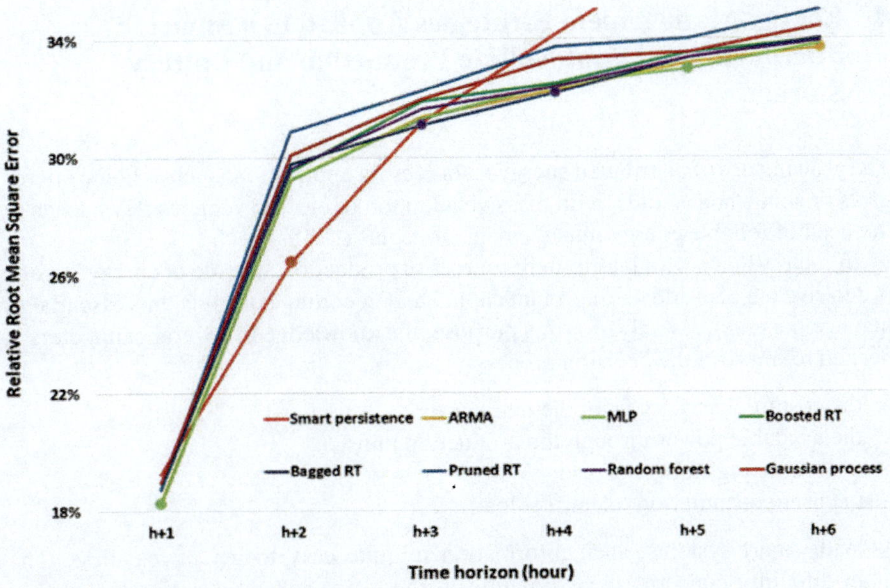

Fig. 5 Zoom of nRMSE values for the eight "best" forecasting models

Table 1 nRMSE versus forecast horizon, the best model is highlighted in blue and the second model is outlined in red

Horizon	h+1	h+2	h+3	h+4	h+5	h+6
Pers	26.60	42.62	54.10	61.39	64.51	63.86
Sma. Pers	19.26	26.46	31.18	34.15	36.92	38.93
ARMA	18.35	29.27	31.38	32.25	33.18	33.69
MLP	18.26	29.26	31.31	32.47	32.98	33.84
RT	24.64	36.88	38.47	39.74	39.95	41.24
Boo. RT	18.75	29.55	31.89	32.51	33.55	33.98
Bag RT	18.76	29.80	31.10	32.17	33.35	34.02
Pru RT	18.72	30.88	32.27	33.76	34.01	35.00
RF	18.97	29.63	31.62	32.38	33.37	33.91
GP	18.97	30.08	31.96	33.29	33.55	34.44
SVR	18.55	38.78	41.03	41.56	41.66	41.60

For the remainder of this study, the ARMA model will be used to forecast the solar irradiation for future time steps and then converted in photovoltaic power.

4 Energy Management Strategies Applied to a Smart Microgrid with Photovoltaic Production and Battery Storage

The growing use of distributed energy resources for applications such as home micro-grids or smart-households, with massive adoption of electric vehicles (EV), leads to the need of reliable energy management strategies (EMSs) [41].

In such systems with intermittent sources of production, the role of energy storage is decisive but also add an important capital and operating cost; it is thus essential to manage the energy wisely. For this purpose, the knowledge of several parameters is needed to improve the decision process:

- the cost of electricity from the main grid;
- the available power through the production units;
- the remaining storage capacity;
- the user consumption (electrical load);

With smart systems, such information is quite easy to get in real time. The main difficulty concerns the estimation of these parameters over time which has an important impact for the optimization of an EMS [42, 43].

In this paragraph, we focus on the evaluation of what the PV forecasting could bring for a real application. This application concerns an accommodation building with 7 rooms and a shared kitchen and living room [44]. The occupancy rate of the building is highly variable through the year. The electrical load also includes an EV, with a capacity of 22 kWh, which is used every day from 9:00 to 16:00 and is considered as fully discharged at 16:00. This EV is charged from 16:00 to 18:00. The case study is presented in Fig. 6.

Fig. 6 Accommodation building and electric vehicle

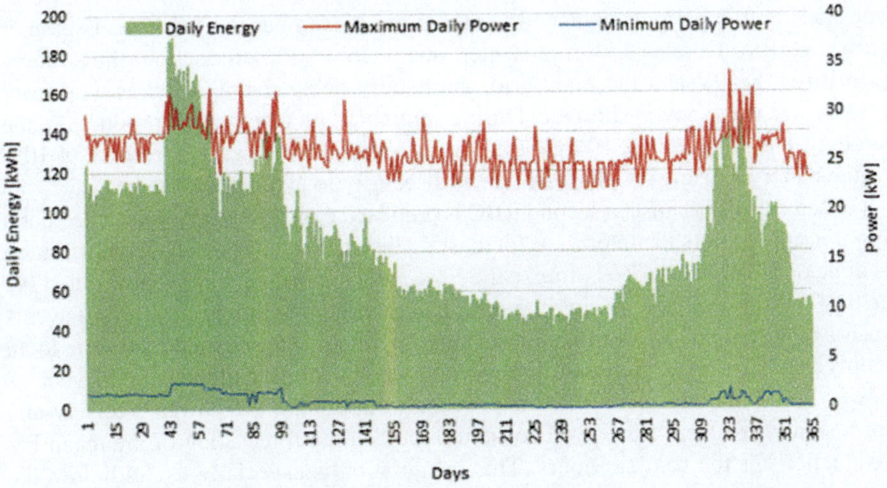

Fig. 7 Load profile of the case study

During the charge, the EV consumes a maximum power of 22 kW and a mean power of 11 kW. If we consider the overall consumption of the case study for a typical year, this leads to a maximum power of 34.2 kW and a mean power of 3.5 kW. Figure 7 presents the load profile for a whole year, highlighting the load consumption over time.

The building is connected to a microgrid which is composed of a PV system (3 × 56 monocrystalline silicon 327 Wp modules connected to 3 × 17 kW inverters) and an energy storage system (24 × 2 V electrochemical lead-acid batteries with a total capacity of 70 kWh DC). This PV microgrid can operate both in grid-connected or in islanded mode. In grid-connected mode, the power limitation for exchange with the main grid is set to 24 kW. It should also be noted that the main grid cannot be used to charge the battery.

The energy is bought and sold from and to the main grid with different prices:

- Energy bought from the main grid: 122.4 €/MWh during "off-peak hours," from 22:00 to 04:00 and 163.1 €/MWh during "peak hours," from 04:00 to 22:00.
- Energy sold to the main grid: constant price of 137.7 €/MWh.

This application allows us to test and compare different EMS in terms of cost efficiency or environmental impact. Here, we want to quantify the economic benefit of a strategy with the use of PV forecasted data. First of all, we define a rule-based control (RBC) strategy fitting our case study. At each time step, the choice depends on the energy cost, the available PV power, and the battery state of charge. The strategy favors the use of the main grid during off-peak hours, when the energy is the less expensive. It also prevents the storage to be full too quickly by allowing the selling of PV power at the right time, when a given PV power level and battery state of charge are reached. At last, some rules are implemented to tackle the seasonal effect, considering two periods: extended winter (from November to March) and summer (from April to October). During winter, due to lower PV production, the strategy prioritizes the PV and the main grid, the battery being mainly used as a security reserve in case of power shortage. During summer, with a higher PV production, the strategy can prioritize the PV and the battery rather than the grid. A reserve of 10% of the battery capacity is still maintain as a security to avoid power shortage.

Based on this premise, a second RBC is proposed considering PV forecasted data. For this application, the interest of forecasting lies in the battery management, which is done on a daily basis. To optimize the EMS, it is important to know how much PV energy will be available during the day. This will help avoiding or limiting events such as PV lost due to a full battery or missing energy (unsatisfied load) due to an empty battery. This is particularly important when dealing with constraints such as power limitation. However, there is no need to know the PV power precisely and for a small time step. Here, we choose to focus on the forecasting of the mean PV production for the next six hours. This information is especially useful before the sunrise, to prepare for peak hours and determine the SoC level that should be kept in the battery. The forecasting method used in this RBC is based on an autoregressive mobile average (ARMA) model presented above. The same strategy will also be simulated with real data instead of ARMA forecasting to estimate how much could be won by improving the forecasting model.

Figure 8 illustrates the behavior of the battery state and the variation of the power flows for a summer week.

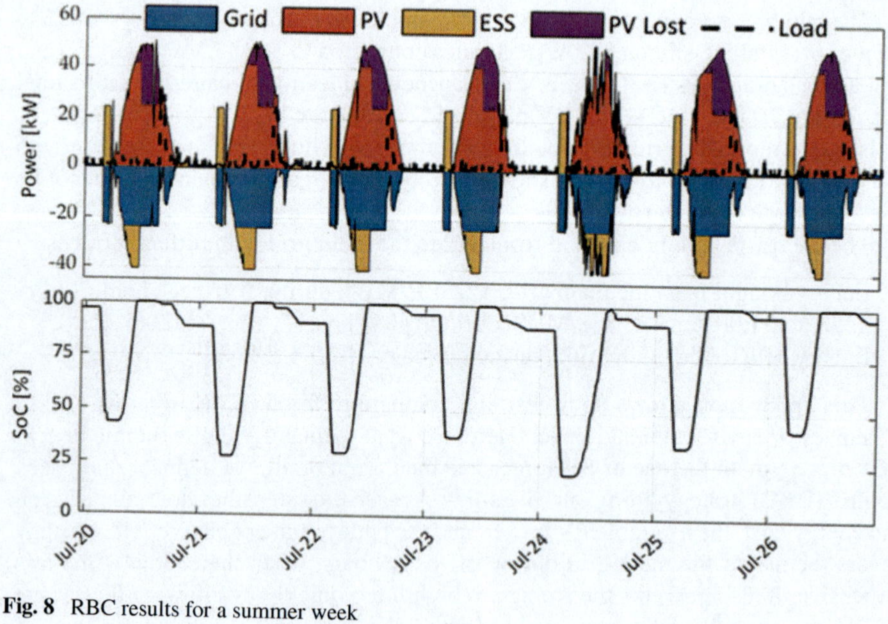

Fig. 8 RBC results for a summer week

To carry out this study, we focus on the economic gain G of the EMS, which is defined by Eq. (11):

$$G = \sum_t \left[\left(P_{PV,g}(t) + P_{b,g}(t) \right) C_{\text{sell}}(t) - P_{g,l}(t) C_{\text{buy}}(t) \right] dt \quad (11)$$

with:

$P_{PV,g}$ the PV power sold to the grid.
$P_{b,g}$ the power from the battery sold to the grid.
$P_{g,l}$ the grid power bought to supply the load.
C_{sell} the electricity selling cost.
C_{buy} the electricity buying cost.

It should be noted that this problem is linear, meaning that if all the power flows are known for the whole period, it is possible to use a mathematical optimization to calculate the optimum. Even if not realistic as a strategy, the calculation of the optimum still has a great interest as it allows us to quantify the performance of our EMS. In this study, we use a linear programing (LP) method to define this optimum.

For a whole year, LP shows a total gain of 6592.7 € and no missing energy. This figure will be used as a reference to calculate the relative performance of the proposed EMS. In Table 2, we present the results for the different EMS version. Even if this study does not focus on the battery degradation, we also present the number of battery equivalent cycles as it will impact the gain if we consider the real operating cost.

Without forecasting, the proposed EMS leads to an economic gain of 5716.6 € for a whole year, representing 86.7% of the optimum. This first result shows that the RBC approach seems relevant to manage the system, with a good compromise between complexity and performance. The addition of PV forecasting with a realistic ARMA model increases the gain to 6113.4 € (+6.9%). The maximum gain with this EMS considering perfect forecasting reaches 6288.2 € (+2.9% compared to ARMA forecasting). Perfect forecasting not being an achievable result, the improvement of the forecasting model would have a very limited effect in this case study. These results confirm the relevance of using an ARMA model to improve the RBC. In order to increase further its performance, it seems more promising to focus on the

Table 2 EMS results for a whole year

	Without forecasting	ARMA forecasting	Perfect forecasting	LP (optimum)
Gain [€]	5716.6	6113.4	6288.2	6592.7
Relative performance [%]	86.7	92.7	95.4	100
Battery cycles [–]	173.2	171.7	166.1	164.1

load forecasting, to anticipate important fluctuations in consumption such as daily, weekly, and seasonal variations.

At last, it should be noted that the number of battery cycle does not vary much from one version of the EMS to another. However, this figure still decreases with the performance of the strategy, meaning that the difference between each version would be higher if we were considering the real operating cost, reinforcing the importance of PV forecasting. Its importance should also be become greater in future, with the increasing cost of electricity and the appearance of more complex contracts with varying costs throughout the day.

5 Conclusion

This paper showed the interest to develop efficient tools to predict solar irradiation and then photovoltaic power and to integrate it into an energy management system.

Several forecasting models were briefly described and then tested on a meteorological site in Ajaccio near the Mediterranean Sea; the predicted data were compared with measured hourly global solar irradiation with a time horizon between 1 and 6 h. It has been shown that the ARMA model was the most suited to our study. It was then introduced in an energy management strategy using a rule-based control and applied to a microgrid including a photovoltaic plant with a battery storage and supplying electricity to a small building and an electric vehicle.

It appears that the addition of PV forecasting with a realistic ARMA model increases the gain of about 7%; this gain could reach 10% with a perfect forecasting reaches.

Today, the electricity tariff for both buying and selling electricity is not well suited to the renewable microgrid development; the increase of the electricity price and the decrease of the battery and photovoltaic cost should make more profitable such systems, also increasing the benefit of solar forecasting.

References

1. Magrini, A., Lentini, G., Cuman, S., Bodrato, A., Marenco, L.: From nearly zero energy buildings (NZEB) to positive energy buildings (PEB): the next challenge—the most recent European trends with some notes on the energy analysis of a forerunner PEB example. Develop. Built Environ. **3**, 100019 (2020). https://doi.org/10.1016/j.dibe.2020.100019
2. Naveen Chakkaravarthy, A., Subathra, M.S.P, Jerin Pradeep, P., Manoj Kumar, N.: Solar irradiance forecasting and energy optimization for achieving nearly net zero energy building. J. Renew. Sustain. Energy. **10**, 035103 (2018). https://doi.org/10.1063/1.5034382
3. Celik, B., Roche, R., Bouquain, D., Miraoui, A.: Decentralized neighborhood energy management with coordinated smart home energy sharing. IEEE Trans. Smart Grid. **9**(6), 6387–6397 (2018). https://doi.org/10.1109/TSG.2017.2710358
4. Enriquez, R., Jimenez, M.J., del Rosaio Heras, M.: Solar forecasting requirements for buildings MPC. Energy Procedia. **91**, 1024–1032 (2016). https://doi.org/10.1016/j.egypro.2016.06.271

5. Khan, A.A., Razzaq, S., Khan, A., Khursheed, F.: Owais: HEMSs and enabled demand response in electricity market: an overview. Renew. Sust. Energ. Rev. **42**, 773–785 (2015). https://doi.org/10.1016/j.rser.2014.10.045
6. Althaher, S., Mancarella, P., Mutale, J.: Automated demand response from home energy management system under dynamic pricing and power and comfort constraints. IEEE Trans. Smart Grid **6–4**, 1874–1883 (2015). https://doi.org/10.1109/TSG.2014.2388357
7. Reinisch, C., Kofler, M.J., Iglesias, F., Kastner, W.: ThinkHome energy efficiency in future smart homes. Eurasip. J. Embed. Syst. **2011**, 1–18 (2011). https://doi.org/10.1155/2011/104617
8. Green Power Labs: predictive energy management services for building managers and operators. https://greenpowerlabs.com/. Accessed 11 Sept 2021
9. Drgoňa, J., Arroyo, J., Cupeiro Figueroa, L., Blum, D., Arendt, K., Kim, D., Perarnau Ollé, E., Oravec, J., Wetter, M., Vrabie, D.L., Helsen, L.: All you need to know about model predictive control for buildings. Annu. Rev. Control. **50**, 190–232 (2020). https://doi.org/10.1016/j.arcontrol.2020.09.001
10. Serale, G., Fiorentini, M., Capozzoli, A., Bernardini, D., Bemporad, A.: Model predictive control (MPC) for enhancing building and HVAC system energy efficiency: problem formulation, applications and opportunities. Energies **11**, 631 (2018). https://doi.org/10.3390/en11030631
11. Lara-Fanego, V., Ruiz-Arias, J.A., Pozo-Vázquez, D., Santos-Alamillos, F.J., Tovar-Pescador, J.: Evaluation of the WRF model solar irradiance forecasts in Andalusia (southern Spain). Sol. Energy **86–8**, 2200–2217 (2012). https://doi.org/10.1016/j.solener.2011.02.014
12. Diagne, M., David, M., Lauret, P., Boland, J., Schmutz, N.: Review of solar irradiance forecasting methods and a proposition for small-scale insular grids. Renew. Sust. Energ. Rev. **27**, 65–76 (2013). https://doi.org/10.1016/j.rser.2013.06.042
13. Heinemann, D., Lorenz, E., Girodo, M.: Forecasting of Solar Radiation in Solar Energy Resource Management for Electricity Generation from Local Level to Global Scale. Nova Science Publishers, New York (2006)
14. Sperati, S., Alessandrini, S., Pinson, P., Kariniotakis, G.: The "Weather intelligence for renewable energies" benchmarking exercise on short-term forecasting of wind and solar power generation. Energies **8**(9), 9594–9619 (2015). https://doi.org/10.3390/en8099594
15. Najac, J.: Wind and photovoltaic production forecasting. In: EDF in'Tech Seminar, INRIA, Grenoble, France, 27/09/2012 (in French) (2012)
16. Diagne, H.M., Lauret, P., David, M.: Solar irradiation forecasting: State-of-the-art and proposition for future developments for small-scale insular grids. In: Processing WREF 2012—World Renewable Energy Forum, May 2012, Denver, United States (2012)
17. Voyant, C., Notton, G., Kalogirou, S., Nivet, M.L., Paoli, C., Motte, F., Fouilloy, A.: Machine learning methods for solar radiation forecasting: a review. Renew. Energy **105**, 569–582 (2017). https://doi.org/10.1016/j.renene.2016.12.095
18. Lauret, P., Voyant, C., Soubdhan, T., David, M., Poggi, P.: A benchmarking of machine learning techniques for solar radiation forecasting in an insular context. Sol. Energy **112**, 446–457 (2015). https://doi.org/10.1016/j.solener.2014.12.014
19. Hornik, K., Stinchcombe, M., White, H.: Multilayer feedforward networks are universal approximators. Neural Netw **2**, 359–366 (1989). https://doi.org/10.1016/0893-6080(89)90020-8
20. Mueller, R.W., Dagestad, K., Ineichen, P., Schroedter-Homscheidt, M., Cros, S., Dumortier, D.: Rethinking satellite-based solar irradiance modeling: the SOLIS clear-sky module. Remote Sens. Environ. **91**, 160–174 (2004). https://doi.org/10.1016/j.rse.2004.02.009
21. Ineichen, P.: A broadband simplified version of the solis clear sky model. Sol. Energy **82**, 758–762 (2008). https://doi.org/10.1016/j.solener.2008.02.009
22. Rigollier, C., Bauer, O., Wald, L.: On the clear sky model of the ESRA—European solar radiation atlas—with respect to the Heliosat method. Sol. Energy **68**, 33–48 (2000). https://doi.org/10.1016/S0038-092X(99)00055-9
23. Chow, T.W.S., Huang, D.: Effective feature selection scheme using mutual information. Neurocomputing **63**, 325–343 (2005). https://doi.org/10.1016/j.neucom.2004.01.194

24. Wiens, T.S., Dale, B.C., Boyce, M.S., Kershaw, G.P.: Three-way k-fold cross-validation of resource selection functions. Ecol. Model. **212**, 244–255 (2008). https://doi.org/10.1016/j.ecolmodel.2007.10.005
25. Sfetsos, A., Coonick, A.H.: Univariate and multivariate forecasting of hourly solar radiation with artificial intelligence techniques. Sol. Energy **68**, 169–178 (2000). https://doi.org/10.1016/S0038-092X(99)00064-X
26. Voyant, C., Paoli, C., Muselli, M., Nivet, M.L.: Multi-horizon solar radiation forecasting for mediterranean locations using time series models. Renew. Sust. Energ. Rev. **28**, 44–52 (2013). https://doi.org/10.1016/j.rser.2013.07.058
27. Kalogirou, S.A.: Applications of artificial neural-networks for energy systems. Appl. Energy **67**, 17–35 (2000). https://doi.org/10.1016/S0306-2619(00)00005-2
28. Mellit, A.: Artificial intelligence techniques for modelling and forecasting of solar radiation data: a review. Int. J. Artif. Intell. Soft Comput. **2008**, 52–76 (2008). https://doi.org/10.1504/IJAISC.2008.021264
29. Rasmussen, C.E.: Gaussian processes in machine learning. In: Bousquet, O., Luxburg, U von., Rätsch, G. (eds.) Advanced Lectures on Machine Learning, pp. 63–71. Springer Berlin Heidelberg (2004). https://doi.org/10.1007/978-3-540-28650-9_4
30. Vapnik, V.: The Nature of Statistical Learning Theory. Springer Science and Business Media (2013). https://doi.org/10.1007/978-1-4757-3264-1
31. Burrows, W.R.: CART regression models for predicting UV radiation at the ground in the presence of cloud and other environmental factors. J. Appl. Meteorol. **36**, 531–544 (1997). https://doi.org/10.1175/1520-0450(1997)036%3c0531:CRMFPU%3e2.0.CO;2
32. Aggarwal, S.K., Saini, L.M.: Solar energy prediction using linear and non-linear regularization models: a study on AMS (American Meteorological Society) 2013–14 solar energy prediction contest. Energy **78**, 247–256 (2014). https://doi.org/10.1016/j.energy.2014.10.012
33. Persson, C., Bacher, P., Shiga, T., Madsen, H.: Multi-site solar power forecasting using gradient boosted regression trees. Sol. Energy **150**, 423–436 (2017). https://doi.org/10.1016/j.solener.2017.04.066
34. Pedro, H.T.C., Coimbra, C.F.M., David, M., Lauret, P.: Assessment of machine learning techniques for deterministic and probabilistic intra-hour solar forecasts. Renew. Energy **123**, 191–203 (2018). https://doi.org/10.1016/j.renene.2018.02.006
35. Breiman, L.: Bagging predictors. Mach. Learn. **24**, 123–140 (1996). https://doi.org/10.1023/A:1018054314350
36. Huang, J., Perry, M.: A semi-empirical approach using gradient boosting and k-nearest neighbors regression for GEFCom2014 probabilistic solar power forecasting. Int. J. Forecast. **32**, 1081–1086 (2016). https://doi.org/10.1016/j.ijforecast.2015.11.002
37. Breiman, L.: Random forests. Mach. Learn. **45**, 5–32 (2001). https://doi.org/10.1023/A:1010933404.324
38. Ibrahim, I.A., Khatib, T.: A novel hybrid model for hourly global solar radiation prediction using random forests technique and firefly algorithm. Energy Convers. Manag. **138**, 413–425 (2017). https://doi.org/10.1016/j.enconman.2017.02.006
39. Voyant, C., Soubdhan, T., Lauret, P., David, M., Muselli, M.: Statistical parameters as a means to a priori assess the accuracy of solar forecasting models. Energy **90-Part 1**, 671–679 (2015). https://doi.org/10.1016/j.energy.2015.07.089
40. Fouilloy, A., Voyant, C., Notton, G., Motte, F., Paoli, C., Nivet, M.L., Guillot, E., Duchaud, J.L.: Solar irradiation prediction with machine learning: forecasting models selection method depending on weather variability. Energy **165-Part A**, 620–629 (2018). https://doi.org/10.1016/j.energy.2018.09.116.
41. Zhu, Y., Chen, Y., Tian, G., Wu, H., Chen, Q.: A four-step method to design an energy management strategy for hybrid vehicles. In: Proceedings of the 2004 American Control Conference, vol. 1, pp. 156–161 (2004).
42. Restrepo, M., Cañizares, C.A., Simpson-Porco, J.W., Su, P., Taruc, J.: Optimization and rule-based energy management systems at the Canadian renewable energy laboratory microgrid facility. Appl. Energy **290**, 116760 (2021). https://doi.org/10.1016/j.apenergy.2021.116760

43. Olivares, D.E., Cañizares, C.A., Kazerani, M.: A centralized energy management system for isolated microgrids. IEEE Trans. Smart Grid **5–4**, 1864–1875 (2014). https://doi.org/10.1109/TSG.2013.2294187
44. Ouédraogo, S., Faggianelli, G.A., Pigelet, G., Duchaud, J.L., Notton, G.: Application of optimal energy management strategies for a building powered by PV/Battery system in Corsica island. Energies **13–17**, 4510 (2020). https://doi.org/10.3390/en13174510

Prediction of Air Quality Index Using Machine Learning Techniques and the Study of Its Influence on the Health Hazards at Urban Environment

J. V. Bibal Benifa, P. Dinesh Kumar, and J. Bruce Ralphin Rose

Abstract Air quality across the globe is degrading at a faster rate due to the industrialization and urbanization which leaves no access to fresh air for breathing in major industrialized regions. Specifically, the industrial regions in and around India contribute a major part in the depletion of air quality in the south Asian region. Further, air pollution plays a vital role on the harmful diseases in the metropolises in India. The pollutant levels have been monitored in real-time to initiate the control measures to keep the air quality index (AQI) in the satisfactory level. Some of the major pollutants are particulate matter (PM2.5 and PM 10), sulfur dioxide (SO_2), nitrogen dioxide (NO_2), ozone (O_3), carbon monoxide (CO), ammonia (NH_3). With the evolution of technology, it is possible to predict the future air quality using various machine learning (ML) techniques. In the present work, the air quality data of major industrialized cities such as Delhi, Bangalore, Chennai, Ahmedabad, and Lucknow have been collected for the past five years (2015–2019) and predictive models are built. The prediction accuracy of traditional ML–models such as decision tree (DT), linear regression (LR), random forest (RF), and gradient boosting (GB) is compared with new hybrid models such as LR + DT and GB + DT. The predicted data are compared with the actual data of the subsequent years. The performance of the regression models is evaluated through mean absolute error (MAE), root mean square error (RMSE), and correlation coefficient (r). The hybrid models have proven efficiency while compared to the traditional methods in prediction of air quality of the subsequent years. Consequently, the health statistics (death rate) released by Institute of Health Metrics and Evaluation (IHME) is correlated with air pollution-based diseases. The presented ML-based work can be used in real-time to predict the future air pollution for the better planning and control of the air pollution-based diseases around the globe.

J. V. Bibal Benifa · P. Dinesh Kumar
Department of Computer Science and Engineering, Indian Institute of Information Technology, 686635 Kottayam, India

J. Bruce Ralphin Rose (✉)
Department of Mechanical Engineering, Anna University Regional Campus, 627007 Tirunelveli, India
e-mail: bruce@auttvl.ac.in

M. Lahby et al. (eds.), *Computational Intelligence Techniques for Green Smart Cities*, Green Energy and Technology, https://doi.org/10.1007/978-3-030-96429-0_12

Keywords Air quality index · Machine learning (ML) · Pollutants · Cardio vascular disease (CVD) · Chronic respiratory disease (CRD)

1 Introduction

Air pollution is caused by various harmful compounds and suspended particles which are emitted by industries and vehicles that react with atmospheric gasses to form the toxic compounds [1]. Air pollution has some severe effects on human beings when the air in the surroundings has a higher value of AQI [2]. Pollution is primarily caused by the emissions from vehicles and industries that are also a major cause of greenhouse effect because of the emission of carbon dioxide (CO_2) [1]. Air pollution is also a major threat for the environment as it has a huge impact on climate change and biodiversity. For the past two decades, climate change has been an argument in global organizations as it causes ozone layer depletion which protects the ultraviolet (UV) rays from entering into the planet's atmosphere.

Nearly, 21 out of 30 world's top most polluted cities are located in India because of the high population and economical barriers [3]. As per WHO report in 2016, nearly, 140 million people in India are breathing air which is 10 times more polluted than the safe limits prescribed by WHO [4]. Pollution caused by industries accounts for about 51% in India, whereas the other forms of pollution are caused by vehicles (27%), crop burning (17%), and fireworks (5%). Global warming is also a key phenomenon which increases the earth's temperature due to the rise of air and water temperatures that are mainly caused by the greenhouse gasses which traps the radiation emitted by the sun and not reflecting it back into space. This prevents the escaping of greenhouse gasses from the planet's surface to the atmosphere [4]. It causes acid rain as well which causes skin cancer for human beings and depletion of soil nutrients because of the reaction with acid rain [5].

1.1 Impact of Air Pollution on Health-Indian Scenario

The air pollution has a great impact on the health of human beings. The various diseases caused by air pollution as per WHO is ischemic heart disease, stroke; COPD, lung cancer, and respiratory infections. Indian Council of Medical Research (IHME) has reported the statistics of death with reasons in various states in India. In the state of Delhi (1990–2016), from the reported deaths in the age group of 15–39, 40–69, and 70 and above, 14.9%, 35.6%, and 43.2%, respectively, are due to cardiovascular disease [6]. In the age group of 40–69 and 70 and above, 6.3% and 9.5%, respectively, are due to the chronic respiratory disease (CRD). In the age group of 15–39, 40–69, and 70 and above, 9.1%, 19.8%, and 9%, respectively, are due to cancer.

Chennai is located in the state of Tamil Nadu in India, and IHME has released the disease burden profile of this city from 1990 to 2016 [7]. In the state of Tamil Nadu

(1990–2016), from the reported deaths in the age group of 15–39, 40–69, and 70 and above, 16.6%, 40.4%, and 39.6%, respectively, are due to cardiovascular disease. In the age group of 40–69 and 70 and above, 6.4% and 8.3%, respectively, are due to the chronic respiratory disease (CRD). In the age group of 15–39, 40–69, and 70 and above, 4.6%, 10.9%, and 5.3%, respectively, are due to cancer.

Ahmedabad is located in the state of Gujarat in India. IHME has released the disease burden profile from 1990 to 2016 [8]. In the state of Gujarat (1990–2016), from the reported deaths in the age group of 15–39, 40–69, and 70 and above, 14.7%, 36.6%, and 37.8%, respectively, are caused by cardiovascular disease. In the age group of 15–39, 40–69, and 70 and above, 2.6%, 12%, and 17.1%, respectively, are due to the chronic respiratory disease (CRD). In the age group of 15–39, 40–69, and 70 and above, 5.7%, 12.5%, and 5.8%, respectively, are caused by cancer.

Lucknow is located in the state of Uttar Pradesh in India. IHME has released the disease burden profile of Uttar Pradesh from the year 1990–2016 [9]. In the state of Uttar Pradesh, from the reported deaths (during 1990–2016) in the age group of 15–39, 40–69, and 70 and above, 9%, 23.7%, and 24%, respectively, are caused by cardiovascular disease. In the age group of 40–69 and 70 and above, 18.8% and 21.1%, respectively, are due to the chronic respiratory disease (CRD). In the age group of 15–39, 40–69, and 70 and above, 6.4%, 12.7%, and 6.2%, respectively, are due to cancer. AQI index primarily depends on the distribution of pollutants present in the atmospheric air. Hence, in order to adopt time-series forecasting models, the emissions rate of each pollutant has to be separately modeled for specific regions. To overcome such difficulties, time-independent supervised learning methods are employed in this chapter to present the detailed insights about the role of ML algorithms in AQI prediction.

Bangalore is located in the state of Karnataka in India. IHME has released the disease burden profile of Karnataka from 1990–2016 [10]. In the state of Karnataka (1990–2016), from the reported deaths in the age group of 15–39, 40–69, and 70 and above, 13.9%, 37.2%, and 36.8%, respectively, are due to cardiovascular disease. In the age group of 40–69 and 70 and above, 10.6% and 13.9%, respectively, are due to the chronic respiratory disease (CRD). In the age group of 15–39, 40–69, and 70 and above, 6.5%, 13.7%, and 7%, respectively, are due to cancer. Lee et al. (2014), in his research, proposed that exposure to air pollution containing $PM_{2.5}$ is closely associated with cardiovascular disease [11]. There are different reasons behind cardiovascular disease such as unhealthy lifestyle; however, air pollution is also one of the key reasons behind it.

1.2 Distribution Levels of Toxic Contents

Particulate matter (PM2.5 and PM 10) levels in India are found to be much higher when compared with other pollutant levels as highlighted in Fig. 1. Similarly, the distribution of toxic content levels (SO_2 and CO) at two highly populated Indian urban environments such as New Delhi and Lucknow is presented in Fig. 2. These

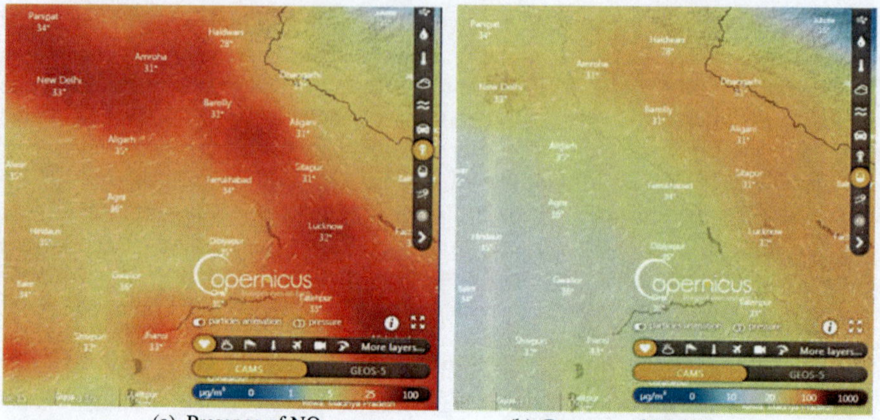

(a). Presence of NO_2 (b). Presence of PM2.5

Fig. 1 Distribution of pollutant levels at New Delhi and Lucknow on 08/06/2021 at 10.40 pm (Courtesy: www.windy.com)

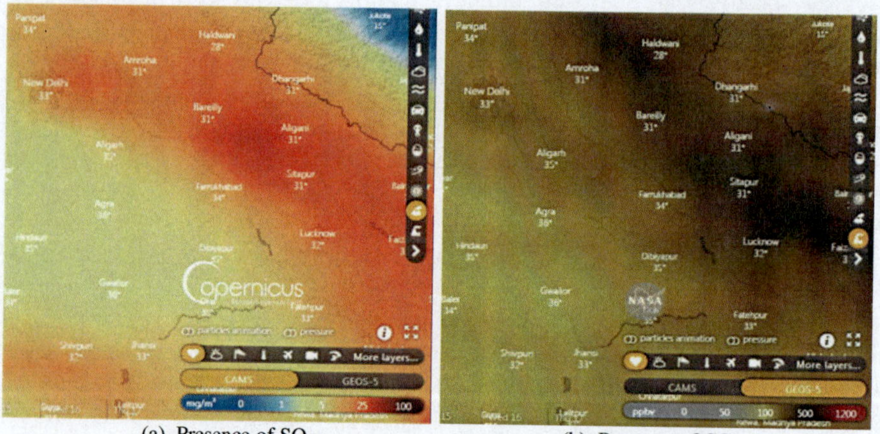

(a). Presence of SO_2 (b). Presence of CO

Fig. 2 Distribution of toxic content levels at New Delhi and Lucknow on 08/06/2021 at 10.40 pm (Courtesy: www.windy.com)

particulate matters are released as a by-product in the combustion process from power plants, and sometimes, the other pollutants may react with atmospheric gasses to form a complex compound which is also a source of particulate matter [5]. The PM has the ability to stay in the air for a longer duration, and it has adverse effects when inhaled by human beings than other pollutants [12]. Primarily, it affects the lungs; later, it affects the entire cardiovascular system to cause fatal diseases such as heart attack and hardening of arteries [13]. Though it influences all the age groups, the aged people (>65) are highly vulnerable to the primary effects. In 2021, a recent study states that about 30.7% of deaths are caused because of the air pollution which accounts for about 2.5 million deaths [1].

According to the WHO report, 7 million premature deaths are directly linked to air pollution in the year 2012 [14]. Among this reported deaths, the different contributors are classified as follows: 40%–ischemic heart disease; 40%–stroke; 11%–chronic obstructive pulmonary disease (COPD); 6%–lung cancer; and 3%–acute lower respiratory infections in children that are caused due to air pollution [3]. The above-mentioned problems and conditions urged to control the pollution by continuous monitoring of air quality by acquiring the detailed information about the pollutant levels. Further, the most commonly available pollutants in the atmospheric air are known as particulate matter (PM2.5), PM 10, SO_2, NO_2, O_3, CO, and NH_3 [15]. The intensity of these pollutants is measured to determine the air quality by calculating the AQI of a particular place which is commonly taken on a daily basis.

The expression used for calculating the AQI is presented in Eq. (1),

$$I = \frac{I_{\text{high}} - I_{\text{low}}}{C_{\text{high}} - C_{\text{low}}} \times (C - C_{\text{low}}) + I_{\text{low}} \tag{1}$$

where "I" is the air quality index, "C" is the pollution concentration, "C_{low}" is the concentration breakpoint that is $\leq C$, "C_{high}" is the concentration breakpoint that is $\geq C$, "I_{low}" is the breakpoint corresponding to "C_{low}," "I_{high}" is the breakpoint corresponding "C_{high}."

"C_{low}," "C_{high}," "I_{low}," and "I_{high}" are computed from the US EPA pollutant breakpoint. The computed AQI range is defined as follows: good [0–50], satisfactory [51–100], moderate [101 200], poor [201–300], very poor [301–400], and severe [401–500] [16]. With the assistance of artificial intelligence (AI)-based machine learning and deep learning methods, air quality monitoring is performed in real-time basis. Moreover, regression models like random forest and decision tree seem to be better for the prediction of air quality as reported by recent works [1, 4, 5].

In the present chapter, a novel AQI prediction strategy is discussed using hybrid methods and the results of such hybrid methods are compared with traditional methods such as random forest (RF), decision tree (DT), linear regression (LR), and gradient boosting (GB) with validations. Subsequently, an illustrative study is presented based on the reports of Indian Council for Medical Research (ICMR) relating to the harmful diseases that are caused by air pollution in the urban environment. The motivation of the chapter presented herein is to apply the ML methods effectively to forecast the future air quality at different cities. Meanwhile, the polluted air causes severe health hazards that should be related with appropriate AQI for each location. Hence, the AQI data are correlated with the decease statistics published by the health department as applicable. If the predicted score is high enough to signify the abnormal death rate due to air pollution, then it enables the implementation of effective measures to reduce the air pollution.

2 Related Works

Ameer et al. [1] presented a comparative analysis of air quality prediction in smart cities that uses regression models such as RF, DT, GB, and artificial neural networks (ANNs) [1]. Here, the datasets are collected through the sensors which are used to monitor the air quality and weather conditions. The performance is evaluated using root mean squared error (RMSE) and mean absolute error (MAE) in which RF regression had produced lower value of errors than other models. Sumit (2019) has predicted the AQI using linear regression, K-nearest neighbor (KNN), DT, and support vector machine (SVM) models [4]. The dataset used herein contains various weather and pollutant records. The dataset is pre-processed and trained using ML methods whereas accuracy metrics have been calculated in which DT outperforms with 99% accuracy. Krishna [17] analyzed and compared the AQI of the different Indian states in which the optimized Bayesian network has delivered the highest accuracy of 99.63% [17].

Džaferović and Karađuzović-Hadžiabdić [18] analyzed the air quality for a city in which the highest R-squared value has been observed in RF with lowest RMSE value as well [18]. The stacked ensemble technique provides a better prediction accuracy than the individual models when compared to gradient descent variants and neural networks [19]. The PM2.5 levels are constantly monitored due to adverse conditions in the smart cities and predicted to monitor and control the concentration in the particular regions [1]. The ensemble models are used to predict the life index that include hybrid ensemble model SVM and ANN, stacking and voting techniques [20]. RF, linear regression, DT, and SVM have delivered a prediction score of above 0.90 in both training and testing dataset [21]. Liu et al. [22] reported the AQI prediction of Beijing city in China, in which the SVR model outperforms the RF model.

Regression models such as RF and DT seem to be the most preferred model for AQI prediction, and the stacking technique is widely used for various ensemble techniques [1, 17, 22]. RF yields better accuracy meanwhile reduces overfitting and training time of the data [23]. DT also offers good performance, because at each split of training, it tries to achieve the maximum information gain and as the depth increases, it learns more about the training data [24].

3 Proposed Work and Methodology

The high-level workflow of the present work is highlighted in Fig. 3. It contains different phases such as data collection, exploratory data analysis, selection of appropriate features for training, model building using traditional and hybrid ML techniques, AQI prediction, and performance analysis. Among the various phases of the work, the performance analysis is elaborately discussed in Sect. 4.

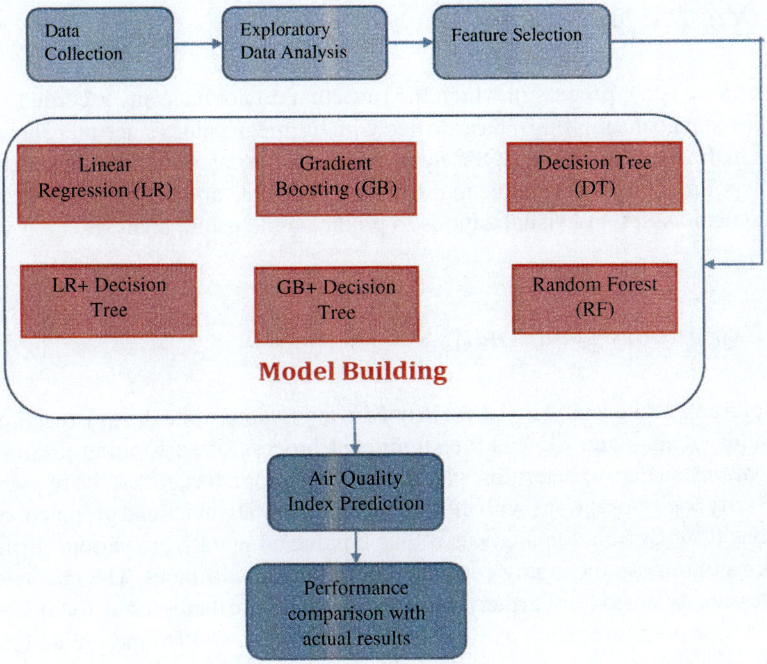

Fig. 3 The high-level work flow of the proposed system

3.1 Data Collection and Exploratory Data Analysis

The essential pollutant data at different urban environments are collected from Central Pollution Control Board (CPCB), India portal which contains the details of daily pollutant levels related to PM2.5, SO_2, NO, NO_2, O_3, CO, NOX, NH_3, ozone, benzene, toluene, Eth-Benzene, MP-Xylene, O-Xylene, relative humidity, wind speed, wind direction, and air temperature [25]. The class labels are assigned as good [0–50], satisfactory [51–100], moderate [101–200], poor [201–300], very poor [301–400], and severe [401–500], respectively. The class labels concerning to the present scenario can be calculated with available data. However, the reliable and robust ML-based prediction models are essential for the future AQI computation through the present/historical data. The data are organized in such a way that it contains the information for the entire day. The data belonging to cities such as Chennai, Bangalore, Lucknow, Delhi, and Ahmedabad are downloaded from the CPCB portal. Other than these pollutant attributes, it also contains some meteorological data such as wind direction, velocity, and pressure. From these data sources, the dataset is retrieved for the period from 2015 to 2019.

3.2 Data Pre-processing

Data cleaning is the process in which the rows that do not have any recorded values are removed, and the missing data are filled with the mean values since there are many outliers in the data. Since, the AQI is not available in the dataset, it is being calculated with the pollutant levels available in the dataset. The data are then processed to view the statistical values and visualizations to predict some useful analysis.

3.3 Exploratory Data Analysis

The exploratory data analysis (EDA) is conducted to determine the key insights and distribution of the data. EDA is the significant process of performing preliminary investigations on data to determine patterns, to spot anomalies, to test the hypothesis, and to verify the assumptions with the help of summary statistics and graphical representations [26]. Correlation analysis is also conducted among the various attributes or features to remove the highly correlated or redundant attributes. The uncorrelated attributes are selected for further investigation. The attributes used for this work are the major pollutants such as PM2.5, SO_2, NO_2, O_3, CO for the prediction and analysis with regression models. After selecting the set of features or attributes, the correlation among the attributes is highlighted in Fig. 4.

In Fig. 4, the lighter shades represent the positive correlation while darker shades represent negative correlation. It is a good practice to ignore correlated attributes during feature selection. Here, it is observed that all the attributes considered are less correlated with the other attributes. A box plot (or box-and-whisker plot) shown in Fig. 5 highlights the distribution of quantitative data in such a way that it facilitates the comparison between the variables. The box shows the quartiles of the dataset while the whiskers extend to show the rest of the distribution. The box plot is a standardized way of displaying the distribution of data based on the five number summary such as minimum value, first quartile, median, third quartile, and maximum value. In the simplest box plot, the central rectangle spans to the first quartile to the third quartile (the interquartile range or IQR). A segment inside the rectangle shows the median and "whiskers" above and below the box that show the locations of the minimum and maximum pollutant concentrations at various cities.

The linearity of the variables is to be verified using a plot distribution graph and look for skewness of features. Figure 6 highlights the skewness of the distribution, and here, all independent variables are found to be right skewed/positively skewed. The attributes used for this work are the major pollutants such as PM2.5, SO_2, NO_2, O_3, CO for the prediction and analysis with regression models. The data have some non-recorded daily data and also have missing values which are filled through pre-processing techniques.

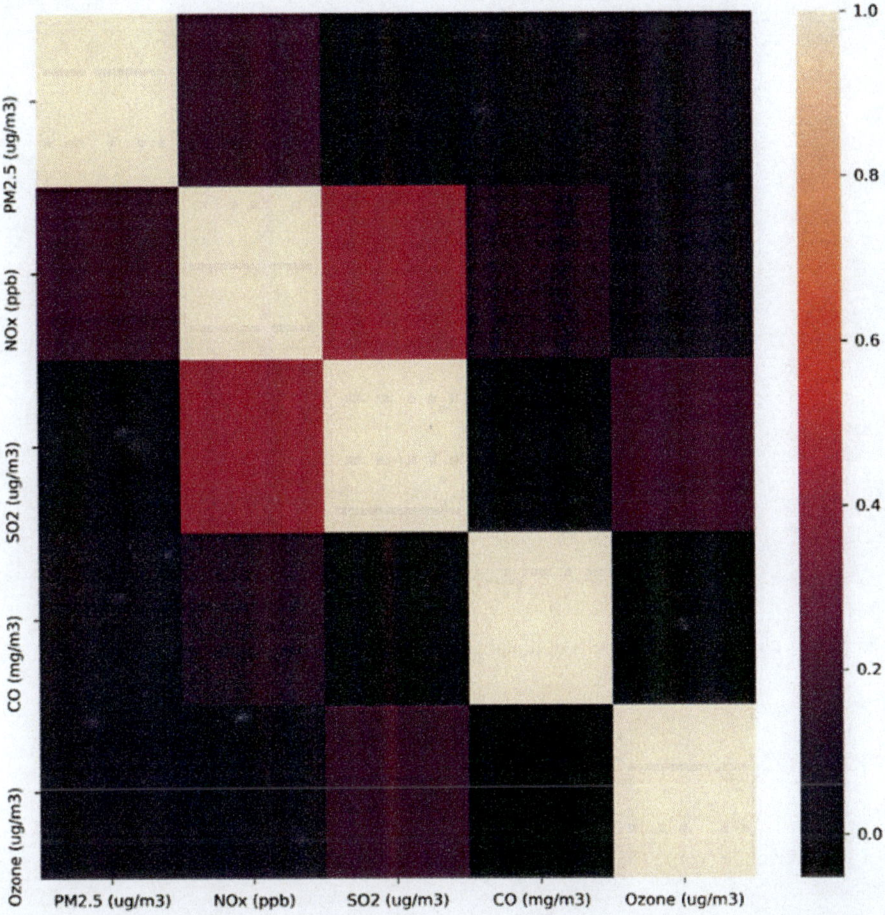

Fig. 4 Correlation between selected attributes

3.4 Model Building

Regression models such as RF, decision tree, linear regression, gradient boosting, and hybrid models are used for model building and to predict the AQI values, and the error analysis methods are used to evaluate the performance metrics of the models. The hybrid models are formed by combining a few ML techniques. The dataset collected for various features in the period of 2015–2019 is used for model building process. The air pollution data concerning to a complete month (January 2015) have passed to the constructed ML model as test data to predict the data corresponding to the subsequent months, and accordingly, the AQI is predicted. The predicted AQI is then compared with the actual sample dataset to measure the accuracy.

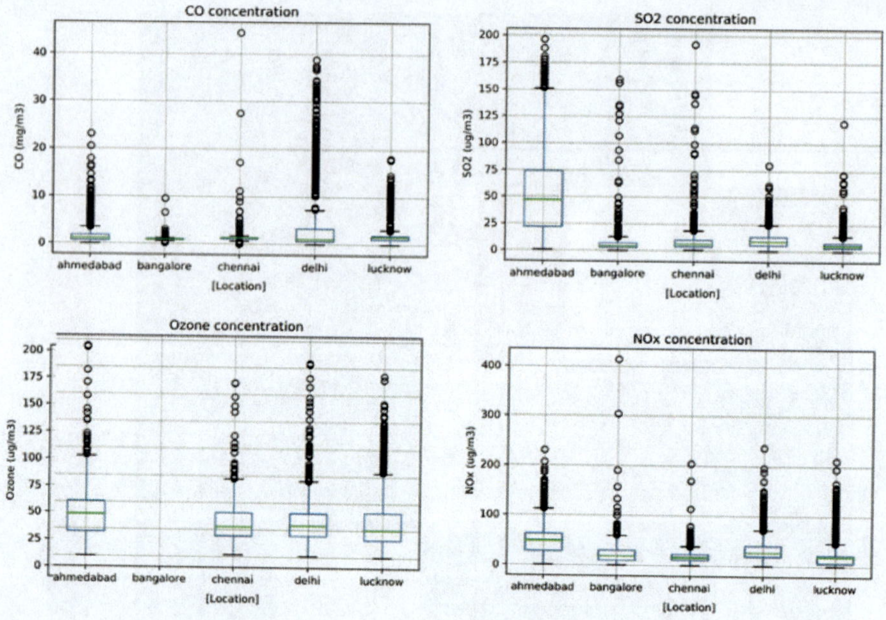

Fig. 5 Box plot distribution details of pollutant in various cities

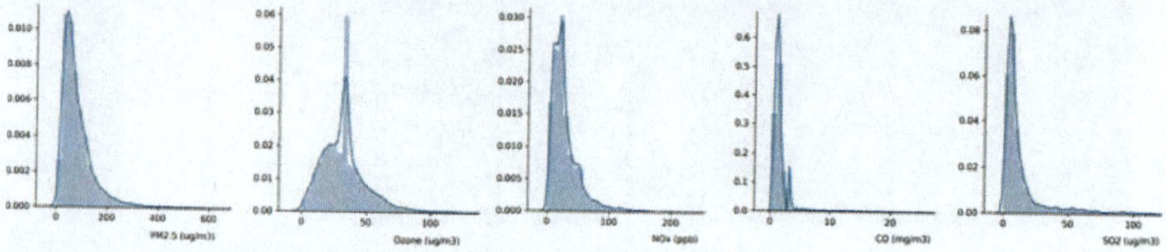

Fig. 6 Distribution and skewness of the pollutant data

(a) ***Decision Trees (DTs)***

DT is a supervised learning algorithm that is used for both classification and prediction problems [27]. The objective of DT is to construct a set of rules that could produce a better predictive model with the set of decisive rules that have been created from the extracted features of data. Moreover, the DT can be created on both linear and non-linear datasets. Entropy is used to calculate the disorderliness to determine the root and child nodes present in the DT. The mathematical expression for computing the entropy is presented in Eq. (2),

$$H(X) = \sum_{i=1}^{n} -p(x_i) \log_2 p(x_i) \tag{2}$$

In Eq. (2), "p" refers to the frequentist probability of an element and entropy is measured from 0 to 1. High entropy value refers to low level of purity whereas low entropy value refers to high purity. Information gain as in Eq. (3) can be defined as the amount of information gained about a random variable or signal from observing another random variable. It can be considered as the difference between the entropy of parent nodes and weighted average entropy of child nodes.

$$IG(S, A) = H(S) - \sum_{i=0}^{n} p(x).H(x) \tag{3}$$

The steps for creating DT are presented in the Algorithm 1.

Algorithm 1. Decision Tree Creation
Input: air pollution dataset with computed AQI
Output: decision rules to predict future AQI

1. Calculate the information gain of each feature
2. Considering that all rows do not belong to the same class, split the dataset S into subsets using the feature for which the information gain is maximum
3. Make a decision tree node using the feature with the maximum information gain
4. If all rows belong to the same class, make the current node as a leaf node with the class as its label
5. Repeat for the remaining features until we run out of all features, or the decision tree has all leaf nodes
6. Predict the AQI from the decision rules

(b) ***Random Forest (RF)***

RF is a supervised and ensemble learning algorithms which are used for both classification and regression models [28]. It constructs multiple DT with the training set in different training time and delivers an output of the class mode which has better classification and regression. It uses classification and decision trees (CARTs) algorithm where each CART is being built through random vectors that will not be calculated for decision rules; rather, it is used only to obtain better outcome variables. The various steps involved for the creation of CART DT are presented in Algorithm 2.

Algorithm 2. DT Creation using Classification and Regression Trees
Input: air pollution dataset with computed AQI
Output: decision rules to predict future AQI

(continued)

(continued)

1. For each ordered variable X,
i. Convert it to an unordered variable X' by grouping its values in the node into a small number of intervals
ii. If X is unordered, then set $X' = X$
2. Perform a chi-squared test of independence of each X' variable versus Y on the data in the node and compute its significance probability
3. Choose the variable $X*$ associated with the X' that has the smallest significance probability
4. Find the split set $\{X* \in S*\}$ that minimizes the sum of gini indexes and use it to split the node into two child nodes
5. If a stopping criterion is reached, exit. Otherwise, apply steps 2–5 to each child node
6. Prune the tree with the CART method and predict the AQI from the decision rules

Since, the RF creates multiple decision trees and the results will not rely on a single DT which provides better prediction results. It makes the predictions by combining the sequences of base models. The combination of several individual models into a final model is represented by Eq. (5). Here, gini impurity is a measure of how often a randomly chosen element from the dataset would be incorrectly labeled as expressed in Eq. (4).

$$\mathrm{Gini(E)} = 1 - \sum_{j=1}^{c} \mathrm{p}_j^2 \tag{4}$$

$$g(x) = f_0(x) + f_1(x) + \ldots + f_n(x) \tag{5}$$

where $f(x)$ corresponds to the result of a single model.

(iii) ***Linear Regression (LR)***

It is a supervised learning algorithm which uses independent variables to build a relationship with the dependent variable [29]. It gives a target prediction based on the independent variable that is used for forecasting purposes. Simple linear regression model is represented in Eq. (6),

$$y = mx + c \tag{6}$$

In Eq. (6), the value of "x" is used to predict the value of "y" whereas "m" and "c" values determine the best fit values of the training model.

Multilinear regression is represented in Eq. (7),

$$y = b_0 + b_1x_2 + \ldots + b_nx_n \tag{7}$$

In multilinear regression, there will be more than one independent variable such as $x_1, x_2,.., x_n$ and its corresponding regression coefficients $b_1, b_2, \ldots ,b_n$.

(iv) ***Gradient Boosting (GB)***

GB is a boosting algorithm (Algorithm 3) uses loss function and weak learners to train the model, and it is trained with a selective model to reduce the loss which makes it a more precise and effective regression model [30]. It trains the model in a sequential and addictive manner, and the gradients are used in the loss function to measure fitting of data with the model. It converts the weak model to a strong model by boosting its performance by building new trees where each tree learns from its previous mistakes.

Algorithm 3. Gradient Boosting Method
Input: air pollution dataset with computed AQI
Output: prediction of future AQI

Fit estimator F^1
For i in $[1, M]//M$ weak estimators
$Loss^i = \sum_{j=1}^{n} (Y_i - F^i(X_j))^2$//loss in i^{th} iteration
Calculate negative gradient: $\frac{\partial L^i}{\partial X_j} = -\frac{2}{n} * (Y_j - F^i(X_j))\forall i$
Fit a weak estimator H^i on $(X, \frac{\partial L}{\partial X})$
//ρ changes the step size
Prediction: $F^m(X) = F^i(X) + \rho * H^i(X) = F^1 + \rho * \sum_{i=1}^{m} H^i(X)$

(e) ***Hybrid Model 1- Gradient Boosting and Decision Tree***

Hybrid model (ensemble) is a combination of a number of different models which tends to be more flexible than individual models with less bias and variance. It can be done in two ways,

- Bagging—It is the process of training individual models in a parallel manner by which each individual model would be trained with random data.
- Boosting—It is the process of training individual models in a series so that the next model learns about the errors generated by the previous model. Ensemble methods such as bagging and boosting deliver better predictions than any other individual models.

In the ensemble method, the gradient boosting and decision tree models are applied sequentially (boosting technique) to obtain better prediction results. Gradient boosting is trained with the dataset and tested with updated values [31]. Subsequently, updated dataset is trained with decision tree model which learns from the errors of the previous model and produces an improved result.

Algorithm 4. Gradient Tree Boosting
Input: air pollution dataset with computed AQI
Output: prediction of future AQI

(continued)

(continued)

1. Initialize $f_0(x) = \arg\min_\gamma \sum_{i=1}^{N} L(y_i, \gamma)$
2. For m = 1 to M:
(a) For I = 1, 2, … , N compute
$r_{im} = -\left[\frac{\partial L(y_i, f(x_i))}{\partial f(x_i)}\right]_{f=f_m-1}$
(b) Fit a regression tree to the targets rim giving terminal regions
$R_{jm}, j = 1, 2, \ldots., J_m$
(c) For $j = 1, 2, \ldots., J_m$ compute
$\gamma_{jm} = \arg\min_{\gamma} \sum_{x_i \in R_{jm}} L(y_i, f_{m-1}(x_i) + \gamma)$
(d) Update $f_m(x) = f_{m-1}(x) + \sum_{j=1}^{J_m} \gamma_{jm} I(x \in R_{jm})$
3. Output $\hat{f}(x) = f_M(x)$

(f) Hybrid Model 2- Linear regression and Decision Tree Model

In this ensemble method, the linear regression (LR) and DT methods are applied sequentially to get an optimal ensemble model [32]. Linear regression equation is trained with the dataset which produces the output data with updated weights. Subsequently, the updated dataset is trained with DT which learns the error made by LR and produces better prediction results than the individual models.

Algorithm 5. Linear Regression and Decision Tree
Input: air pollution dataset with computed AQI
Output: prediction of future AQI

Select the samples;
Do stepwise linear regression on samples in current node;
Estimate the residual sum square (RSS_{node});
If (number of samples > predefined minimum node size)
For each predictor variable x_i

> Sort x_i in ascending order;
> **For** each value d in the above sorted list
> Using d as the threshold value, partition the samples
> Do stepwise linear regression on each sample and calculate the corresponding RSS;
> Estimate the cumulative RSS
> **End For**
> Estimate the least cumulative RSS achievable from splitting current node on x_i,

End For
Estimate the least RSS of all possible splits of current node ($MinRSS_{splitting}$);
Estimate the improvement from splitting the current node
If (improvement > predefined minimum improvement)
Split current node into two new nodes using the variable and "d" that give the $MinRSS_{splitting}$;
End If
End If

4 Experimental Results

4.1 Experimental Environment

The work presented herein is implemented using Jupyter Notebook which is installed on a Windows 10 operating system that contains 8 GB RAM and an Intel Core i7 processor. Here, the air quality dataset has been collected from CPCB core data Website for the years 2015–2019 which is pre-processed and fed into the ML models to get the prediction results. The performance of the ML models is also measured with mean absolute error (MAE), root mean squared error (RMSE), and correlation coefficients.

4.2 Performance Analysis

The values used to predict the AQI magnitude are obtained by splitting the dataset into training and test datasets, respectively. The predicted values are evaluated with MAE, RMSE, and correlation coefficient (r). MAE is defined as the average of absolute differences between the predicted and observed values. The method of calculation of MAE value is expressed in Eq. (8) as

$$MAE = \frac{1}{n}\sum_{i=1}^{n}|y_i - y| \tag{8}$$

where y_{i}—Actual value, y—Predicted value, n—Number of observations.

RMSE is the measure of differences among the observed and predicted values that provides the deviation of predicted values from the actual values [33]. It is obtained by getting the mean of the squared differences and applying a root over the final value [1]. RMSE is calculated by the expression presented in Eq. (9) as

$$RMSE = \sqrt{\frac{1}{n}\sum_{i=1}^{n}|y_i - y|^2} \tag{9}$$

Correlation coefficient (r) is the relative measure between actual and predicted value which ranges from 0 to 1. The values that are closer to 1 are said to have a good association and model performance between predicted and actual values [33]. Correlation coefficient value is calculated using the Eq. (10),

$$r = \frac{\sum_{i=n}^{n}(x_i - \overline{x})(y_i - \overline{y})}{\sqrt{\sum_{i=1}^{n}(x_i - \overline{x})^2 \sum_{i=1}^{n}(y_i - \overline{y})^2}} \tag{10}$$

4.3 Experimental Results and Discussion

The results of both the individual and the ensemble model predictions of major industrialized cities in India are included in this discussion. The results of AQI values with respect to various cities are shown in Fig. 7. The AQI values are grouped with the location to visualize the air quality data for the past five years (2015–2019). The mean value of AQI for the past five years at the specific urban locations in India is displayed in Fig. 8.

The performance values are tabulated for various regression models RF, DT, GB, and LR as presented in Table 1. The values of test data and predicted data have been visualized through the scatterplots shown in Fig. 9. From the Table 1, it is observed that the lowest errors of 0.836 and 4.073 in MAE and RMSE, respectively, with highest correlation coefficient value of 0.993. The GB and LR models have indicated higher MAE values than RF and DT.

The scatterplot of overlapping of predicted and observed values of the regression models are shown in Fig. 9.

In the above stated regression model, the LR and GB have higher MAE values so that it can be trained with other models sequentially. Therefore, an ensemble model is created by combining any of the two individual models such as GB, DT, and LR. The performance metrics of these ensemble models are shown in Table 2. The

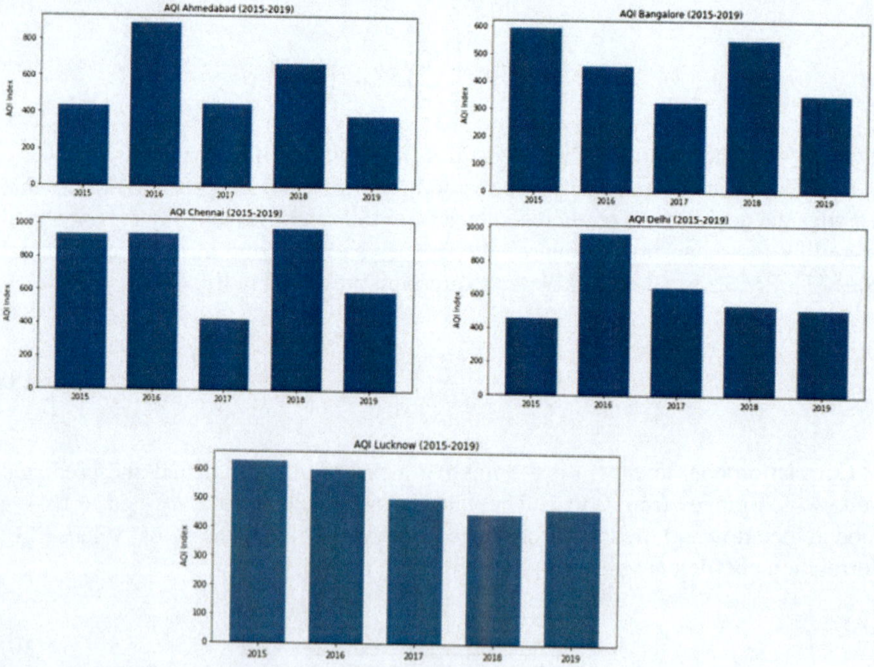

Fig. 7 Estimated AQI of various cities (2015–2019)

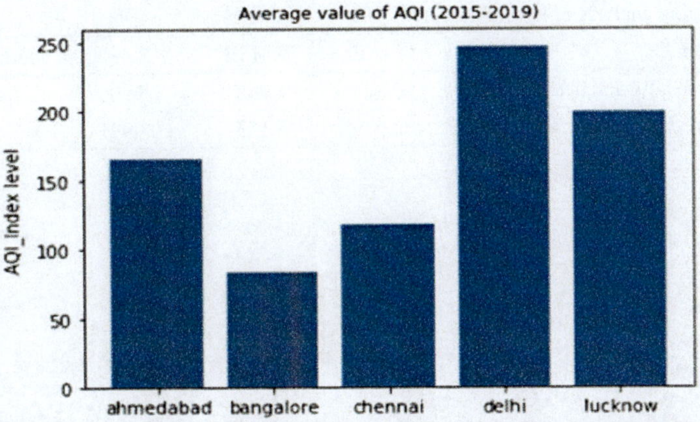

Fig. 8 Mean value (AQI) for the year 2015–2019

Table 1 Errors analysis of traditional ML methods

Regression model	MAE	RMSE	Correlation coefficient
Random forest	0.836	4.073	0.9993
Decision tree	0.959	5.291	0.9989
Linear regression	13.626	22.758	0.9796
Gradient boosting	3.006	5.082	0.9989

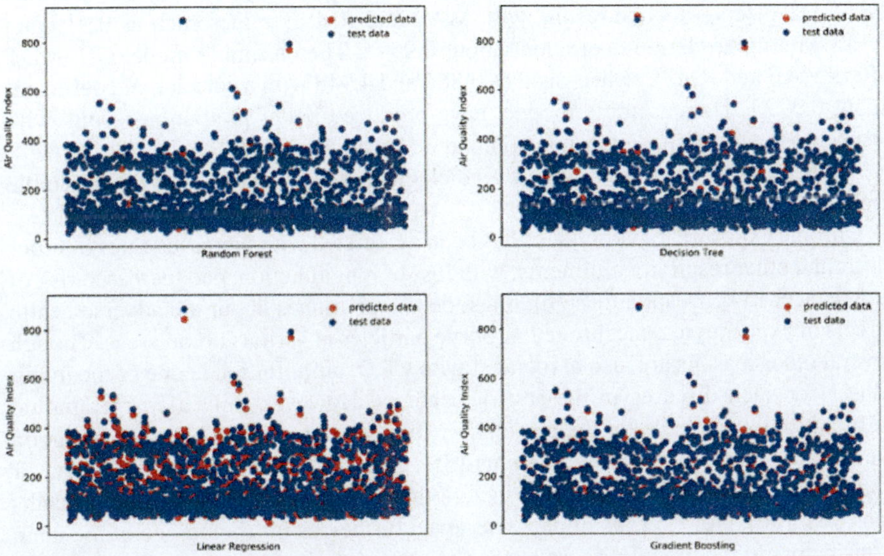

Fig. 9 Predicted AQI of Indian cities using traditional ML techniques

Table 2 Errors analysis of hybrid model

Ensemble model	MAE	RMSE	Correlation coefficient
Gradient boosting and decision tree	0.885	4.948	0.9990
Linear regression and decision tree	0.919	5.153	0.9991

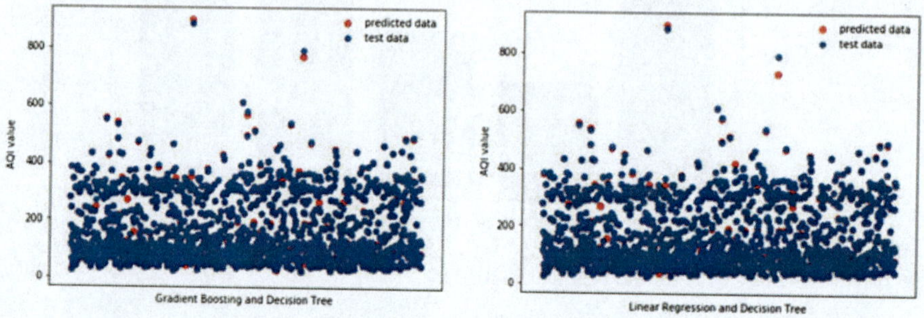

Fig. 10 Predicted AQI of Indian cities using hybrid ML techniques

ensemble model GB with DT has significant reduction in the MAE and RMSE values such as 0.0 and 2.121, respectively. The decrease in MAE and RMSE is about 0.343 and 0.134, respectively, and an increase in correlation coefficient value of 0.0001 is obtained when it is compared with DT and GB, respectively. The ensemble model LR + DT offers effective results with MAE and RMSE values such as 0.919 and 5.153 with a correlation coefficient about 0.9991. The ensemble model GB + DT offers MAE and RMSE values such as 0.885 and 4.948 with a correlation coefficient about 0.9990. The ensemble models have produced better prediction results with respect to the error analysis when compared to the individual models.

The scatterplot of overlapping of predicted and observed values of the ensemble models are shown in Fig. 10.

Chronic exposure to particulates has been associated with increased rates of bronchitis and other respiratory ailments, with loss of lung function, and increased risk of lung cancer [34]. Despite these findings, debate continues about the adverse health effects of exposure to concentrated airborne particles at various urban areas. Though several other reasons are also associated with CRD, air pollution is one of the major causes for such diseases in the stated locations. However, only a sparse amount of literature is available on the correlation studies between CRD and air pollution metrics at specific urban environments [35]. Ritchie & Roser [36] stated that air pollution is one of the leading risk factors that is responsible for 5 million deaths each year [36]. However, its impact goes even further, as it is being one of the main contributors to the global disease burden. As the air pollution continues to increase at various industrialized cities in India, there is a chance that the death rates would be augmented in the absence of swift remedial measures. Hence, considerable steps with regulatory measures should be accelerated to reduce the air pollution at the industrial cities to mitigate the major airborne health hazards in the near future.

5 Conclusions

In the presented chapter, hybrid ML models are proposed to get the closer values of AQI with less prediction error as compared to the traditional ML techniques. The datasets used for the presented work have been collected from the CPCB database which monitors and maintains the real-time air quality data across the major cities in India. ML algorithms are trained over the data obtained for the years 2015–2019, and the prediction models are being developed. Subsequently, one month data in relevance to the pollutants level are utilized to predict the subsequent month and so on. The presented ML and hybrid models have been evaluated using mean absolute error (MAE), root mean squared error (RMSE), and correlation coefficient (r). The hybrid models have also shown a significant difference in the error values when compared with the traditional ML models. From the two hybrid models compared herein, the gradient boosting with the decision tree model outperforms with MAE, RMSE values of 0.885 and 4.948, respectively, as compared to other methods. The prediction results are correlated with the death statistics that was released by IHME with respect to the corresponding cities at specific time periods. It is also observed that the death rate due to air pollution related diseases is less in the mildly polluted regions as compared to highly polluted areas. So, air pollution has a greater influence with few airborne diseases and it requires larger sampling analysis to ensure the degree of impact. In future, time-series models will be developed to predict the AQI in proportion to the pollutant levels at individual zones of interest for best results and custom measures.

References

1. Ameer, S., Shah, M.A., Khan, A., Song, H., Islam, S.U., Asghar, M.N.: Comparative analysis of machine learning techniques for predicting air quality in smart cities. IEEE Access **7**, 128325–128338 (2019). https://doi.org/10.1109/ACCESS.2019.2925082
2. Pang, Y., Huang, W., Luo, X.-S., Chen, Q., Zhao, Z., Tang, M., Hong, Y., Chen, J., Li, H.: In-vitro human lung cell injuries induced by urban PM2.5 during a severe air pollution episode: variations associated with particle components. Ecotoxicol. Environ. Saf. **206** (2020). https://doi.org/10.1016/j.ecoenv.2020.111406.
3. World Air Quality Report. https://www.iqair.com/world-most-polluted-cities/world-air-quality-report-2019-en.pdf
4. Upadhyay, S.: Comparative analysis of machine learning regression algorithms on air pollution dataset. Int. J. Sci. Res. Comput. Sci., Eng. Inf. Technol. 125–136 (Jul 2020)
5. Nasir, H., Goyal, K., Prabhakar, D.: Review of air quality monitoring: case study of India. Indian J. Sci. Technol. **9**(44). https://doi.org/10.17485/ijst/2016/v9i44/105255
6. Delhi Disease Burden Profile, 1990–2016. http://www.healthdata.org/sites/default/files/files/Delhi-Disease_Burden_Profile%5B1%5D.pdf
7. TamilNadu Disease Burden Profile, 1990–2016. http://www.healthdata.org/sites/default/files/files/TamilNadu-Disease_Burden_Profile%5B1%5D.pdf
8. Gujarat Disease Burden Profile, 1990–2016. http://www.healthdata.org/sites/default/files/files/Gujarat_-_Disease_Burden_Profile%5B1%5D.pdf

9. Uttar Pradesh Disease Burden Profile, 1990–2016. http://www.healthdata.org/sites/default/files/files/Uttar_Pradesh_Disease_Burden_Profile%5B1%5D.pdf
10. Karnataka Disease Burden Profile, 1990–2016. http://www.healthdata.org/sites/default/files/files/Karnataka_Disease_Burden_Profile%5B1%5D.pdf
11. Lee, B.J., Kim, B., Lee, K.: Air pollution exposure and cardiovascular disease. Toxicol Res. **30**(2), 71–75 (2014). https://doi.org/10.5487/TR.2014.30.2.071
12. Xing, Y.F., Xu, Y.H., Shi, M.H., Lian, Y.X.: The impact of PM2.5 on the human respiratory system. J. Thorac. Dis. **8**(1), E69–E74 (2016). https://doi.org/10.3978/j.issn.2072-1439.2016.01.19
13. Chen, C.H., Wu, C.D., Chiang, H.C., et al.: The effects of fine and coarse particulate matter on lung function among the elderly. Sci. Rep. **9**, 14790 (2019). https://doi.org/10.1038/s41598-019-51307-5
14. World Health Organization. https://www.who.int/mediacentre/news/releases/2014/air-pollution/en/
15. Technical Report by Working group of WHO, Health Aspects of Air Pollution with Particulate Matter, Ozone and Nitrogen Dioxide. https://www.euro.who.int/__data/assets/pdf_file/0005/112199/E79097.pdf
16. Janarthanan, R., Partheeban, P., Somasundaram, K., Navin Elamparithi, P.: A deep learning approach for prediction of air quality index in a metropolitan city. Sustain. Cities Soc. **67** (2021). https://doi.org/10.1016/j.scs.2021.102720
17. Chaitanya, A.K., Prasad, K.V.: A comparative study on prediction of Indian air quality index using machine learning algorithms. J. Crit. Rev. **7**(13), (2020). https://doi.org/10.31838/jcr.07.13.05
18. Džaferović, E., Karađuzović-Hadžiabdić, K.: Air quality prediction using machine learning methods: a case study of Bjelave neighborhood, Sarajevo, BiH. Advan. Technol., Syst., Appl. V: Papers Sel. Techn. Sci. Div. Bosnian-Herzegovinian Am. Acad. Arts Sci. **2020**(142), 423–434 (2020). https://doi.org/10.1007/978-3-030-54765-3_29
19. Ganesh, S.S., Arulmozhivarman, P., Tatavarti, R.: Forecasting air quality index using an ensemble of artificial neural networks and regression models. J. Intell. Syst. **28**(5), 893–903 (2019). https://doi.org/10.1515/jisys-2017-0277
20. Erdogan, Z., Namli, E.: A living environment prediction model using ensemble machine learning techniques based on quality of life index. J. Ambient Intell. Humanized Comput. https://doi.org/10.1007/s12652-019-01432-w
21. Liang, Y.-C., Maimury, Y., Chen, A.H.-L., Juarez, J.R.C.: Machine learning-based prediction of air quality. Appl. Sci. **10**, 9151 (2020)
22. Liu, H., Li, Q., Yu, D., Gu, Y.: Air quality index and air pollutant concentration prediction based on machine learning algorithms. Appl. Sci. **9**(19), 4069 (2019). https://doi.org/10.3390/app9194069
23. He, H., Luo, F.: Study of LSTM air quality index prediction based on forecasting timeliness. IOP Conference Series: Earth and Environmental Science, vol. 446, p. 2. Environmental Analysis and Pollution Control Engineering
24. Halsana, S.: Air quality prediction model using supervised machine learning algorithms. Int. J. Sci. Res. Comput. Sci., Eng. Inf. Technol. https://doi.org/10.32628/CSEIT206435
25. Dataset. https://app.cpcbccr.com/ccr/#/caaqm-dashboard-all/caaqm-landing
26. Amatoa, F., Laib, M., Guignard, F., Kanevski, M.: Analysis of air pollution time series using complexity-invariant distance and information measures. arXiv:1909.11484v1 [stat.AP] (21 Sep 2019)
27. Maryam Sarkhosh, Ali Asghar Najafpoor, Hosein Alidadi, Jamal Shamsara, Hanieh Amiri, Tittarelli Andrea, Fatemeh Kariminejad, "Indoor Air Quality associations with sick building syndrome: An application of decision tree technology", Building and Environment, Volume 188, (2021). https://doi.org/10.1016/j.buildenv.2020.107446
28. Font, A., Tremper, A.H., Lin, C., Priestman, M., Marsh, D., Woods, M., Heal, M.R., Green, D.C.: Air quality in enclosed railway stations: quantifying the impact of diesel trains through deployment of multi-site measurement and random forest modelling. Environ. Pollut. **262** (2020). https://doi.org/10.1016/j.envpol.2020.114284

29. Shams, S.R., Jahani, A., Kalantary, S., Moeinaddini, M., Khorasani, N.: The evaluation on artificial neural networks (ANN) and multiple linear regressions (MLR) models for predicting SO_2 concentration. Urban Clim. **37** (2021). https://doi.org/10.1016/j.uclim.2021.100837
30. Cheng, B., Ma, Y., Feng, F., Zhang, Y., Shen, J., Wang, H., Guo, Y., Cheng, Y.: Influence of weather and air pollution on concentration change of PM2.5 using a generalized additive model and gradient boosting machine. Atmos. Environ. **255**. https://doi.org/10.1016/j.atmosenv.2021.118437
31. Zhang, Y., Haghani, A.: A gradient boosting method to improve travel time prediction. Transp. Res. Part C: Emerg. Technol. **58**(Part B) (2015). https://doi.org/10.1016/j.trc.2015.02.019
32. Huang, C., Townshend, J. R. G.: A stepwise regression tree for nonlinear approximation: applications to estimating subpixel land cover. Int. J. Remote Sens. **24**(1):75–90
33. Kumar, A., Goyal, P.: Forecasting of air quality in Delhi using principal component regression technique. Atmos. Pollut. Res. **2** (2011)
34. Schwartz, J.: Particulate air pollution and chronic respiratory disease. Environ. Res. **62**(1), 7–13 (1993). https://doi.org/10.1006/enrs.1993.1083. PMID: 8325268
35. Dávid, A., Kégel, E., Rudnai, P., Sárkány, E., Kertész, M.: A levegöszennyezettség mértéke és a gyermekek légúti megbetegedése közötti összefüggés vizsgálata Dorogon, [Correlation between air pollution and respiratory morbidity in children at Dorog]. Orv Hetil. **131**(10):513–7 (11 Mar 1990). Hungarian. PMID: 2179805
36. Ritchie, H., Roser, M.: Air pollution. Our World Data (2017)

Deep Learning for Green Smart Environment

Tuan Nguyen, L. C. Ngoc, Tung Nguyen Son, Duc Ha Minh, and T. Ha Phuong Dinh

Abstract Urban waste management has always been a challenging problem due to the increasingly abundant amount of mixed domestic waste without household waste segregation. The remarkable advancement in deep learning helps computer vision systems gain splendid achievements in image classification and image recognition, including image-based waste identification and classification. We separate three significant categories of domestic waste: recyclable waste (plastic, paper, glass-metal), biodegradable waste, and non-recyclable waste. With the deep learning techniques, our model achieves an 87.50% prediction accuracy on the test dataset. The automatic waste classification helps address complicated problems such as recycling contamination and then mitigates the risk of incorporating many types of waste in a bin. The automated waste sorting model also provides an optimal solution for users by segregating garbage into suitable compartments. This solution characterizes the application of a convolutional neural network to optimize innovative waste management: classifies main categories of waste at its source to support the recycling system.

Keywords Waste classification · Deep learning · Computer vision · ResNet50

1 Problem

The risks associated with waste have become an increasingly severe problem in environmental protection. According to a World Bank [1] report, the worlds' cities generated 2.01 billion tons of solid waste in 2016, amounting to a footprint of 0.74 kg per person per day. With rapid population growth and urbanization, annual waste generation is expected to increase by 70% from 2016 to 3.40 billion tons in 2050. Urbanization and economic growth increase lead waste disposal becoming a

T. Nguyen (✉) · T. N. Son · D. H. Minh · T. H. P. Dinh
National Economics University, Hanoi, Vietnam
e-mail: nttuan@neu.edu.vn

L. C. Ngoc
Ha Noi University of Science, Hanoi, Vietnam

M. Lahby et al. (eds.), *Computational Intelligence Techniques for Green Smart Cities*, Green Energy and Technology, https://doi.org/10.1007/978-3-030-96429-0_13

significant concern for the entire society. Urbanization, economic expansion, and population explosion have posed more significant pollution treatment and waste management challenges. Currently, waste is often not appropriately recycled and is usually either dumped in a landfill or incinerated. Besides resource-intensive methods, landfills threaten the environment and pose a range of health concerns, particularly to people living near the landfill and workers who collect waste. Using a waste incinerator to deal with problems may raise concerns that fine particulate emissions will cause heavy air pollution. Sameh Wahba [2], World Bank Director for Urban and Territorial Development, Disaster Risk Management and Resilience, said that "poorly managed waste is contaminating the world's oceans, clogging drains and causing flooding, transmitting diseases, increasing respiratory problems from burning, harming animals that consume waste unknowingly, and affecting economic development, such as through tourism." As a result, improving solid waste management is an urgent priority.

Waste needs to be classified based on different components recycled in various ways to help safeguard the environment and human health. The waste management sector follows a generally accepted hierarchy. The earliest known usage of the "waste management hierarchy" appears to be Ontario's Pollution Probe in the early 1970s. The hierarchy started as the *3 Rs—reduce, reuse, recycle*—but now a fourth "R" is frequently added—*recovery* and ended by waste disposal. However, given the current pace of trash growth, waste management is a challenge facing society. Developing methods for managing waste efficiently, economically, and environmentally is essential for supporting sustainable urban development. Therefore, classification is a core component of the waste management system. The purpose of trash classification is to increase the reuse of waste, minimize landfill capacity, and lower waste disposal costs. Therefore, waste classification affects the economic, environmental, and waste treatment benefits.

However, the existing waste classification method is insufficient and exacerbates several urgent problems. It is why scientists are developing and researching automated classification methods that increase the efficiency of the recycling process. Waste classification has piqued the interest of many researchers to apply computer vision in solving traditional manual classification methods. Using deep learning technology in the field of waste classification can improve the efficiency of recycling factories. The "smart bins system" could automate the waste classification process.

2 Related Works

Traditional waste classification methods have focused on physical characteristics, including weight, form (solid, liquid, aqueous, or gaseous), and waste generating processes. However, image classification has become a major research area in recent years, thanks to the release of large publicly available datasets, especially ImageNet (https://www.image-net.org). The increase in data rates and the availability of enormous datasets have led to the rapid development of large deep neural network models.

Recently, computer vision approaches have been used to localize, identify, and classify waste on the streets. This enables the street cleaning equipment to locate the areas with the highest waste rates and concentrate its efforts there. Internal cleaning robots these days have also used computer vision techniques to recognize the type of garbage that the robot is about to clean on the floor. Finding and sorting trash are essential features for a floor cleaning robot, as it allows the robot to detect and avoid difficult-to-clean things. Waste classification has spurred many researchers to apply computer vision in solving traditional manual methods. Waste classification using deep learning has the potential to improve the efficiency of recycling systems. This application has a positive influence on the environment as well as energy saving. Zhou et al. [3] have demonstrated how waste classification can be utilized to aid thermal conversion in waste-to-energy studies.

Recently, Salimi et al. [4] have created a trash robot capable of detecting and sorting rubbish into organic and non-organic waste. This robot would go to public locations and autonomously scan and dispose of rubbish. Meanwhile, Y Chu et al. [5] introduced a deep learning neural network system to automatically classify residential garbage in metropolitan regions automatically. This model used a convolutional neural network-based algorithm to extract features from the data and a multi-layer perceptron (MLP) method to merge image features and other feature information to classify the image waste into either recyclable or other wastes. Research by Balakrishnan Ramalingam et al. [6] proposes a debris classification and detecting model for automatic floor mop robot using convolutional neural network—CNN and support vector machine—SVM. The proposed technique can detect and classify floor fragments with 95.50% accuracy, according to test results. Moreover, CNN network takes only 71 ms for the entire debris detection and sorting process, which implies that the suggested approach is suitable for deployment in real-time cleaning applications.

Much research on the waste classification problem is now being conducted. To classify waste on the "TrashNet" dataset, Aral et al. [7] used transfer learning models derived from various well-known CNN models for image classification, including DenseNet121, DenseNet169, InceptionResnetV2, MobileNet, and Xception. According to the findings of the experiments, DenseNet121 transfer learning model had the greatest accuracy, with a 95% yield. Aghilan et al. [8] have lowered the time complexity by reducing the number of parameters from seven million to two million (utilizing another lightweight transfer learning model—MobileNetV2). However, the accuracy rate only fluctuates around 80%. The application of multiple CNN models for automatic trash classification was once again examined by Ruiz et al. [9]. The best performance results were found utilizing a ResNet-based architecture, which had an average accuracy of 88.66% on the TrashNet dataset. For Vietnam's garbage classification task, Vo et al. [10] gathered 5904 pictures from Vietnam sorted into three categories: organic, inorganic, and medical waste to create the waste-set dataset. This research developed a deep neural network model named DNN-TC, which enhances the ResNext model for better prediction performance. The experimental results reveal that DNN-TC yields 94% and 98% accuracy in the TrashNet and waste-set datasets.

The common limitation of the above studies is that the training datasets are mostly gathered from the Internet and do not truly follow up with actual waste samples since they cannot scan all possible scenarios when users throw garbage into the container (in terms of angle, light, etc.). Furthermore, how to divide labels for classification of the previous research, such as in the study of author Vo et al. [10], does not have practical effects in post-production waste treatment. Because of that, we reconstructed the training dataset and separated it into more specific labels:

- *Recyclable waste*: Split into three small separate groups of paper-cardboard, glass-metal, and plastic.
- *Organic waste*
- *Non-recyclable waste.*

After building the dataset in accordance with the actual, we select the feasible tuning neural networks to build a classification model based on three criteria: accuracy, mass volume (based on the number of parameters during the training process), and compatibility when integrating with hardware devices.

3 Convolutional Neural Network (CNN)

3.1 Overview

Under deep learning viewpoint, waste classification is an image classification problem. Toward this problem, convolutional neural networks (CNNs or ConvNets) are the most common approach. CNN is inspired by the structure of neural networks and the neural activities of cells in the brain. The connection pattern between neurons is like the organizational structure of the cerebral cortex (responsible for the management of animal vision). The first convolutional neural network is LeNet-5 which was introduced by Bengio et al. [11]. LeNet-5 can recognize handwritten digits with high accuracy, evaluated based on the public dataset MNIST (http://yann.lecun.com/exdb/mnist). CNNs are widely applied in the fields of image and video recognition, recommendation systems, image classification, object detection, medical image analysis, natural language processing, financial time series analysis, etc.

3.2 Compare Convolutional Neural Network (CNN) to Artificial Neural Network (ANN)

Both CNN and traditional artificial neural network (ANN) are composed of self-optimizing network nodes through training. Each neuron still receives input and operates (such as a dot product followed by a non-linear function). The entire network still represents a single score function through weights from the raw image matrix to the output layer. The last class will contain loss functions corresponding to the classes to be classified or identified. Techniques used to develop traditional ANNs are still applied to CNNs.

There are two main reasons why CNNs perform better than traditional neural networks in pattern classification tasks. Firstly, for complex computer vision problems, traditional neural network training takes much time and requires extremely powerful computation. Secondly, using a CNN can reduce overfitting of the model. Overfitting occurs when the model is too complex to simulate the training data and contains too many parameters, which is inefficient in predicting tasks. In a multi-layer perceptron (MLP) network, model complexity can be considered the number of hidden layers and the number of units in the hidden layers.

One of the most significant limitations of traditional ANNs is their tendency to overstate the complexity required to compute the image data. Popular benchmark datasets such as MNIST of handwritten digits are suitable for most forms of ANNs, due to the relatively small image size of only 28×28. With this dataset, a single neuron in the first hidden layer will contain 784 weights ($28 \times 28 \times 1$), where note that MNIST is normalized to only black and white values, which is manageable for most network architectures ANN. If processing a colored input image with a more substantial size of 64×64, the number of weights on a single neuron of the first layer will increase significantly. The most notable difference between CNN and traditional ANN is that CNN is mainly used in pattern recognition and classification from image data. This allows the model to encode image-specific features into the network architecture, making the network more suitable for image-centric tasks while reducing the parameters required for model establishment. In the areas mentioned above, convolutional neural networks have achieved breakthrough results in the past decade. The most crucial factor that makes CNNs more efficient than MLP networks is the significant reduction in the number of parameters in the training model. This breakthrough has pushed researchers and developers to approach the model to solve complex tasks that were not possible with classical ANNs. Another critical problem that CNNs have solved is that the extracted features are not spatially independent. For example, in the face recognition problem, the sole goal is to detect the object feature regardless of the position in the input data.

The CNN network is a modified version of the multi-layer perceptron (MLP) network. Perceptrons distributed at multiple layers lead to neurons (network nodes) in one layer that is connected to all neurons in the next layer (fully connected). The fully connected of these networks makes the model vulnerable to overfitting with too much data. Typical techniques to correct too many parameters involved in the calculation process include reducing model weights through weights adjustment and removing unnecessary connections. Dropout, CNNs take a different approach toward regularization: take advantage of the hierarchical pattern in increasingly complex data by using smaller and simpler samples embossed in the model's filter. Therefore, in terms of network connectivity and complexity, CNNs are more efficient than MLPs when handling image data modeling and analysis tasks.

CNNs use relatively little preprocessing compared to other image classification algorithms. This means that the network learns to optimize filters through automatic learning, whereas these filters are designed manually in traditional algorithms. The independence combined with human intervention in image feature extraction is a great advantage of this network.

3.3 The Basic Architecture of Convolutional Neural Network

Convolutional neural networks are widely used in computer vision. Through element-wise multiplication, critical features from the image are extracted and transferred into convolutional layers. Each convolutional layer consists of many units. The result of each unit is a convolutional transformation from the previous layer through cross-correlation with the kernel. A convolutional neural network has three essential processes included convolution, pooling, and fully connected (Fig. 1).

Convolution: This process happens through the element-wise multiplication between the input matrix and the kernel to return new values in a new layer. This process could happen continuously in the first layers of the network combined with non-linear activation functions. The objective is to extract two-dimensional features through the classes.

Pooling: The depth of the network is determined by the number of channels. Because the depth often grows exponentially, later layers after feature extraction will need many parameters. To reduce the computation, the model needs to reduce the dimensions of the input matrix or reduce the number of floor units. Since each unit would be a representative result of applying a kernel to find a particular feature, reducing the number of units would not be feasible. By finding a representative value for each spatial region that the filter passes through, reducing the input matrix size would not change the main contours of the image but reduce the size of the image.

Fully connected: After reducing the size, the matrix could be flattened into a vector. This process happens at the end of the network and uses a non-linear activation

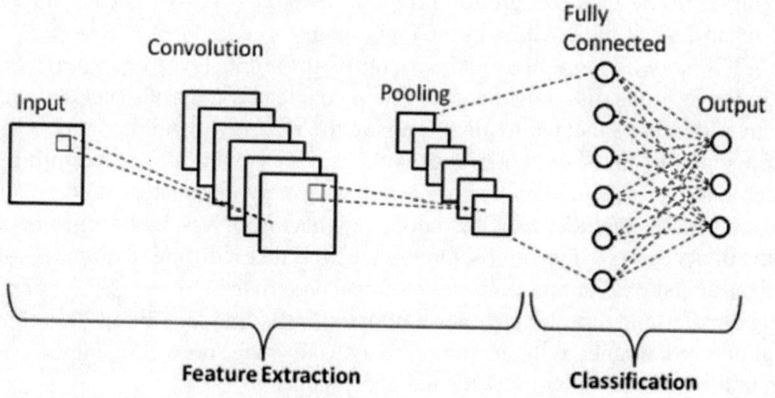

Fig. 1 Basic convolutional neural network architecture (Phung and Rhee [12])

function. The last fully connected layer applies the Softmax activation function to calculate the probability distribution of each group.

3.4 Properties of Convolutional Neural Network

A convolutional neural network mitigates the limitations imposed by the MLP architecture by exploiting the spatially strong local correlations present in images. CNN has some distinguishing characteristics.

Sliding connectivity: CNNs do not connect to the entire image data but only to each region (local region). Each local region has the same size as the kernel of image size. The kernel slide in the direction of the image from left to right and top to bottom and calculate the convolution values. Then, it populates these values to the feature map.

3D neuron blocks: Layers are arranged in 3 dimensions included width, height, and depth. Each neuron inside a convolutional layer is connected to only a small region of the layer before called the receptive field. Both locally and fully connected, distinct layer types are stacked to form the CNN architecture (Krizhevsky et al. [13]). For example, in Fig. 2, AlexNet network consists of 9 layers that extract features of images and perform the objective tasks.

Local connectivity: CNNs exploit spatial locality by implementing a model of local connectivity between neurons of adjacent layers. Thus, the architecture ensures that the learned filters produce the most robust response to the spatially localized input pattern. Furthermore, such multi-layer stacking leads to non-linear filters that become increasingly "global" (responsive to a larger region of pixel space).

Synthesis: In the composite layers of CNN, the feature map is divided into rectangular subregions. The feature subregions representing the image's features are sampled independently down one unique value, usually average pooling or max

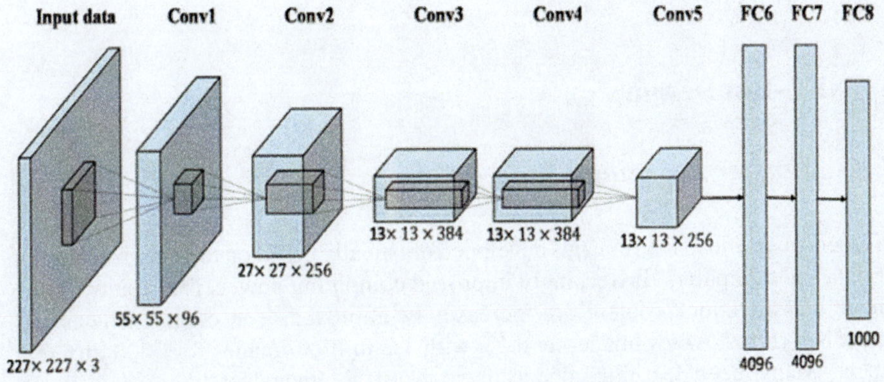

Fig. 2 Structure of 3D neural blocks of AlexNet network (Butt et al. [14])

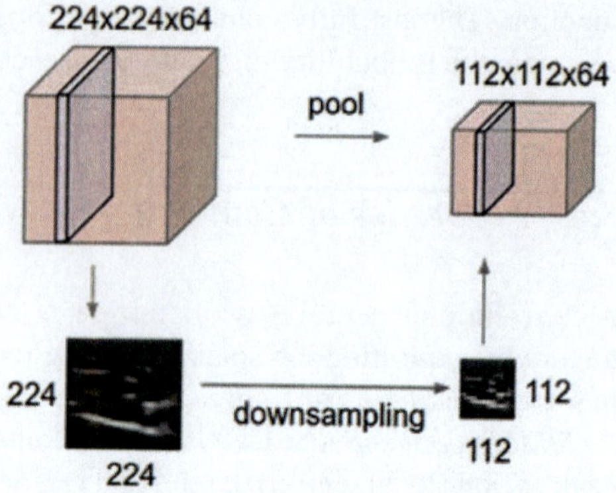

Fig. 3 Synthesis of CNN. (*Source* https://cs231n.github.io/convolutional-networks)

pooling. The pooling operation reduces the size of the feature map and provides a degree of local translation invariance for the features in the image (Myburgh et al. [15]) (Fig. 3). In addition, it allows CNNs to be more robust with other features variations in their position.

More complex in depth: CNNs could "learn" more abstract features as the input data propagates toward deeper layers. In image classification, vertical and horizontal edges could be detected in the first layer, followed by simple contour recognition in the second layer. Then, specific features were extracted, such as facial expressions in subsequent layers as shown in Fig. 4

Share weights: In CNN, each filter is duplicated across the entire visual field. These copied units share the same weight and offset vectors, forming a feature map (Fig. 5). This means that all neurons in each convolutional layer respond to the same feature in their field. They grant translational equivalence with a stride.

Together, these properties allow CNN to achieve better generalizability. In addition, weight sharing reduces the number of learned parameters, thus reducing memory requirements to run the network. Therefore, it allows training more extensive networks.

4 NEU-Bin Design

4.1 Transfer Learning

In recent years, deep learning has developed drastically based on the massive amount of data and computers' increasingly improved computing power. The results for the image classification problem are increasingly improved. The common dataset is ImageNet (https://www.image-net.org), with 1.2 million images for 1000 different labels. Many deep learning models have won the ImageNet large scale visual recognition challenge—ILSVRC (https://goo.gl/1A8drd): AlexNet [KSH12], ZFNet [ZF14], GoogLeNet [SLJ+15], ResNet [HZRS16], VGG [SZ14]. In general, these

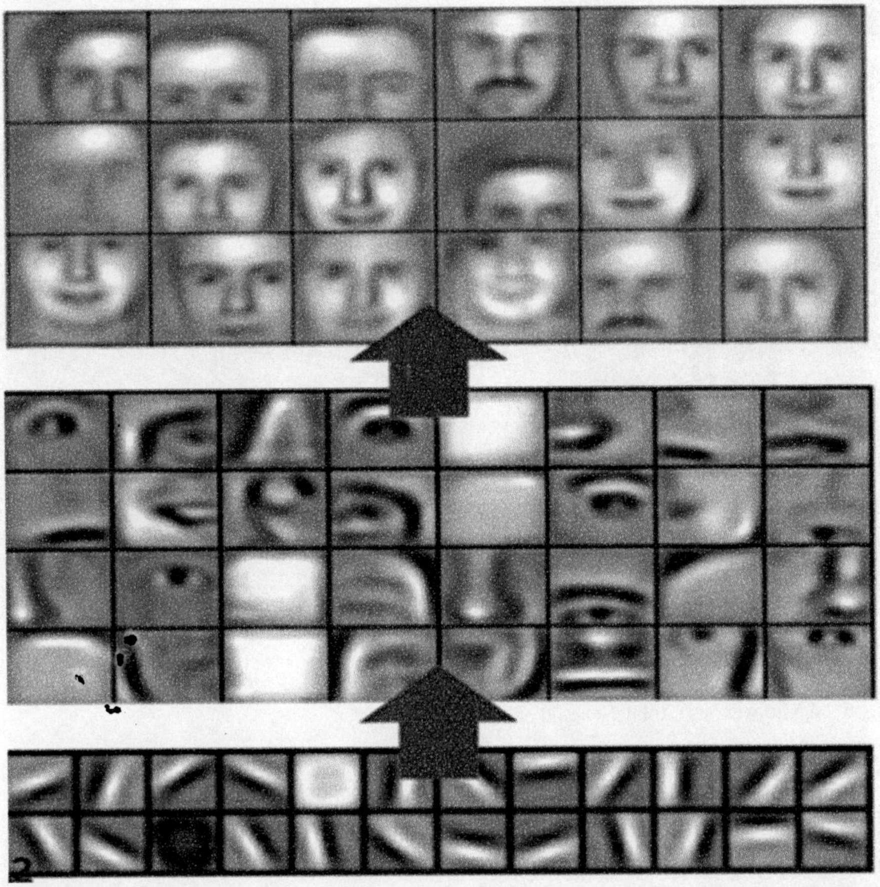

Fig. 4 Lower level features progressively combine to form higher-layer features in deep learning (Bengio [16])

models are multi-layer neural networks. The front layers are usually convolutional layers. The last layer is fully connected and usually a Softmax regressor. The pre-final layers' output could be considered the feature vector, and Softmax regression is the classifier used. The feature extractor and classifier are trained together through coefficient optimization in deep neural networks, which makes these models achieve good results.

However, these models all include many layers of weights. Therefore, training based on more than a million photos takes much time. For problems classifying other image data with a small training dataset, we may not need to rebuild the neural network and train it from scratch. Instead, we could use the trained models mentioned above and change the architecture of the network accordingly. This method of using pre-trained models is known as transfer learning. Transfer learning is a research area in machine learning that focuses on storing knowledge gained while solving a

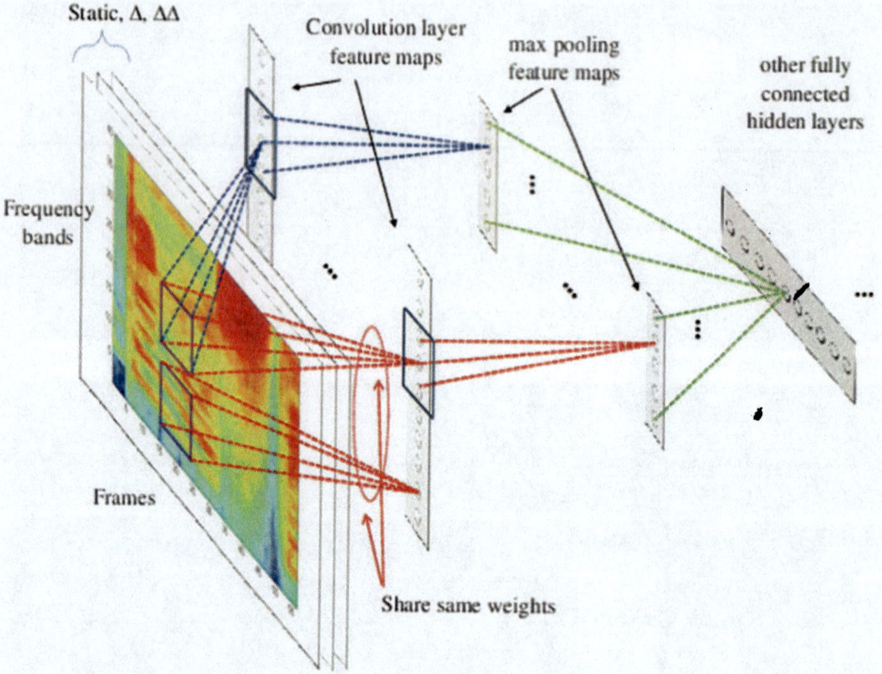

Fig. 5 Shared weights (Abbdel-Hamid et al. [17])

problem and applying it to another but related problem. In short, transfer learning transfers knowledge learned from the source dataset to the target dataset. All layers except the output layer could be considered feature extractors. This is based on the observation that objects in image data often have similar characteristics. Then, we train another classifier based on the extracted feature vector. This approach could significantly increase classification accuracy compared to using manual features because deep neural networks could extract high-level features of an image.

With the transfer learning method, the deep learning neural network is layered with the initial layers preserving the essential features of the image, such as edges, contours, and layers that are then extracted. Based on that, we can freeze some blocks and update the last layers of the model. One of the problems when updating the model using a tweak is that the parameter at the layers in the non-freeze mode must be updated to solve the new problem. When there is a new task, the algorithm creates a new neural network and shares representative features between tasks. However, this approach is not suitable due to space constraints and complexity, such as the number of linear networks and new tasks to learn.

Why use Transfer Learning?

Three main points have made transfer learning more popular in deep learning.

Data are not enough. Deep learning models need numerous data. It consumes data resources to "learn" (training) on that. These days, the data augmentation technique with transfer learning is often an efficient solution to almost all problems. The reason is that, in some cases, we might not have enough data to train our machine learning models. Working with insufficient data would result in lower performance; starting with a pre-trained model would help data scientists build better models.

Insufficient resources: Learning on large data sets is very resource-intensive. Transfer learning contributes to reducing the training time. In our experience, proper implementation of a deep learning model on a task required much data. However, it is common to be in scenarios where there is insufficient labeled training data. It is cumbersome attempting to label data for a deep learning model from scratch. In tasks where domain expertise may be required to create a quality dataset, it is pretty time-consuming to gain (that needed) domain knowledge—this is an excellent example of where deep transfer learning may come into play.

Improve the quality: There are many cases where transfer learning improves a target task's prediction quality compared to retraining from scratch. The reason could be that the source network was trained with the numerous data and learned better generalization. The pre-trained model with the target task while the network still had the source task "knowledge" for dynamic separation of multi-task learning.

When using Transfer Learning?

Some cases apply transfer learning but do not find it compelling. The reason for that is that transfer learning is only suitable for some specific situations as follows learning (mainly based on our experience).

Only apply transfer learning between two models with the same domain. Pre-trained model A and the model to be trained B do not share the same data domain; the features learned from the feature extractor of A will not be helpful in the classification of model B. For example, if you want to build a sound application that wakes up Google's virtual assistant in Vietnamese when saying "Wake up Google," you already have pre-trained model A for speech to text tasks but trained in English. As such, you should not perform transfer learning in this case. Transfer learning only works if the initial and target problems of both models are similar enough. A network is already pre-trained on a similar task, which is usually trained on massive amounts of data. Instead of building a model from scratch to solve a similar problem, we use the model trained on another problem as a starting point.

The training data of the pre-trained model A is larger than that of model B. If we transfer the coefficients from a pre-trained model trained on the small size data, then the features learned from model A could not generalize to solve the target task. Imagine, learning from a pre-trained model is like teaching a child (model) to learn. It is more effective to let that child has access to a vast and rich source of knowledge than to rely on a small, concentrated source of knowledge.

Pre-trained-model A is a good-quality model. This is an obvious requirement because only good models extract "good" features (garbage in garbage out).

The approach uses the pre-trained model and retrains the last few layers based on the specific data. This common technique is called *fine-tuning*. With this technique, the target model copies all model designs with their parameters from the source model except the output layer and fine-tunes these parameters based on the target dataset. In contrast, the output layer of the target model needs to be trained from scratch. Furthermore, in fine-tuning technique, the weights of some of the layers are updated and trained as fully connected layers at the end of the model structure. Therefore, this technique is a bit more resource-intensive due to the training of several classes earlier.

Each neural network consists of many layers; after training, each layer will be adjusted to detect specific features in the input data. For example, in a convolutional neural network tasked with image classification, the first layers will detect general features such as edges, corners, circles, or patches of color. In short, the early layers in the neural network detect general features, while the deeper layers detect specific features.

4.2 Residual Network 50 (ResNet50)

When designing increasingly deep networks, it is vital to understand how adding layers increase the complexity and performance of the network. More important is the ability to design networks where adding layers to the network inevitably increase the representation rather than making a slight difference.

A residual neural network (ResNet) is built from the structure of pyramidal cells in the cerebral cortex. ResNet utilizes skip connections to jump over some layers. Typical ResNet models are implemented with double or triple-layer skips that contain nonlinearities (ReLU) and batch normalization in between (He Kaiming et al. [18]).

ResNet utilizes skip connections to mitigate the accuracy saturation problem or to avoid the problem of vanishing gradients. Adding more layers to a suitable model lead to higher training error could lead to accuracy saturation (He et al. [18]). During training, the weights adapt to mute the upstream layer and amplify the previously skipped layer. In the simplest case, only the adjacent layer's connection weights are adapted, with no explicit weights for the upstream layer. This works best when a single non-linear layer is stepped over, or the intermediate layers are all linear. If not, then an explicit weight matrix should be learned for the skipped connection.

Skipping effectively simplifies the network; using fewer layers in the initial training stages spurs learning by reducing the impact of vanishing gradients. The network gradually restores the skipped layers as it learns the feature space. At the end of the training process, when all layers are expanded, it learns faster. Although a neural network without residual parts explores more of the feature space, it becomes more vulnerable to perturbations and necessitates extra training data to recover.

The main idea of residual network is that each layer added should have a component that is an identity function. This means that if we train the newly added class to be a homogeneous map $f(x) = x$, the new model will be at least as efficient as the original model. Because the added layer can better match the training data, resulting in smaller training errors. Better yet, the identity function should be the simplest in a

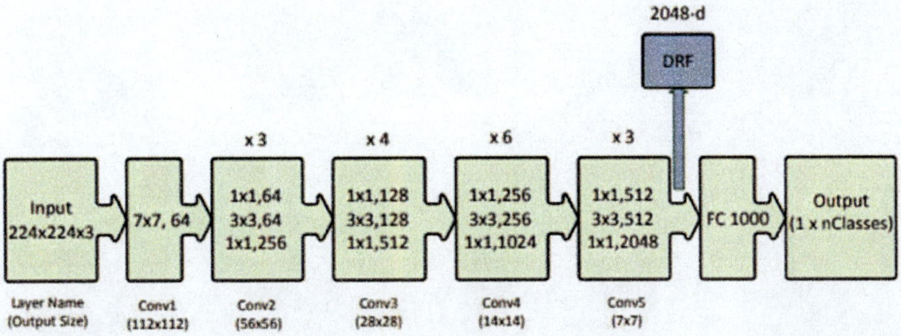

Fig. 6 The structure of ResNet50 (Mahmooh [19])

class instead of the null function $f(x) = 0$. To ensure that adding layers increase the representation of the network, larger function classes must contain smaller classes. ResNet50 is a variant of the ResNet model that has 48 convolutional layers along with one max pooling layer and one average pooling layer. ResNet50 has 3.8×10^9 floating-point operations. In a convolutional neural network, the convolutional layer transforms the imported image using a sequence of $3 \times$ three filters, extracting specific features from the input data. Our model is built from a pre-trained ResNet50 model, based on an ImageNet dataset with a size of 256×256 and classified into 1000 labels. The structure of ResNet50 is depicted in Fig. 6.

4.3 Data Augmentation

Data augmentation is a common technique used to increase the amount of training data based on the existing data and apply some image transformations such as zoom in (or out), random crop, and rotate to generate a new set of additional images. Especially, it is used in deep learning to control overfitting, improve model accuracy, and generalization. The randomness of this process helps the model not have to be equipped with too much local training data, saving time, and effort in data collection. In our experiments, we use the ImageDataGenerator class1 from Keras to provide several transformations for generating new training data, such as rotation, zooming, translation, randomly flipping images horizontally, and filling new pixels with their nearest surround pixels. Besides, depending on the purpose of the transformation, modelers could apply a variety of methods. For example,

- *normalize*: Normalize each image to a normal distribution by subtracting the mean and dividing all the corresponding pixels in each channel by the variance.
- *zoom range*: This method is about a range of image magnification values: [lower, upper]. The magnification value of an image will be randomly generated within the zoom range. The smaller this magnification value, the larger the image will be.

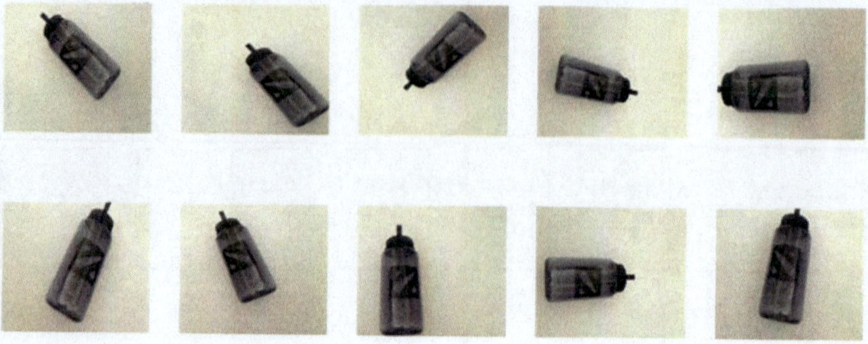

Fig. 7 Examples of data augmentation for waste dataset

- *rotation*: Random rotation of an image. Usually set from 10 to 20 degrees.
- *brightness range*: Range to adjust the brightness of the image. Brightness will be a random value from [min value, max value].

When initializing the data generator with pre-trained models, we perform the data transformation steps in the data pipeline identical to the pipeline applied on the pre-trained models. Then, the features made from the base network have a good classification effect. Figure 7 shows new data generated from this technique.

4.4 Model

When performing transfer learning to build the model, we froze the first layers of the ResNet50 pre-trained model. These are classes that detect general features that are common across all networks. Then, the deeper classes are refined with the collected data, and adding new classes are added to classify the new categories included in the training dataset. When there is a significant difference between the source and the target, or the training dataset has few specific features, we unfreeze the classes relative to the pre-trained model. Next, we add a new classifier and refine the unfrozen classes with the new data. This process is called "full model fine-tuning" or "whole model tuning," this type of transfer geometry requires large amounts of training data.

NEU-Bin model is formed by freezing the first layers of the ResNet50 network to detect general image features. Then, we added an average pooling layer to reduce the number of image parameters but still retain essential features. A fully connected layer connects all the units of the previous layer with the units of the current layer. Dropout layer whose task is to limit overfitting for the model. Overfitting occurs when the model is too complex to simulate the training data and contains too many parameters, inadvertently extracting some residual variation (noise) as if that variation represented the model structure. Finally, the output layer is also a fully connected

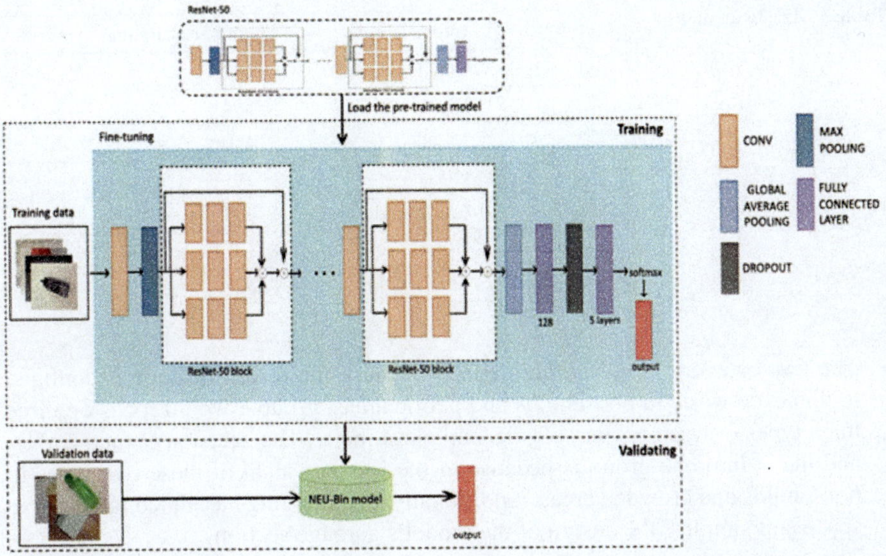

Fig. 8 NEU-Bin pipeline

layer with a unit number of 5 equivalent to 5 labels of NEU-Bin with an activation function of Softmax. It is used to classify objects belonging to 5 classes.

The process of building NEU-Bin consists of two stages, as illustrated in Fig. 8:

- ***Phase 1***: Since the layers of the pre-trained model have been trained on the ImageNet dataset, we freeze the classes of the ResNet50 model and only update the weights of added layers. When the loss function becomes more stable, and the network reaches a higher level of accuracy with the added layers, we continue to the next phase.
- ***Phase 2*****:** At this stage, we unfreeze the last few layers of the pre-trained model and continue training with these layers along with the newly added adjustment layers.

5 Methodology

5.1 Dataset

The overall dataset includes 4664 images combined from selected data from TrashNet and Waste-set. Given the nature of the data and the purpose of research toward solutions contributing to environmental protection and support appropriate supply for recycling plants, the data are classified into three main categories.

Table 1 TrashNet dataset

	Label	Number of images
1	Plastic	482
2	Paper	594
3	Metal	410
4	Glass	501
5	Cardboard	403
6	Others	137
	Total	2527

- The first category is *recyclable wastes*. Because the research scope is confined to domestic waste in households and public areas, group 1 would focus on three main types of waste: paper, plastic, and glass-metal. The decision to merge glass and metal into one group is because of the less prevalent of these two types in households and crowded areas. Additionally, this merging has aimed to conserve space and simplify the design of the model's garbage sections.
- The second category is *organic wastes*. This type of waste mainly focuses on leftovers or expired food such as tangerine peel, longan seed, and rancid meat.
- The third category is *non-recyclable wastes*. This waste cannot be recycled or takes a significant amount of time, effort, and resources to recycle, such as plastic bags, broken glass, and cigarette butts.

TrashNet

In 2016, Yang and G. Thung [20] released the TrashNet dataset (Table 1), spanning six classes: glass, paper, cardboard, plastic, metal, and other trash. Currently, the dataset consists of 2527 images: 501 glasses, 594 paper, 403 cardboard, 482 plastic, 410 metal, 137 trash. All the images were taken using an Apple iPhone 7 Plus, an Apple iPhone 5S, and an Apple iPhone SE. The original dataset has a capacity of 3.5 GB with 2527 images resized to 512×384 with three RGB color channels (Red, Green, and Blue). Figure 9 illustrates some examples of each label in the dataset. Research on the waste classification problem is now being conducted, using the TrashNet dataset to evaluate their proposed methods.

Waste-set

In addition, we gathered more waste images and created the waste-set dataset, which represents actual waste images in Vietnam.

This dataset consists of three prominent labels: recyclable waste, organic waste, and "other" waste. Recycled waste includes three labels: plastic, paper-cardboard, and metal-glass. Cameras captured the 2137 images in the waste-set dataset from smartphone devices at the street, public locations, and households in Hanoi, Vietnam. Table 2 presents a description of the waste-set dataset, and Fig. 10 illustrates the dataset.

Fig. 9 Examples in the TrashNet dataset

Table 2 Waste-set dataset (collected)

	Label	Description	Number of images
1	Plastic	Plastic bottles, plastic boxes, pens, plastic household items	523
2	Paper and cardboard	Newspapers, hardcovers, flyers, boxes, envelopes, etc.	527
3	Metal and glass	Glass bottles, water cans, building materials, keys, etc.	526
4	Organic waste	Fruits, vegetables, leaves, nuts, etc.	311
5	Others	Masks, candy shells, plastic bags, foam boxes, napkins, etc.	250
	Total		2137

The research divided the test dataset, including TrashNet and waste-set, with the ratio of 60% used for training, 20% for validation, and 20% for testing.

5.2 Experimental Settings

The training phase was implemented in Python 3.7 and performed on the TensorFlow framework, an end-to-end open-source machine learning platform. This chapter

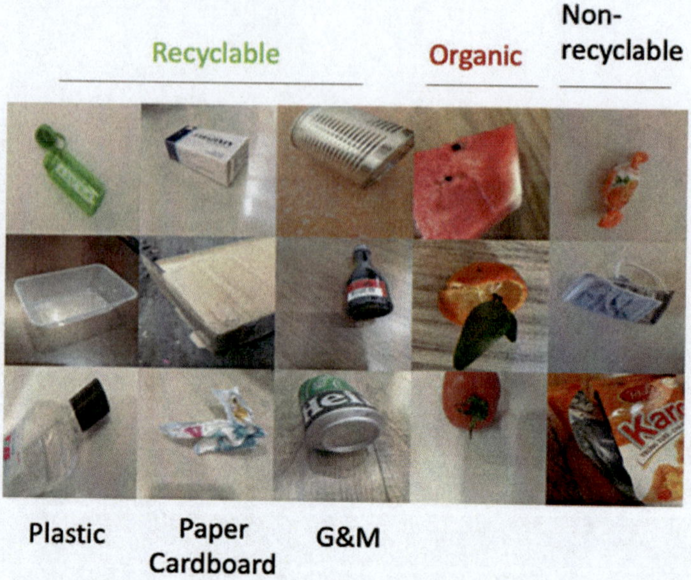

Fig. 10 Some examples in the waste-set dataset

conducted various trash classification tasks, including ResNet50, DenseNet121, MobileNetV2, VGG16, and InceptionV3. All the models mentioned above were adjusted with appropriate parameters during the training phase to offer the best results possible. For NEU-Bin, we pre-trained and fine-tuned the ResNet50 by replacing the final fully connected layers. In the first training stage, as mentioned in Sect. 4.4, the root mean square propagation (RMSprop) is utilized as an optimizing algorithm with the initial learning rate $\alpha = 0.0001$ for the loss to converge faster. After 20 epochs, since the loss on the validation dataset appears to stabilize and the network achieved a considerable level of accuracy with the added layers, the process proceeds to the second stage. In stage 2, we unfreeze the last few layers of the pre-trained model and continue training with these layers along with the added adjustment layers for another 20 epochs using Adam optimizer with learning rate $\alpha = 0.00001$ to achieve the smallest potential loss. Additionally, this study used batch size = 32, and the model's performance on the validation set is evaluated for each epoch during training processing.

6 Result

6.1 Experimental Result

Table 3 displays the accuracy and number of parameters involved in the training process of many popular pre-trained models, including ResNet50, DenseNet121, MobilenetV2, VGG16, and InceptionV3, when trained on the validation dataset.

ResNet50 achieved the highest precision prediction rate of 87.50% on the test dataset, which is more than 1.00% higher than the second-highest, DenseNet121, with 86.50%, and more than 5.20% compared to the model giving the lowest accuracy rate is VGG16 with 82.30%. DenseNet121 has a slightly lower precision prediction rate than ResNet50, but its size is considerably less when the total parameter of DenseNet121 is only equivalent to 0.3 of ResNet50's parameters; therefore, the time complexity of the DenseNet model may be less than ResNet50. Since the goal of this study is toward the waste treatment process once the garbage is collected, the accuracy is critical; even minor classification errors might result in significant problems. For example, if the recyclable container has been carried to the recycling plant but found to be mixed with organic waste, the process would have to go through another manual sorting step, which would be costly and time-consuming. As a result, we chose to adopt the model with the highest accuracy performed on the validation set—ResNet50 as our primary model to construct NEU-Bin.

Figure 11 displays the loss function and prediction rate over 20 epochs (only training the added layers while the rest of the layers froze). The loss function's

Table 3 The accuracy and size of each pre-trained model

Model	Accuracy (%)	Number of parameters
ResNet50	87.50	23,696,261
DenseNet121	86.50	7,103,429
MobileNetV2	83.40	2,340,293
VGG16	82.30	14,747,845
InceptionV3	82.50	21,934,245

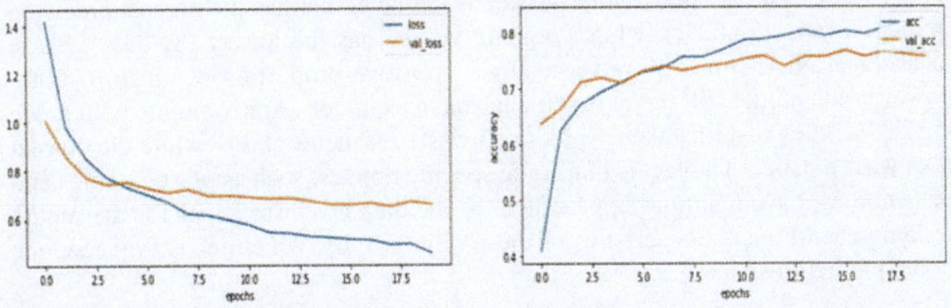

Fig. 11 Loss and accuracy before fine-tuning

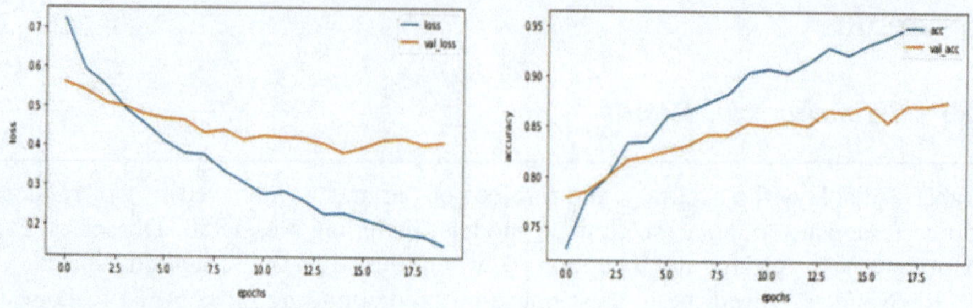

Fig. 12 Loss and accuracy after fine-tuning

drop on the training and test dataset is significant and relatively close. Similarly, the increasing rate after each epoch on the training dataset is negligible compared to the test dataset. However, the correct classification rate after 20 epochs is still relatively low, averaging around 75% in the training and 72% in the test. After releasing some last layers of the pre-trained model and continued training as in Fig. 12, the loss function declines but slower. Accuracy increased significantly on the training and testing datasets, rising from 75 to 92% on the training dataset and 72 to 84% on the testing dataset.

By experimenting with various models, we explore that ResNet is the suitable transfer learning model for the waste classification task. In this case, we could develop a ResNet-based neural network and adjust parameters according to the characteristic of the Vietnam garbage dataset.

6.2 *Confusion Matrix*

Figure 13 depicts the confusion matrix on the test dataset of the NEU-Bin model, with the horizontal axis indicating the network's predicted label and the vertical axis representing the actual label of the object that the model predicts. The highest prediction rates are mainly distributed on the diagonal from the top left corner to the lower right corner of the matrix. The average of all values on that diagonal is also the accuracy of the model. "Other waste" is the label that has the highest precision prediction ratio (93.62%), while "organic waste" has the lowest (81.03%). Thus, on average, NEU-Bin achieved a relatively positive proportion of up to 87.50%. However, the model still predicts some labels inaccurately. For example, with a rate of 12.37%, the most mistakenly predicted waste is "metal-glass," while the correct label was "plastic." The fact is that some specific objects, such as plastic bottles and glass bottles, have a similar appearance, or the data are insufficient for the model to comprehend the characteristics of these objects fully, which leads to inaccurate model predictions.

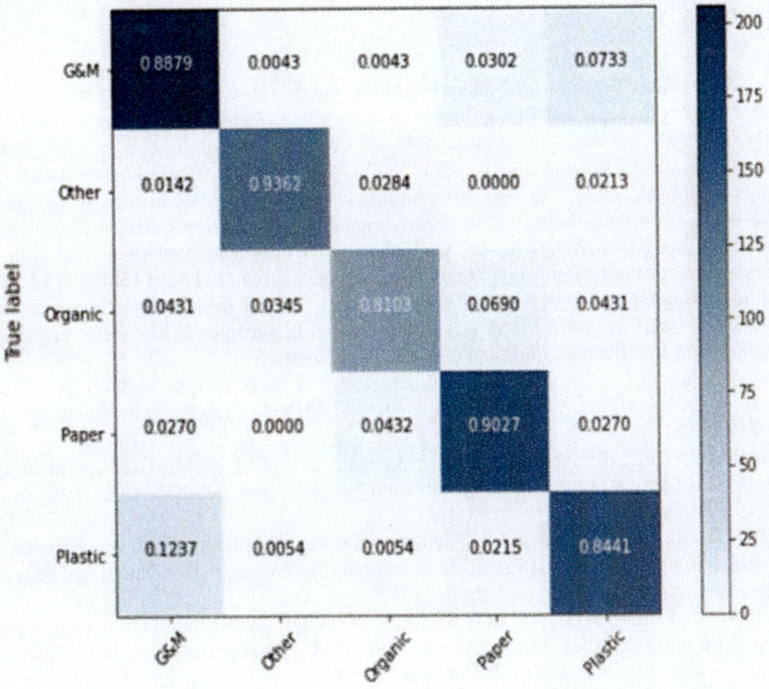

Fig. 13 Confusion matrix

7 Conclusion

The "green smart city" promises to solve many urgent problems associated with the urbanization process, including the issue of sorting domestic waste. Concretely, "smart" bins could automate the waste classification process. It supports users not to have to worry about which bin to throw garbage in. This solution could limit manual waste classification and save labor costs, reducing the adverse effects of not sorting waste. Therefore, the cities could be "greener."

On the downside, with an accuracy of 87.5%, the model still has inaccurate predictions for some types of waste with a similar appearance. In addition, there is another limitation that we try to solve, which is "trash-in-trash." It is the condition when different types of waste are mixed at one user's disposal. Therefore, a reasonable approach is to put all the "trash-in-trash" into the "other" compartment. In this method, carefully sorting will significantly reduce the amount of waste.

In future, the waste treatment system will automate the entire process, then minimize manual waste classification. The classification quality of the intelligent trash model depends on the quality of the camera. In factories, garbage sorting systems can integrate conveyors inside for a larger scale.

References

1. The World Bank.: What a waste: an updated look into the future of solid waste management. In: World Bank (2018). Available online: https://www.worldbank.org/en/news/immersive-story/2018/09/20/what-a-waste-an-updated-look-into-the-future-of-solid-waste-management
2. Ijjasz-Vasquez, E., Wahba, S., Kaza, S.: Here's what everyone should know about waste. In: Blogs.worldbank.org (2018). Available online: https://blogs.worldbank.org/sustainablecities/here-s-what-everyone-should-know-about-waste. Accessed 28 Sep 2021
3. Zhou, H., Long, Y., Meng, A., Li, Q., Zhang, Y.: Classification of municipal solid waste components for thermal conversion in waste-to-energy research. Fuel **145**, 151–157 (2015)
4. Salimi, I., Bayu Dewantara, B.S., Wibowo, I.K.: Visual-based trash detection and classification system for smart trash bin robot. 2018 International Electronics Symposium on Knowledge Creation and Intelligent Computing (IES-KCIC) (2018)
5. Chu, Y., Huang, C., Xie, X., et al.: Multilayer hybrid deep-learning method for waste classification and recycling. Comput. Intell. Neurosci. **2018**, 1–9 (2018). https://doi.org/10.1155/2018/5060857
6. Ramalingam, B., Lakshmanan, A., Ilyas, M., et al.: Cascaded machine-learning technique for debris classification in floor-cleaning robot application. Appl. Sci. **8**, 2649 (2018). https://doi.org/10.3390/app8122649
7. Aral, R.A., Keskin, S.R., Kaya, M., Haciomeroglu, M.: Classification of trashnet dataset based on deep learning models. 2018 IEEE International Conference on Big Data (Big Data) (2018). https://doi.org/10.1109/bigdata.2018.8622212
8. Aghilan, M., Kumar, M.A., Aafrid, T.S.M. et al.: Garbage waste classification using supervised deep learning techniques. In: Papers.ssrn.com (2020). Available online: https://papers.ssrn.com/sol3/papers.cfm?abstract_id=3563564
9. Ruiz, V., Sánchez, Á., Vélez, J.F., Raducanu, B.: Automatic image-based waste classification. From Bioinspired Syst. Biomed. Appl. Mach. Learn. 422–431 (2019) https://doi.org/10.1007/978-3-030-19651-6_41
10. Vo, A.H., Hoang Son, L., Vo, M.T., Le, T.: A novel framework for trash classification using deep transfer learning. IEEE Access **7**, 178631–178639 (2019). https://doi.org/10.1109/access.2019.2959033
11. Lecun, Y., Bottou, L., Bengio, Y., Haffner, P.: Gradient-based learn-ing applied to document recognition. Proc. IEEE **86**, 2278–2324 (1998). https://doi.org/10.1109/5.726791
12. Phung, R.: A high-accuracy model average ensemble of convolutional neural networks for classification of cloud image patches on small datasets. Appl. Sci. **9**, 4500 (2019). https://doi.org/10.3390/app9214500
13. Krizhevsky, A., Sutskever, I., Hinton, G.E.: ImageNet classification with deep convolutional neural networks. Commun. ACM **60**, 84–90 (2017). https://doi.org/10.1145/3065386
14. Butt, M.M., Latif, G., Iskandar, D.N.F.A., et al.: Multi-channel convolutions neural network based diabetic retinopathy detection from fundus images. Procedia Comput. Sci. **163**, 283–291 (2019). https://doi.org/10.1016/j.procs.2019.12.110
15. Myburgh, J.C., Mouton, C., Davel, M.H.: Tracking translation invariance in CNNs. Artif. Intell. Res. **1342**, 282–295 (2020). https://doi.org/10.1007/978-3-030-66151-9_18
16. Bengio, Y., Courville, A., Vincent, P.: Representation learning: a review and new perspectives. IEEE Trans. Pattern Anal. Mach. Intell. **35**, 1798–1828 (2013). https://doi.org/10.1109/tpami.2013.50
17. Abdel-Hamid, O., Mohamed, A., Jiang, H., et al.: Convolutional neural networks for speech recognition. IEEE/ACM Trans. Audio, Speech, Lang. Process. **22**, 1533–1545 (2014). https://doi.org/10.1109/taslp.2014.2339736
18. He, K., Zhang, X., Ren, S., Sun, J.: Deep residual learning for image recognition. In: Openaccess.thecvf.com (2016). https://openaccess.thecvf.com/content_cvpr_2016/html/He_Deep_Residual_Learning_CVPR_2016_paper.html

19. Mahmood, A., Ospina, A.G., Bennamoun, M., et al.: Automatic hierarchical classification of kelps using deep residual features. Sensors **20**, 447 (2020). https://doi.org/10.3390/s20020447
20. Yang, M., Thung, G.: Classification of Trash for Recyclability Status (2006)

Machine Learning and Fuzzy Technique for Environmental Time Series Analysis

Dung Truong, Ngoc C. Le, Hung Nguyen The, and Minh-Hien Nguyen

Abstract In this chapter, we will revise the application of machine learning techniques in some Environmental Time Series problems. In Particular, we will consider the application of fuzzy techniques combined with machine learning methods. Especially, we will apply a multi layer perceptron equipped with fuzzy layers, say ANFIS model, and a fuzzy deep learning neural network, say Fuzzy Auto Encoder. These approach has shown the improvement in the results of some real problems. Two study cases including the eutrophication on Han river (Ly et al, Sci Total Environ 797:149040 (2021), [1]) and the air quality monitoring (Nghiem et al, Air Qual Atmos Health 14(1): 7-18 (2021) [2]) in Hanoi will be considered.

Keywords Environmental time series · Machine learning · Deep learning · Fuzzy method

1 Introduction

One important thing for the Green Smart Cities is to provide a good habitat environment for all citizens. The pollution includes a lot of factors made by humans and nature. From nature, there is a lot of climate each year, which makes the element become bad. But the impact factor is from Human life and industry makes a lot of important pollution. It's clear that humans have bringed terrible pollution, which predicts the apocalypse for Earth.

There are a lot of elements that can cause pollution at the moment, but the drop of air and water quality, which is vital for human life, are getting a lot of attention. These pollutions are calculated by the concentration of the chemical factors inside.

D. Truong (✉) · N. C. Le · H. N. The
School of Applied Mathematics and Informatics, Hanoi University of Science and Technology, Hanoi 100000, Vietnam
e-mail: ttdung997@gmail.com
URL: http://sami.hust.edu.vn/

D. Truong · M.-H. Nguyen
Grooo international, Hanoi 100000, Vietnam

M. Lahby et al. (eds.), *Computational Intelligence Techniques for Green Smart Cities*, Green Energy and Technology, https://doi.org/10.1007/978-3-030-96429-0_14

The chemical factors bring the drop off the heath and death to the human and other creatures. More than that, chemical factors can bring some bad natural phenomenon, which make the abnormal developed of species, destroy the ecosystem [3].

Air pollution is one of the world's biggest health hazards to people everywhere. In 2020, there were about seven million premature deaths annually by air pollution; in the Western Pacific Region alone, around 2.2 million people die each year; and in Vietnam, around 60,000 deaths.[1] Due to fossil fuel emission alone, air pollution has a bad effect on the global economy, which burns about $2.9 trillion per year[2] [3]. World Health Organization (WHO) have point out that the emission of pollutants, such as particulate matter (PM), sulphur oxides (SOx), nitrogen oxides (NOx), carbon 56 monoxide (CO) and carbon dioxide (CO2), are generated in greater quantity recent year. Several scientific studies have shown that the poor air quality, which comes from emission of pollutants, has a big impact on people's health and its consequences overtime. More than that, PM2.5 is confirm to lead to the dead, which destroy human lung, causing serious diseases such as stroke, lung cancer, chronic obstructive pulmonary disease, heart disease and respiratory infections such as pneumonia [4].

WHO has defined the Air Quality Index (AQI) [2] with the aim to evaluate air quality and found a way to protect people's health though try to drop the AQI. There are six factors, including CO, SO2, PM10, PM2.5, O3, and NO2, that have a big impact on AQI and human life. For example, O3 causes lung diseases, especially for children and older adults who are active outdoors are the most sensitive groups, while CO causes heart disease. PM2.5 was confirmed to be the important factor which makes the patient become bad, push to the death.The COVID-19, which attacks people's respiratory and cardiovascular systems, was deemed a pandemic. The World air quality report has shown that if we have better air quality, we can save about 7–33% death from the pandemic [4].

Besides, water pollution, which reduces the freshwater, has a big impact on human life [1], just like the air. People using freshwater for multi-proposes, such as drinking water sources, industrial use, agricultural irrigation, and recreation. However, the quality of water are dropping down significant, due to the human activities, such at rapid urbanization, industrialization, overpopulation, and climate from the nature [5, 6]. For example, eutrophication, caused mainly by a surplus of nitrogen and phosphorus, stimulates the proliferation of specific phyto-plankton and macrophytes while inhibiting other aquatic species.

Eutrophication leads to somes unwanted environmental consequence, including freshwater algal blooms, which causes the bad outcome on environment and public health, such as oxygen depletion, heavy odor, and toxin. Therefore, severe and harmful algal blooms in eutrophic streams and lakes on a global scale have drawn much attention recently [1].

Some of chemistry factors that cause eutrophication including chemical oxygen demand (COD), biological oxygen demand (BOD), total organic carbon (TOC),

---

[1] https://www.who.int/vietnam/health-topics/air-pollution.

[2] https://www.iqair.com/world-most-polluted-cities/world-air-quality-report-2020-en.pdf.

total suspended solids (TSS), total phosphorus (TP), dissolved total phosphorus (DTP), phosphate (PO4-P), total nitrogen (TN), dissolved total nitrogen (DTN) [7–9]. Besides that, chlorophyll a (chl-a), which provides nutrients for the rise of algal blooms, can be used to evaluate the harm level of eutrophication. The literature has proven that excessive nutrient loads are the key driving factor for algal blooms in freshwater, which is the relative between the chemistry factors and chl-a. For example, in the US, an examination of 1264 lakes demonstrated a significant correlation between chl-a with total nitrogen (TN) and total phosphorus (TP) [10]. In Asia, there are many papers that have confirmed the same. In China, a survey of 15 shallow lakes has revealed significant correlations between chl-a, a primary surrogate for algal biomass, and TN (R2 of 0.25–0.29, $p < 0.001$) and TP (R2 of 0.51–0.66, $p < 0.001$) [11]. The results of researches in Han river found a strong correlation (R2 of 0.96) between algal density and TP [9]. In addition, many studies have revealed that physical parameters such as temperature, light [8, 12], flowrate [9, 13], and water residence time [14, 15], also have pronounced effects on algae proliferation. In general, warm temperatures and adequate light intensity make algae more develop [8, 12]. In contrast, a high flowrate tends to adversely effect on algal density [15].

Besides that, human activities have a big impact on eutrophication through dam construction, flood protection, land use, and water extraction, ...This kind of work can significantly alter the traits of freshwater ecosystems, ultimately influencing the distribution, composition, and density of algae in a given region [16, 17]. Therefore, it's hard to know what is the main factor between nutrients, hydrology, climate, and the influence of anthropogenic activities, for the freshwater algal blooms problem. Trophic state index (TSI) has been widely used to determine trophic states of freshwater bodies, not only for lentic systems but also for streams and rivers [18, 19]. Determination of TSI-Chla was given in the SI [20]:

$$TSI - Chla = 9.82 * \ln chl - a + 30.6 \tag{1}$$

The consequences of air and water pollution are large, and much research has been done to develop early warning strategies to prevent the pollutions. The measurements of chemical factors are collected automatically by the IoT sensors. Then, the data will be transferred to the data warehouse ready for analysis. These data are time series naturally leading to the pollution predictions. A time series is a sequence of collected data over times, normally in time equi-length, say hourly, daily, weekly, monthly, quarterly, or annually, ...Time series can be decomposed into four components, each expressing a particular aspect of the movement of the values of the time series, say the long-term trend, the seasonal and periodical components, and the non-regular component including other nonrandom sources of variations of the series.

In the pollution problem, time-series data mining can point out the factor causing the rising pollution level then may be some solutions can be proposed. However, the data can contains some missing value and lead to difficulty in the analysis process. This missing can be due to the malfunctions of the sensors, lost in transmitting data process, softwares errors, or many other reasons from the system. These missing need to be filled up and the filling process may be very important in the environment data related tasks. In this chapter, we summary the results in [1] and [2] as the data analysis respects in two most environmental polluted factors, say water and air.

2 Related Work

Basically, the time-series dataset can be processed by traditional methods including chemical analysis or statistics based prediction models. Within the algal bloom problem, numerous numerical models have been applied to solve the problem. MIKE 21 software, nutrient simulation models, Environmental Fluid Dynamics Code (EFDC), Water Quality Analysis Simulation Program (WASP), Better Assessment Science Integrating Point and Nonpoint Sources (BASINS), Hydrological Simulation Program—FORTRAN (HSPF) [6, 13, 18, 21, 22], which have been applied to process the water data series.

With the air quality, statistical model examples are the AutoRegressive Moving Average (ARMA) [23], AutoRegressive Integrated Moving Average (ARIMA) [24], Threshold Autoregressive (TAR) model [25], Hidden Markov Model (HMM) [26]. Regression models have been used widely in ambient air data. In [27], the authors applied a linear regression method to predict PM2.5 emissions in the Northeast United States from 2002 to 2013 based on fine-resolution aerosol optical depth. Oteros et al. [28] used different factors of pollen concentration and took into account the extreme weather events in the Mediterranean climate characteristics to establish a multivariate regression model. It is presented in [29] a real-time approach based on multiple linear regression for air quality prediction. These models all faced challenges in prediction tasks with highly nonlinear time series data.

For real practical problem, there are significant challenges to build traditional mathematical models for prediction. First, the processes are time-consuming and require updated high-frequency data, thus making them costly [30]. More than that, the statistical methods are based on theoretical assumptions and prior knowledge of the data, which need a lot of background knowledge. Modern Machine learning (ML) methods, like Deep Neural Network, require no prior physical information or linear assumptions. Having the ability to unravel the complex nonlinear response of structures, deep learning models can prediction using learning techniques based on historical data. They can also perform operations based on preset rules and algorithms to adaptively learn model parameters and obtain hidden relationships lying inside the data. The trained model is then used to predict the future development trends and patterns of time series data, suitable for the quality problem [1].

In eutrophication, ML has been successfully applied for water-related data because it could handle the nonlinearity, unstable and interdisciplinary features of water quality parameters [31, 32]. For example, in [33], the authors considered eight parameters, including air temperature, precipitation, solar radiation, total nitrogen, total phosphorus, the ratio of total nitrogen to total phosphorus, and chl-a as input variables for algal bloom prediction. In [1], an extreme learning machine model exhibited better prediction and generalization performance as compared to multiple linear regression (LR), a conventional neural network with backpropagation (BP), and adaptive neuro-fuzzy inference system (ANFIS) [33]. Other studies have demonstrated that the support vector machines (SVMs) showed superior performance in generating early warnings for the eutrophication of reservoirs [34] and rivers [35]. In [36], deep learning models, such as long-short-term memory (LSTM) and recurrent neural network (RNN) are applied to predict algal bloom, using temperature, pH, BOD, COD, DO, cyanobacteria, water level, and pondage as input variables. The results show that LSTM has the better result, compare to RNN and statistical method, i.e., ordinary least square regression analysis.

In air quality problems, artificial neural networks (ANN) have advantages in short-term prediction. Ni et al. [37] collected CO, NO2, SO2, combined with meteorological data, including rainfall, temperature, relative humidity, wind speed, then using ANN to predict PM2.5 [38]. Is studied in [39] the combined prediction method of shallow nonlinear autoregressive network (NAR) on the basis of BP based on time and space dimension. The authors in [40] developed the geographically-weighted gradient boosting machine to address the spatial non-stationarity of the relationships between PM2.5 concentrations based on aerosol optical depth (AOD) and meteorological conditions. The above results show that these machine learning methods, which are mainly structured with shallow network features, can only obtain accurate results in short-term prediction since they are relatively simple. Therefore, if we want to achieve accurate long-term prediction performance, we should use more complex architectures.

Comparing with the shallow network, the so-called deep learning networks have shown the outstanding ability for the complex time series relation. Recurrent neural network (RNN) [41] for time series prediction has attracted extensive attention from researchers because it could capture the high nonlinearity of time series data. In [42], the authors predicted hourly/daily/monthly average solar radiation by an RNN, and an adaptive learning rate for the RNN was proposed. The approach was found promising when compared to the multilayer perceptron. As an improved version of the RNN, long short-term memory (LSTM), replaced it as a popular time-series data prediction technology [43]. The gated recurrent unit (GRU) [44] inherits the advantages of LSTM to automatically learn features and efficiently model long-term dependent information, and it also exhibits a significant increase in computational speed. Besides, it has been introduced in [45] the spatio-temporal ambient air quality index by using RNN and LSTM. LSTM has been used in the research in New Delhi [46]. The research focused on PM 2.5 and used data in 2009 in the southeast of the USA. The chemist factors, including pH, BOD, COD, DO and the nature index, such as temperature, wind, rainflow has been used to predicted water quality by RNN in recent year [32].

Although deep learning method show the best performance for many problems and artificial neural networks are accredited for their good performance in handling time series analysis, they are widely considered as "black-box" models because the model structure hyper parameters hardly reflect physical meanings while they are the most crucial influencing factors in enhancing forecasting accuracy and reliability [1].

On the other hand, missing-value data filling is another vital problem. With a normal dataset, there are many ways to deal with the missing data, such as simple removal, replacing with the average, or split and replace combining. But in time-series dataset, replacement can lead to loss of the trend of the dataset hence the lowering accuracy in predictions. Deep learning has been applied to deal with the program. For example, ANN was used to fill-up data for rainfall dataset [47]. Basis function neural network (RBFNN) layer uses Gaussian transfer functions rather than sigmoid functions employed by Multi-layer perceptrons. It has been applied with the aim of demonstrating its successful performance.

3 Proposed Methodology

3.1 Fuzzy Technique and Membership Function

The term fuzzy was introduced with the 1965 proposal of fuzzy set theory by Lotfi Zadeh [48]. Fuzzy logic had however been studied since the 1920s, as infinite-valued logic notably by Lukasiewicz and Tarski [49]. Fuzzy logic is based on the observation that people make decisions based on imprecise and non-numerical information. Fuzzy models or sets are mathematical means of representing vagueness and imprecise information (hence the term fuzzy). These models have the capability of recognizing, representing, manipulating, interpreting, and utilizing data and information that are vague and lack certainty.

While variables in mathematics usually take numerical values, in fuzzy logic applications, non-numeric values are often used to facilitate the expression of rules andfacts [49]. A linguistic variable such as age may accept values such as young and its antonym old. Because natural languages do not always contain enough value terms to express a fuzzy value scale, it is common practice to modify linguistic values with adjectives or adverbs. For example, we can use the hedges 'rather' and 'somewhat' to construct the additional values 'rather old' or 'somewhat young'.

Fuzzification operations can map numerical or categorical input values into fuzzy membership functions. The contra ones, de-fuzzifying operations, can be used to map a fuzzy output membership function into a "crisp" output value that can be then used for decision or control purposes. Fuzzy logic has been applied to many fields, from control theory to artificial intelligence. Original fuzzy set and fuzzy logic have been extended to intuitionistic fuzzy set and logic [50]. Beside the truth value, intuitionistic fuzzy logic introduces falsity value. Then, this is extended more by introducing the neutrality, i.e. to evaluate a proposition, instead of one value as original fuzzy logic or two values as intuitionistic fuzzy logic, we use three values [51].

Membership function of a fuzzy set is a generalization of the indicator function for classical sets. In fuzzy logic, for any set X, a membership function on X is any function from X to the real unit interval [0, 1]. In machine learning and deep learning architecture, the membership function will transfer the raw input data into the fuzzy data, with the large dimension, describing the relationship of each number with the specific threshold of the dataset. Some membership function set which have been used in the can be show below:

1. Bell function

$$f(x) = \frac{1}{1 + |\frac{x-c}{a}|^{2b}} \tag{2}$$

2. Zmf function

$$f(x; a, b) = \begin{cases} 0 \text{ If } x \le a \\ 2(\frac{x-a}{b-a})^2 \text{ If } a \le x \le \frac{a+b}{2} \\ 1 - 2(\frac{x-b}{b-a})^2 \text{ If } \frac{a+b}{2} \le x \le b \\ 1 \text{ If } x \ge b \end{cases} \tag{3}$$

3. Smf function

$$f(x; a, b) = \begin{cases} 0 \text{ If } x \le a \\ 2(\frac{x-a}{b-a})^2 \text{ If } a \le x \le \frac{a+b}{2} \\ 1 - 2(\frac{x-b}{b-a})^2 \text{ If } \frac{a+b}{2} \le x \le b \\ 1 \text{ If } x \ge b \end{cases} \tag{4}$$

4. Gauss function

$$f(x) = e^{\frac{(x-c)^2}{2\sigma^2}} \tag{5}$$

These function will be combine together in a deep learning layer, which named the Fuzzy layer in the deep learning model in the chapter.

3.2 Data Processing Model

Autoencoder Autoencoder [52] includes two-parts, the encoder and decoder, which are connected directly. Autoencoder aims to learn the representation for a dataset. While the encoder layer transfer the original input **x** to some hidden layer H with a function $\mathbf{h} = f(\mathbf{x})$, the decoding ones using this hidden layer to represent as the reconstruction $\mathbf{x'} = g(\mathbf{h})$. The autoencoder formula can be summarized as

$$\mathbf{h} = \sigma(W_1\mathbf{x} + b_1) \tag{6}$$

$$\mathbf{x'} = \sigma(W_2\mathbf{h} + b_2), \tag{7}$$

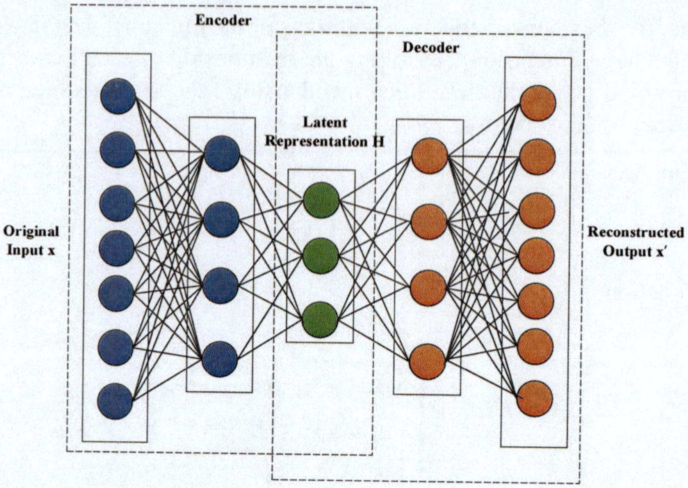

Fig. 1 The general autoencoder architecture

where σ is an activation function and b_i is bias vector. The parameter matrix of the ith layer is $W_i \epsilon \Re^{I_{di} \times O_{di}}$ projecting a I_{di} dimensional input into an O_{di} dimensional output. The optimization problem is to minimize the reconstruction error, which is the gap between the input value **x** and reconstructed value **x'**. There are some popular way to calculate the reconstruction error, such as (Fig. 1):

$$\min \mathcal{L} = \min E(\mathbf{x}, \mathbf{x'}) = \min ||\mathbf{x} - \mathbf{x'}|| \tag{8}$$

$$\mathcal{L}(\mathbf{x} - \mathbf{x'}) = -\sum_{c=1}^{M} \mathbf{x'}_c \log(x_c) \tag{9}$$

There are several types of the autoencoder, including Vanilla, Deep and Regularized Autoencoder [53]. While the Vanilla Autoencoder (VA) is the simplest autoencoder, which has only one hidden layer H, the Deep Autoencoder (DA) has many. Both of them have different applications in practice. Regularized Autoencoder (RA) encourages the model to have other properties, such as rank deficiency and sparsity apart from the ability of reconstructing the input **x**. The RA can be either a spare or denoising autoencoder. Spare autoencoder involves a sparsity penalty $\Omega(H)$ in the core layer H.

Moreover, Autoencoder has many extensions, which combine with other neural networks. For example, an LSTM autoencoder [54] is an implementation of an autoencoder for sequence data using an Encoder-Decoder LSTM architecture. Once fit, the encoder part of the model can be used to encode or compress sequence data that in turn may be used in data visualizations or as a feature vector input to a supervised learning model, or in this case for data distribution learning. On the other hand, the

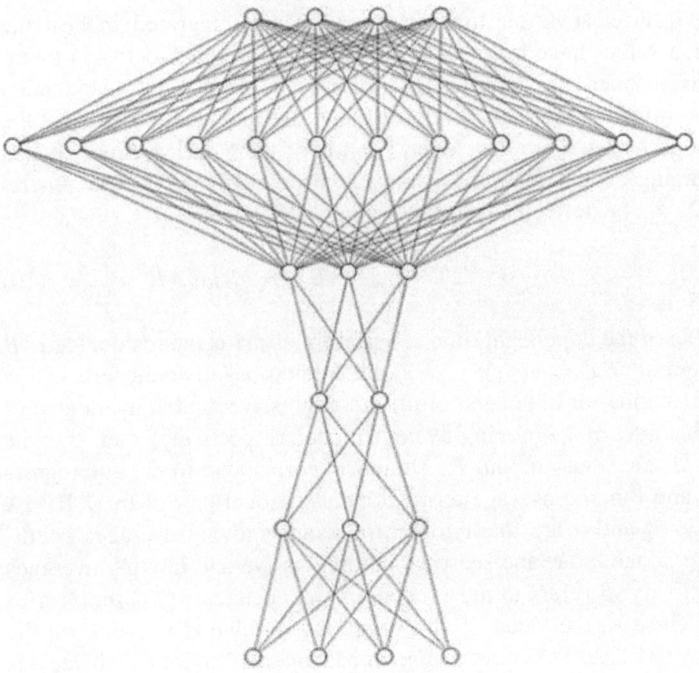

Fig. 2 The fuzzy autoencoder architecture

mathematical technique, such as fuzzy, entropy can be pushed in the Autoencoder, adding some special attribute for the network. Autoencoder have many applications. Among them, the auto reconstruction model can be used for missing data filling.

Fuzzy Autoencoder In deep learning, fuzzy techniques can add some attributes to the network, based on the membership function and values. With Autoenders, this function is believed to help the network understand more about the original data, then can have a better representation. These model constructs are shown in Fig. 2.

In this chapter, Fuzzy autoencoder will be applied along with some extension of Autoencoder, such at bidirectional GRU-Autoencoder, bidirectional LSTM-Autoencoder, Regularized Autoencoder to create the fuzzy extension, apply to fill the missing data from the environment dataset

3.3 Regression Models

Classical Time Series Analysis Methods An autoregressive integrated moving average (ARIMA) model is one of the most effective algorithms fitting the long-term time series for forecasting [55, 56]. The interested evolving variable is regressed on its own prior values while the regression error is a linear combination of error terms whose values have occurred at various times in the past. The I (integrated) in the name implies that the data values have been replaced with the difference between themselves and the previous ones. Seasonal ARIMA (i.e., SARIMA) [57] is an extended version of ARIMA, in which the seasonal components are included to represent the periodicity in time series. A more developed version of the SARIMA model (i.e., SARIMAX) containing exogenous regressors was introduced in [58]. The general form of SARIMAX can be derived as the following.

$$\varnothing_p(B)\phi_p(B^s)(1-B)^D(a-B^s)^D Y_t = c + X_t B + \theta_q(B)\omega_Q(B^S)\epsilon_t \tag{10}$$

where Y_t is the value of the dependent time series at time t; ϵ_t is the residual at t; B is the backshift operator ($BY_t = Y_{t-1}$); p, d, and q are non-negative integers, which stand for the order (number of time lags) of the autoregressive model, the degree of differencing, and the order of the moving-average model, respectively; s refers to the number of periods in each season; and P, D, and Q correspond to the autoregressive, differencing, and moving average terms for the seasonal part of the ARIMA model, respectively; $\varnothing$ and θ are the autoregressive and moving averages coefficients, respectively; ϕ and ω are the seasonal autoregressive and moving averages coefficients, respectively; β refers to the exogenous (independent) (X_t) time-series parameters. In SARIMAX, the value of an exogenous variable is imposed on the model. For example, CO2 and SO2 have a huge impact on the PM2.5, which decides the air quality level. TN, TP is the important factors to evaluate chl-a.

Regression Analysis Methods Several traditional regression methods, such as LR [59], SVR [60], and DTR [61] are applicable for predicting trophic status. In the LR method, the relationship between the dependent variable and regressors is established by the following model, in which the loss function is optimized by the gradient descent method:

$$y = w_0 + \langle w, x\rangle = w_0 w_1 x_1 + w_2 x_2 + \cdots + w_k x_k \tag{11}$$

The SVR method is based on solving the minimum problem $||w||^2$ subjected to

$$||y - \langle w, x\rangle - W_0|| \le \epsilon \tag{12}$$

The Decision Tree Regressor (DTR) is of the tree structure form. It breaks a dataset down into smaller subsets while the associated tree consisting of decision nodes and leaf nodes is developed incrementally. A decision node has two or more branches, each representing the values of the tested attribute. The leaf node represents a decision on the numerical target. The root node, which is the top-most decision node in the tree, corresponds to the best predictor. In this study, we used entropy to split the branches of the tree. The input variables were used for the branching process on the tree, followed by one of the leaf nodes. Then, the value assigned to the leaf node was used as the output variable.

The time-series data is converted to the linear input, by flatten data from n-row to one dimension vector. The date of the row can be select from the continuous date, or by special knowledge. In the environment problem, the data usually have relative with the data on the same period in the year, depend on the cycle of nature.

Fuzzy Deep Learning Methods A Deep Neural Network consists of multiple processing layers including a set of perceptrions. The method consists of algorithms to uncover the hidden relationship and potential structure in the data. Recurrent Neural Network (RNN) type models are known as advanced artificial neural networks (ANN) structures offering more concise and reliable results. Unlike the traditional feed forward architecture, RNNs possess inner-loops within hidden layers. These loops allow RNN combining current with previous data to compute the output.

Thus, RNNs proceed not only single data points but also entire the sequences. However, due to the vanishing gradient phenomenon and long-term dependency, RNN models may be failed in some special cases. LSTM is a specially design to overcome those problems. However, if these dependencies are not so meaningful, the higher complexity of the model may amplify more noise and reduce the accuracy. To address the complexity of LSTM, the GRU was developed featuring a similar structure but it does not contain the output gate and contains fewer weights. Hence, it is relatively simpler and trains faster than LSTM.

Another deep structure, the Adaptive Neuro-Fuzzy Inference System (ANFIS) is one of the most popular fuzzy-based models for examining water quality [31]. ANFIS is an artificial neural network based on the Takagi-Sugeno fuzzy inference system [62]. Beside the input one, there are five layers consisting of the fuzzification, the rule, the normalized, the defuzzification, and the output layers [62]. ANFIS may possess either linear or nonlinear parameters. Combining neural networks and fuzzy principles, it is benefit from both being in a single framework.

In [1], the simplest variant of ANFIS, which includes five layers, was employed. For the fuzzy layer, the Gaussian membership functions have been used, due to the distribution of the nature data. This function can be calculate by:

$$\mu_{ij} = e^{\frac{x_i - m_{ij}}{2\sigma_{ij}}}, \tag{13}$$

where $\mu_{ij}(x_i)$ is the membership function of the jth linguistic variable, x_i is the ith input of some particular input layer node, $m_i j$ and $\sigma_i j$ are the mean and the standard deviation. In the rule layer, for each jth rule in the four rules, its firing strength was computed as

$$\omega_j = \prod_{i=1}^{5} \mu_{ij} \tag{14}$$

In Layer 3, the computed firing strength was normalized as

$$v_j = \frac{w_j}{\sum_{k \neq q} w_k} \tag{15}$$

Then, in Layer 4, the Takagi-Sugeno inference was performed and the consequence parameters are (p, q, r). In the last layer, the center of area (COA) was used for the defuzzification. COA was adopted because it showed the best results.

$$TS_j = v_j(b_{j0} + \sum_{i=1}^{5} b_{ji}x_i) \tag{16}$$

3.4 Metrics and Residuals

The MAE (mean absolute error) was used to further compare the accuracies of the different algorithms.

$$MAE = \frac{1}{n}\sum_{j=1}^{n} |y_j - \hat{y}_j| \tag{17}$$

where n is the number of samples, $\hat{y}_j$ is the prediction and y_j is the true value of the response. In this sense, MSE is used to be the metric for both filling missing data problems and regression output.

In the classification problem, positive and negative labels were assigned to the algal bloom phenomenon. The different metrics used to measure the performance of a classifier or predictor were determined as follows:

$$\text{Precision} = \frac{\text{TP}}{\text{TP} + \text{FP}} \tag{18}$$

$$\text{Recall} = \frac{\text{TP}}{\text{TP} + \text{FN}} \tag{19}$$

$$\text{F1-score} = \frac{2 * \text{Precision} * \text{Recall}}{\text{Precision} + \text{Recall}} \tag{20}$$

where true positive (TP) is the number of positive cases in both prediction and reality; false positive (FP) is the number of positive cases in prediction but negative in reality; true negative (TN) is the number of negative cases in both prediction and reality; and false negative (FN) is the number of negative cases in prediction but positive actually.

3.5 Study Sites

Eutrophication Analysis and Algal Bloom Prediction in Han River The experiment in [1] was conduct in the Han River, the second-longest river in South Korea, with a length of approximately 500 km. Located in the central part of the Korean

Peninsula, the Han River plays a critical role in supplying water for agriculture, navigation, recreation, as well as a drinking source for approximately half of the Korean population (about 24 million people). The Han River watershed (34,428 km squares) consists of two major branches: the north and south Han River, which join Paldang Reservoir [63]. Sampling sites were selected downstream from Paldang Reservoir, which flows through the metropolitan city of Seoul. It was previously reported that the system had been severely affected by discharges from wastewater treatment plants and residential/commercial activities. Using Carlson's trophic state equation [20] (SI), chl-a concentrations obtained from 2010 to 2020 implied that the lower Han River ranged from mesotrophic to hypereutrophic states, but mostly eutrophic condition. The critical trophic state leads to severe algal bloom in this regulated river. For example, the chl-a concentration cells during summer in 2015, 2018, and 2019 were reported to reach 25 mg/m3, 32.3 mg/m3, and 34.8 mg/m3, respectively, a concern alert warning level declared by the National Institute of Environmental Research [13, 15]. Diatoms were found to be dominant during all seasons, except for summer when green algae and blue-green algae were more prevalent. In addition to Microcystis aeruginosa, which was found the most predominant blue-green species i.e., up to 80–90%, and other species were also observed, such as Anabaena, Aphanizomenon, and Oscillatoria [13, 15]. Considering that different cell numbers in each harmful alga could pose different levels of the warning system and the difficulty in separately detecting the biomass, chl-a is still regarded as a typical primary proxy for algal bloom [6] (Fig. 3).

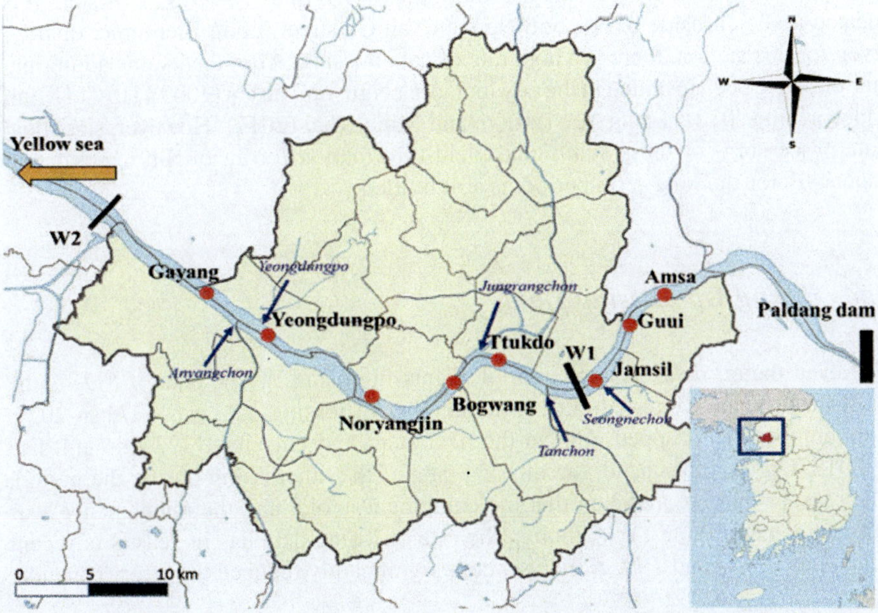

Fig. 3 Water monitoring stations along the mainstream of the Han River

Data collected on a monthly basis between January 2011 and April 2020 were acquired from the Korean Ministry of Environment.[3] Surface water was sampled at 40 different stations, where the authors [1] mostly focused on eight mainstream stations: Amsa (M1), Guui (M2), Jamsil (M3), Ttukdo (M4), Bogwang (M5), Noryangjin (M6), Yeongdungpo (M7), and Gayang stations (M8). Data on the following 20 parameters were obtained from the website: chemical oxygen demand (COD), biological oxygen demand (BOD), total organic carbon (TOC), total suspended solids (TSS), total phosphorus (TP), dissolved total phosphorus (DTP), phosphate (PO4-P), total nitrogen (TN), dissolved total nitrogen (DTN), nitrate (NO3-N), amonia (NH4-N), chl-a, temperature (°C), precipitation (mm), flowrate (m3/s), DO, pH, electroconductivity (EC), total coliform (TColi), and fecal coliform (FColi). Of these, dissolved parameters, DTN, DTP, NO3-N, and PO4-P were obtained by filtering water samples through Whatman GF/C glass fiber filters (Maidstone, England). Water quality analyses were conducted following standard methods as in [63, 64]. In all national freshwater monitoring stations, Quality Assurance/Quality Control (QA/QC) had been applied to avoid the uncertainty of data collected in the database. The annual precipitation data were obtained from the Korea Meteorological Administration data.[4]

Air Quality in Ha Noi In Hanoi, Vietnam, ambient air quality is measured by the automatic monitoring station for main air pollutants. In the early years of 2000s five first automatic air quality monitoring stations, namely, Giai Phong (2000), Pham Van Dong (2000), Lac Long Quan (2001), Nguyen Trai (2002), and Nguyen Chi Thanh/Lang (2002), were installed. Air quality data used in [2] is the hourly concentrations of PM_10, $PM_2.5$, SO_2, NO, NO_2, O_3, and CO measured in the Nguyen Van Cu automatic air quality monitoring station in the period of 2010-2018. This station is located at the curbside of No. 556 Nguyen Van Cu street, Long Bien inner district. Even for this station, there is a high rate of data missing. After that, some additional stations have been installed in the city including Nguyen Van Cu (2009), Ho Chi Minh Mausoleum (2012), Trung Yen (2017), and Minh Khai (2017). However, statistical data processing for these stations to yield information for air quality assessment, especially for the long-term period, is still limited.

3.6 Filling Up Missing Data

However, the air datasets contain a lot of missing data [2]. The data of SO_2 is the most losing data average. In 2010, the number of lossing data was 4975 in 2021. This number has dropped a lot in the next year, and only has 1123 missing data in 2018. On the other hand, the missing data of the other rising during the period. The reasons can be predicted that the equipment is old, and the maintenance was not guaranteed (Table 1). Although the rate of losing data has increased in recent years, the $PM1_0$ and $PM_2.5$ indexes have significantly reduced the error. This may

[3] http://water.nier.go.kr/.

[4] data.seoul.go.kr.

Table 1 Number of loss data in the air dataset

Methods	Number of missing data								
–	2010	2011	2012	2013	2014	2015	2016	2017	2018
NO2	1708	632	514	310	181	880	1836	1480	1791
CO	1277	499	444	451	338	1040	1371	890	4810
O3	1561	768	720	319	1651	1498	1712	8760	5646
PM10	1714	1153	824	523	2389	3867	2173	696	735
PM2.5	1695	1124	925	614	2398	3977	2704	834	963
SO2	4975	4309	3786	2523	3622	2323	1613	4227	1123
Total of loss	12,930	8485	7213	4740	10,579	13,585	11,409	16,887	15,068

be from the alternative devices, or the update of the current system, which focuses on collecting the PM data. However, with the modern trend of analysis using more PM indexes, this error is temporarily acceptable. In general, data sets lose a lot of data, it is not possible to use the methods of averaging, binning, and interval division. The use of automatic data filling methods, including autoencoder is necessary.

For dealing with the large number of missing values of the data, we used some autoencoders method, including Bidirectional GRU Autoencoder, Fuzzy bidirectional GRU Autoencoder, Fuzzy bidirectional LSTM Autoencoder and Fuzzy Regularized Autoencoder. Our aim is to find the best method to fill the missing value of the dataset. The methods that get the minimum of MSE. After filling data, it will be applied in the regression problem in the next section,

The fuzzy pattern can be clearly seen when incorporated into the Autoencoder network for significant changes, both positively and negatively. With complex deep learning networks, such as LSTM, GRU, the combination gives an inefficient result. This can be explained because the complex deep learning network already includes many different processing layers. By adding a fuzzy layer, these models increases noise and instability, specified in the network.

In contrast, with simpler models, like the Fuzzy Regularized Autoencoder network, the results are really impressive. Compared with the original deep learning network, the MSE is the same. But with the simple refinement of the network, with fewer layers, the fuzzy layer has played an important role in helping the network to better analyze the data and conduct the reconstruction.

At the final result, the Fuzzy Regularized Autoencoder is applied to filling data. The data which have been filled will be used in the regression model. On the other hand, the missing data of Han river is less than 5%, so that we use the average replacement for them.

Table 2 Percent of loss data in the air dataset

Methods	Number of missing data								
–	2010 (%)	2011 (%)	2012 (%)	2013 (%)	2014 (%)	2015 (%)	2016 (%)	2017 (%)	2018 (%)
NO2	19.50	7.21	5.85	3.54	2.07	10.05	20.90	16.89	20.45
CO	14.58	5.70	5.05	5.15	3.86	11.87	15.61	10.16	54.91
O3	17.82	8.77	8.20	3.64	18.85	17.10	19.49	100.00	64.45
PM10	19.57	13.16	9.38	5.97	27.27	44.14	24.74	7.95	8.39
PM25	19.35	12.83	10.53	7.01	27.37	45.40	30.78	9.52	10.99
SO2	56.79	49.18	43.10	28.80	41.35	26.52	18.36	48.25	12.82

3.7 Regression Model

Correlation and entropy analysis Spearman correlation and entropy analyses of individual variables for TSI-Chla and algal blooms, respectively, are shown in Table 2. Shannon entropy was effectively used in complex and interdisciplinary water quality to reveal the contribution of each variable to the final result; the higher entropy, the more important of the parameter relative to the output. Hence, entropy analysis is one of the most suitable measures of input variables and can help reduce the uncertainty and noise for ML methods. According to Table 2, 14 parameters were further screened out from the initial 20 water quality parameters. The highest entropy value was noted for BOD, followed by nitrogen group i.e., NO3-N, DTN, TN, phosphorus group e.g., TP, DTP, then weather patterns i.e., temperature, and precipitation, which suggested that the level of algal bloom was significantly governed by either environmental factors or anthropogenic activities. The low (linear) correlation coefficients imply that the simple LR model was unsuitable for prediction. Although LR is widely used to explain the underlying mechanisms of water quality and to construct mathematical models for water quality prediction, this traditional approach still suffers from significant drawbacks because the correlations among different variables are very complex owing to their high dependency on spatiotemporal patterns and human interference. For example, an earlier study demonstrated that EC was a robust and convenient parameter to predict the water quality of the Han River, with r values of 0.81 and 0.87 for the log-transformed concentrations of TN and COD, respectively [64]. However, the results did not align well with the extended observation period (i.e., 2011–2020), in which comparatively lower correlations were seen ($r < 0.7$) (Table 4).

With the air dataset, the correlation heatmap has been show in Fig. 4.

The correlation is not really effective, which is hard to use the traditional methods to deal with. CO and NO2 show the relative with PM2.5, which the number is 0.35 and 0.33 respectively. The Tempuration (TEMP) and humidity (HUMD) show the opposite

Table 3 Percent of loss data in the air dataset

Method	Layer detail	MSE
Bidirectional GRU autoencoder [2]	192, 96, 48, 24, 24, 48, 96, 192	50.27840
Fuzzy bidirectional GRU autoencoder	192, 96, 48, 24, 24, 48, 96, 192	127.9801
Fuzzy bidirectional LSTM autoencoder	192, 96, 48, 24, 24, 48, 96, 192	100.4851
Fuzzy regularized autoencoder	192, 48 24, 48, 192	48.00100
Fuzzy regularized deep autoencoder	192, 96, 24, 24,96, 192	48.89200

Table 4 Entropy and corelation with algal bloom of diferent factors

Varable	Entropy	Spearman corelation	Varable	Entropy	Spearman corelation
NH3-N	0.0068	0.0125	pH	0.0066	0.1278
NO3-N	0.0149	−0.1889	DO	0.0019	−0.0375
PO4-P	0.0064	−0.0259	BOD	0.0269	0.1840
fT-N	0.0135	−0.0148	Temperature	0.0133	0.2053
T-P	0.0115	−0.0178	Flowrate	0.0068	0.0935
Dissolved total N	0.0138	−0.1105	Precipitation	0.0098	−0.0361
Dissolved total P	0.0072	−0.0485	EC	0.0078	0.175

Predicting Environmental Quality Using the ML Application

Eutrophication Analysis and Algal Bloom Prediction A combination of only two or three variables could not explain the trophic state of the Han River. For example, neither nutrients (NH3-N, NO3-N, and PO4-P) nor climatic conditions (temperature, flowrate, and precipitation) had a direct impact on the occurrence of algal blooms, as evidenced by the high MAE values of 0.1179 and 0.1187, respectively (Table 3). Interestingly, using a higher number of variables (e.g., nine) did not increase accuracy, probably because the model tends to become more complicated as more input variables (features) are incorporated. If additional parameters do not make an effective contribution to the explanation of the output variable, they might cause noise and induce the model to be unstable. This can be more pronounced, especially when the added parameters have a close relationship with others [65]. For example, the high correlations between DTN and TN ($r = 0.98$) and between DTP and TP ($r = 0.91$) in this study would probably cause multicollinearity in mathematical models. The best performance was observed for the case using seven input variables, including DTP, DTN, pH, DO, BOD, temperature, precipitation, and flow rate, as shown by the lowest MAE value of 0.1085, regardless of the ML algorithms used (Table 3).

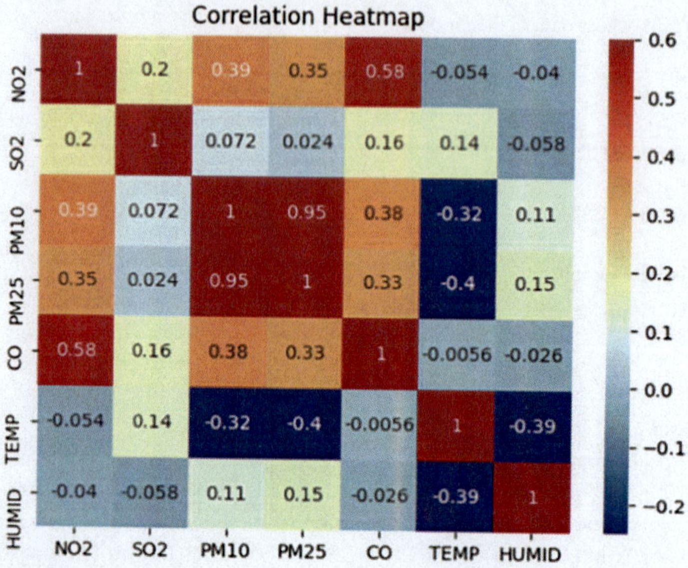

Fig. 4 Correlation heatmap for chemist factors impact air quality

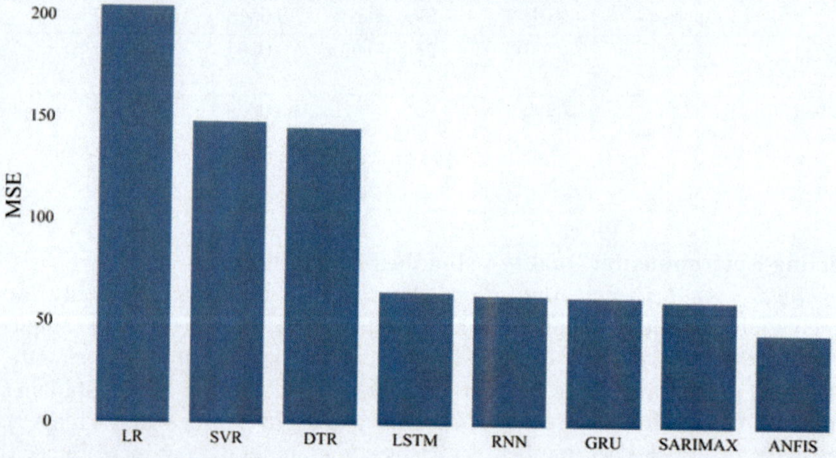

Fig. 5 Air regression results

The ML results suggest that eutrophication and algal proliferation were primarily driven by nutrients, organic contaminants, and environmental factors, without in-depth information of the characteristics of the Han River (Table 5).

Air Quality Analysis and PM2.5 Prediction (Fig. 5)

It's clear to see that the traditional linear method is not good to predict the PM2.5, which has been predicted by the correlation heatmap. For detail, LR shows the highest MSE of the methods, which confirm there are no linear relative between the factors.

Table 5 Entropy and corelation with algal bloom of diferent factors

Input	RNN	LSTM	GRU	SVR
NH3-N, NO3-N, PO4-P	0.1245	0.1284	0.1325	0.1125
NH3-N, NO3-N, PO4-P, TN, TP	0.1179	0.1231	0.1189	0.1247
NH3-N, NO3-N, PO4-P, TN, TP, DTN, DTP	0.1135	0.1214	0.1156	0.1235
DTN,DTP, pH	0.1122	0.1187	0.1178	0.1157
DTN, DTP	0.1145	0.1147	0.1189	0.1187
DTN, DTP, pH, DO, BOD	0.1126	0.1141	0.1196	0.1151
NH3-N, NO3-N, PO4-P, DTN, DTP, pH	0.1178	0.1189	0.1158	0.1146
Temperature, flowrate, precipitation	0.1188	0.1157	0.1174	0.1158
DO, BOD, temperature, flowrate, precipitation	0.1098	0.1121	0.1048	0.1168
DTN, DTP, DO, BOD, temperature, flowrate, precipitation	*0.1085*	*0.1106*	*0.1116*	*0.1335*
NH3-N, NO3-N, PO4-P, TN, TP, BOD, temperature, flowrate, precipitation, EC	0.1236	0.1116	0.1126	0.1154
All variables	0.1125	0.1125	0.1125	0.1174
Input	*Linear*	*DTR*	*SARIMAX*	*ANFIS*
NH3-N, NO3-N, PO4-P	0.1231	0.1758	0.1147	0.1125
NH3-N, NO3-N, PO4-P, TN, TP	0.1257	0.1747	0.1148	0.1118
NH3-N, NO3-N, PO4-P, TN, TP, DTN, DTP	0.1224	0.1825	0.1140	0.1118
DTN,DTP, pH	0.1174	0.1789	0.1139	0.1104
DTN, DTP	0.1185	0.1803	0.1141	0.1101

(continued)

Table 5 (continued)

Input	RNN	LSTM	GRU	SVR
DTN, DTP, pH, DO, BOD	0.1175	0.1759	0.1139	0.1083
NH3-N, NO3-N, PO4-P, DTN, DTP, pH	0.1208	0.1741	0.1142	0.1060
Temperature, flowrate, precipitation	0.1190	0.1720	0.1139	0.1086
DO, BOD, temperature, flowrate, precipitation	0.1185	0.1709	0.1137	0.1084
DTN, DTP, DO, BOD, temperature, flowrate, precipitation	*0.1167*	*0.1705*	*0.1135*	*0.1067*
NH3-N, NO3-N, PO4-P, TN, TP, BOD, temperature, flowrate, precipitation, EC	0.1189	0.1712	0.1147	0.1089
All variables	0.1174	0.1708	0.1140	0.1068

On the other hand, SVR shows the best in the traditional linear group, showing us the classification relative of the input. The deep neural network has found the hidden relative of the input data, which the MSE is low. The results of the network are nearly the same. The fuzzy technique has a big impact on the network, which can make the network understand more about the relative. ANFIS has shown the best performance for the problem again.

The results of two different environment regression problems show the same structure. The ANFIS, which is applied by the fuzzy technique, was identified as the best algorithm for these problems. The neural network, such as RNN, LSTM, can find the hidden relative of the input factors, and get good output. On the other hand, the traditional time-series, or linear methods, depend on the correlation of the input, and need using the knowledge domain to have good results.

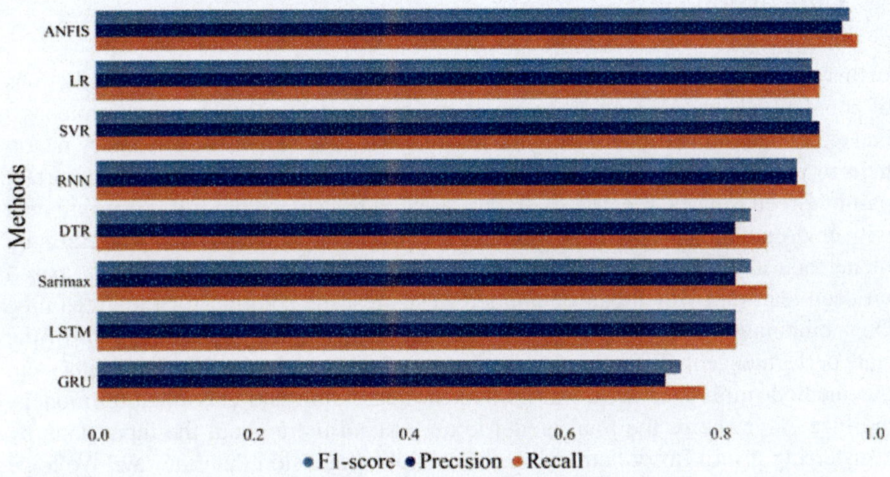

Fig. 6 Algal bloom prediction in an urban river with different methods

3.8 *Classify Model*

With algal bloom prediction in an urban river, the classification will split into different levels, and algal blooms will bloom on the high level. The level is calculated by split the regression model by the technical threshold. Evaluation metrics such as precision, recall, and the F1-score to estimate the qualitative performance of classification models were compared for the different models (Fig. 6).

Among the RNN-like methods, the original RNN showed the best performance, followed by the LSTM and GRU models. For the TSI-Chla classification, the accuracies of the regression models were similar among the different evaluation metrics, with the precision of the RNN, LSTM, and GRU models being 0.93, 0.93, and 0.82, respectively. Of the ML algorithms, the ANFIS algorithm presented the best performance with precision, recall, and F1-score values of 0.96, 0.98, and 0.97, respectively, while the SARIMAX algorithm was the least reliable model for the classification problem. The large discrepancy of the SARIMAX algorithms in the performance between classification and regression could be explained by its intrinsic feature, whereby quantitative values (i.e., the regression between the predicted index and real index) easily fell on different sides of the classification threshold if they were slightly different; thus, even a small error in values may lead to large differences in the class prediction. Similarly, if the gap between the values of the predicted index and actual index is large, they would probably fall on the same side of the trophic threshold, leading to a similar classification of trophic state. The measured values and the calculated data from the different ML algorithms were compared at each mainstream station.

4 Conclusion

In this chapter, we have introduced methods to analyze and get some early predictions of population sources in the urban areas, such as air and water. Their pollution has a bad effect on human life, knowing the reasons, which are about the chemistry factor, help to prevent them early. It will increase human health, which is an important part of Green Smart Cities. Overall, our research has shown the full process to deal with environmental problems, beginning with collecting the data and filling-missing them, then analyzing for the element's quality, and predicting phenomena based on deep learning, which can be applied to suggest the solution for the green city. Our combination between fuzzy and deep learning methods has shown out-of-the state performance in different phases of the total process. In the filling-missing step, our method minimizes MSE values. On the regression and classification models, the fuzzy layer helps the final model to understand more about the input data, by transferring it into fuzzy data, which have triple dimension than the raw. We hope that our chapter will help people to have a basic vision to deal with the population factors, using machine learning and deep learning methods. We believe it is necessary to build a green smart city.

References

1. Ly, Q.V., Nguyen, X.C., Ngoc, C.L., Truong, T.D., Hoang, T.-H.T., Park, T.J., Maqbool, T., et al.: Application of machine learning for eutrophication analysis and algal bloom prediction in an urban river: a 10-year study of the Han River. South Korea. Sci. Total Environ. **797**, 149040 (2021)
2. Nghiem, T.-D., Mac, D.-H., Nguyen, A.-D., Le, N.C.: An integrated approach for analyzing air quality monitoring data: a case study in Hanoi. Vietnam. Air Qual. Atmos. Health **14**(1), 7–18 (2021)
3. World Health Organization: WHO announces COVID-19 outbreak a pandemic (2020). https://www.euro.who.int/en/health-topics/health-emergen-cies/coronavirus-covid-19/news/news/2020/3/who-an-nounces-covid-19-outbreak-a-pandemic
4. World Health Organization: (2018) https://www.who.int/vietnam/news/detail/02-05-2018-more-than-60-000-deaths-in-viet-nam-each-year-linked-to-air-pollution
5. Ly, Q.V., Lee, M.-H., Hur, J.: Using fluorescence surrogates to track algogenic dissolved organic matter (AOM) during growth and coagulation/flocculation processes of green algae. J. Environ. Sci. **79**, 311–320 (2019)
6. Xia, R., Zhang, Y., Wang, G., Zhang, Y., Dou, M., Hou, X., Qiao, Y., Wang, Q., Yang, Z.: Multi-factor identification and modelling analyses for managing large river algal blooms. Environ. Pollut. **254**, 113056 (2019)
7. Huisman, J., Codd, G.A., Paerl, H.W., Ibelings, B.W., Verspagen, J.M.H., Visser, P.M.: Cyanobacterial blooms. Nat. Rev. Microbiol. **16**(8), 471–483 (2018)
8. Ly, Q.V., Maqbool, T., Hur, J.: Unique characteristics of algal dissolved organic matter and their association with membrane fouling behavior: a review. Environ. Sci. Pollut. Res. **24**(12), 11192–11205 (2017)
9. Xin, X., Zhang, H., Lei, P., Tang, W., Yin, W., Li, J., Zhong, H., Li, K.: Algal blooms in the middle and lower Han River: characteristics, early warning and prevention. Sci. Total Environ. **706**, 135293 (2020)

10. Wurtsbaugh, W.A., Paerl, H.W., Dodds, W.K.: Nutrients, eutrophication and harmful algal blooms along the freshwater to marine continuum. Wiley Interdiscipl. Rev. Water **6**(5), e1373 (2019)
11. Lv, J., Hongjuan, W., Chen, M.: Effects of nitrogen and phosphorus on phytoplankton composition and biomass in 15 subtropical, urban shallow lakes in Wuhan. China. Limnologica **41**(1), 48–56 (2011)
12. Descy, J.-P., Leprieur, F., Pirlot, S., Leporcq, B., Wichelen, J.V., Peretyatko, A., Teissier, S., et al.: Identifying the factors determining blooms of cyanobacteria in a set of shallow lakes. Ecol. Inform. **34**, 129–138 (2016)
13. Kim, J., Lee, T., Seo, D.: Algal bloom prediction of the lower Han River, Korea using the EFDC hydrodynamic and water quality model. Ecol. Model. **366**, 27–36 (2017)
14. Cha, Y., Cho, K.H., Lee, H., Kang, T., Kim, J.H.: The relative importance of water temperature and residence time in predicting cyanobacteria abundance in regulated rivers. Water Res. **124**, 11–19 (2017)
15. Kim, M., Lee, J., Yang, D., Park, H.Y., Park, W.: Seasonal dynamics of the bacterial communities associated with cyanobacterial blooms in the Han River. Environ. Pollut. **266**, 115198 (2020)
16. Wang, C.X., Yanhua, C., Zucong, Z., Maoheng, Y.: Chun: Response of the nitrogen load and its driving forces in estuarine water to dam construction in Taihu Lake. China. Environ. Sci. Pollut. Res. **27**, 31458–31467 (2020)
17. Domingues Rita, B., Barbosa Ana, B., Sommer, U., Galvão Helena, M.: Phytoplankton composition, growth and production in the Guadiana estuary (SW Iberia): unraveling changes induced after dam construction. Sci. Total Environ. **416**, 300–313 (2012)
18. Chalar, G., Arocena, R., Pacheco, J.P., Fabián, D.: Trophic assessment of streams in Uruguay: a trophic state index for benthic invertebrates (TSI-BI). Ecol. Indic. **11**, 362–369 (2011)
19. Lopes, O.F., Rocha, F.A., de Sousa, L.F., da Silva, D.M.L., Amorim, A.F., Gomes, R.L., da Silva, A.L.S., Junior, J., Mota, R.: Influence of land use on trophic state indexes in northeast Brazilian river basins. Environ. Monit. Assess. **191**, 77 (2019)
20. Carlson, R.E.: A trophic state index for lakes. Limnol. Oceanogr. **22**, 361–369 (1977)
21. da Silva Burigato Costa, C.M., da Silva, M.L., Almeida, A.L., Leite, I.R., de Almeida, I.K.: Applicability of water quality models around the world a review. Environ. Sci. Pollut. Res. **26**, 36141–36162 (2019)
22. Zhang, Z., Wang, J.: Phytoplankton, dissolved oxygen and nutrient patterns along a eutrophic river-estuary continuum: observation and modeling. J. Environ. Manage. **261**, 110233 (2020)
23. Benmouiza, K., Cheknane, A.: Small-scale solar radiation forecasting using ARMA and nonlinear autoregressive neural network models. Theor. Appl. Climatol. **124**, 945–958 (2016)
24. Kocak, C.: ARMA (p, q) type high order fuzzy time series forecast method based on fuzzy logic relations. Appl. Soft Comput. **58**, 92–103 (2017)
25. Aero, O.: Ogundipe, Adeyemi & #x201C;Fiscal deficit and economic growth in Nigeria: ascertaining a feasible threshold. Publ. Soc. Sci, Electr (2018)
26. Guo, H., Pedrycz, W., Liu, X.: Hidden Markov models-based approaches to long-term prediction for granular time series. IEEE Trans. Fuzzy Syst. **26**, 2807–2817 (2018)
27. Tang, C.-H., Coull, B.A., Schwartz, J., Di, Q., Koutrakis, P.: Trends and spatial patterns of fine-resolution aerosol optical depth derived PM2. 5 emissions in the Northeast United States from 2002 to 2013. J. Air Waste Manage. Assoc. **67**(1), 64–74 (2017)
28. Oteros, J., García-Mozo, H., Hervás, C., Galán, C.: Biometeorological and autoregressive indices for predicting olive pollen intensity. Int. J. Biometeorol. **57**, 307–316 (2013)
29. Donnelly, A., Misstear, B., Broderick, B.: Real time air quality forecasting using integrated parametric and non-parametric regression techniques. Atmos. Environ. **103**, 53–65 (2015)
30. Jianfeng, Z., Zhu, Y., Zhang, X., Ye, M.Y., Yang, J.: Developing a long short-term memory (LSTM) based model for predicting water table depth in agricultural areas. J. Hydrol. **561**, 918–929 (2018)
31. Tiyasha, T.T., Minh, Y., Mundher, Z.: A survey on river water quality modelling using artificial intelligence models: 2000–2020. J. Hydrol. **585**, 124670 (2020)

32. Zhou, Y.: Real-time probabilistic forecasting of river water quality under data missing situation: deep learning plus post-processing techniques. J. Hydrol. **589**, 125164 (2020)
33. Yi, H.-S., Sangyoung, P., An, K.-G., Kwak, K.-C.: Algal bloom prediction using extreme learning machine models at artificial weirs in the Nakdong River, Korea. Int. J. Environ. Res. Publ. Health **15**, 2078 (2018)
34. Park, Y., Cho, K.H., Park, J., Cha, S.M., Kim, J.H.: Development of early-warning protocol for predicting chlorophyll-a concentration using machine learning models in freshwater and estuarine reservoirs. Korea. Sci. Total Environ. **502**, 31–41 (2015)
35. Shen, J., Qin, Q., Wang, Y., Sisson, M.: A data-driven modeling approach for simulating algal blooms in the tidal freshwater of James River in response to riverine nutrient loading. Ecol. Modell. **398**, 44–54 (2019)
36. Lee, S., Lee, D.: Improved prediction of harmful algal blooms in four Major South Korea's Rivers using deep learning models. Int. J. Environ. Res. Publ. Health **15**(7), 1322 (2018)
37. Ni, X.Y., Huang, H., Du, W.P. Relevance analysis and short-term prediction of PM 2.5 concentrations in Beijing based on multi-source data. Atmos. Environ. **150**, 146–161 (2017)
38. Zhan, Y., Luo, Y., Deng, X., Chen, H., Grieneisen, M.L., Shen, X., Zhu, L., Zhang, M.: Spatiotemporal prediction of continuous daily PM2.5 concentrations across China using a spatially explicit machine learning algorithm. Atmos. Environ. **155**, 129–139 (2017)
39. Shang, Z., Deng, T., He, J., Duan, X.: A novel model for hourly PM2.5 concentration prediction based on CART and EELM. Sci. Total Environ. **651**, 3043–3052 (2019)
40. Du, P., Wang, J., Hao, Y., Niu, T., Yang, W.: A novel hybrid model based on multi-objective Harris hawks optimization algorithm for daily PM2. 5 and PM10 forecasting. Appl. Soft Comput. **96**, 106620 (2020)
41. Wang, Y., Wang, Y., Lui, Y.W.: Generalized recurrent neural network accommodating dynamic causal modeling for functional MRI analysis. NeuroImage **178**, 385–402 (2018)
42. Yadav, A.P., Kumar, A., Behera, L.: RNN based solar radiation forecasting using adaptive learning rate. In: International Conference on Swarm. Evolutionary, and Memetic Computing, pp. 442–452. Springer, Cham (2013)
43. Lin, H., Shi, C., Wang, B., Chan, M.F., Tang, X., Ji, W.: Towards real-time respiratory motion prediction based on long short-term memory neural networks. Phys. Med. Biol. **64**(8), 085010 (2019)
44. Zhao, R., Wang, D., Yan, R., Mao, K., Shen, F., Wang, J.: Machine health monitoring using local feature-based gated recurrent unit networks. IEEE Trans. Ind. Electron. **65**(2), 1539–1548 (2017)
45. Tong, W., Li, L., Zhou, X., Hamilton, A., Zhang, K.: Deep learning PM 2.5 concentrations with bidirectional LSTM RNN. Air Qual. Atmos. Health **12**(4), 411–423 (2019)
46. Krishan, M., Jha, S., Das, J., Singh, A., Goyal, M.K., Sekar, C.: Air quality modelling using long short-term memory (LSTM) over NCT-Delhi. India. Air Qual. Atmos. Health **12**(8), 899–908 (2019)
47. Nkuna, T.R., Odiyo, J.O.: Filling of missing rainfall data in Luvuvhu River Catchment using artificial neural networks. Phys. Chem. Earth Parts A/B/C **36**(4–15), 830–835 (2011)
48. Zadeh, L.A.: Outline of a new approach to the analysis of complex systems and decision processes. IEEE Trans. Syst. Man Cybern. **1**, 28–44 (1973)
49. Pelletier, F.J.: Review: Petr Hájek. Metamathematics of fuzzy logic. Bulletin Symbol, Logic (2000)
50. Atanassov, K.T.: Intuitionistic fuzzy sets. In: Intuitionistic fuzzy sets, pp. 1–137. Physica, Heidelberg (1999)
51. Cuong, B.C., Kreinovich, V.: Picture Fuzzy Sets-a new concept for computational intelligence problems. In: 2013 Third World Congress on Information and Communication Technologies (WICT 2013), pp. 1–6. IEEE (2013)
52. Alain, G., Bengio, Y.: What regularized auto-encoders learn from the data-generating distribution. J. Mach. Learn. Res. **15**(1), 3563–3593 (2014)
53. Goodfellow, I., Bengio, Y., Courville, A.: Deep learning. MIT press (2016)

54. Srivastava, N., Mansimov, E., Salakhutdinov, R.: Unsupervised Learning of Video Representations using LSTMs (2016)
55. Faruk, D.Ö.: A hybrid neural network and ARIMA model for water quality time series prediction. Eng. Appl. Artif. Intell. **23**, 586–594 (2010)
56. Lv, X., Zhang, J., Liang, P., Zhang, X., Yang, K., Huang, X.: Phytoplankton in an urban river replenished by reclaimed water: features, influential factors and simulation. Ecol. Indic. **112**, 106090 (2020)
57. Wang, S., Li, C., Lim, A.: Why Are the ARIMA and SARIMA not Sufficient. arXiv:1904.07632
58. Fathi, M.M., Awadallah, A.G., Abdelbaki, A.M., Mohammed, H.: A new Budyko framework extension using time series SARIMAX model. J. Hydrol. **570**, 827–838 (2019)
59. David, F.A.: Statistical Models: Theory and Practice, Cambridge University Press (2009)
60. Drucker, H., Burges, Chris, J.C., Linda, K., Alex, S.J., Vladimir, V.N.: Support vector regression machines. Adv. Neural Inf. Process. Syst. **9**, NIPS 1996, 155–161. MIT Press (1997)
61. Breiman, L., Friedman, J., Richard, O.A., Charles, S.J.: Classification and Regression Trees: Wadsworth & Brooks/Cole Advanced Books & Software (1984)
62. Jyh-Shing, J.R.: ANFIS: adaptive-network-based fuzzy inference system. IEEE Trans. Syst. Man Cybern. **23**(3), 665–685 (1993)
63. Chang, H.: Spatial analysis of water quality trends in the Han River basin. South Korea. Water Res. **42**(13), 3285–3304 (2008)
64. Kim, J.-W., Ki, S.J., Moon, J., Yoo, S.K., Ryu, A., Won, J., Choi, H., Kim, J.H.: Mass load-based pollution management of the Han River and its tributaries. Korea. Environ. Manage. **41**(1), 12–19 (2008)
65. Bermingham, M.L., Pong-Wong, R., Spiliopoulou, A., Hayward, C., Rudan, I., Campbell, H., Wright, A.F., Wilson, J.F., Agakov, F., Navarro, P., Haley, C.S.: Application of high-dimensional feature selection: evaluation for genomic prediction in man. Sci. Rep. **5**(1), 1–12 (2015)

Calculation of the Energy of a Two-Circuit Solar System with Thermosiphon Circulation Based on the Internet of Things

Yedilkhan Amirgaliyev, Murat Kunelbayev, and Talgat Sundetov

Abstract The article describes the calculation of the energy of a two-circuit solar system with thermosiphon circulation based on the Internet of Things. A new design of a flat solar collector has been developed, as well as a two-circuit solar installation with thermosiphon circulation. A controller capable of controlling the current temperature of the solar thermal system has been developed to control the solar system. For this purpose, the proposed measurement system uses temperature sensors (DS18B20 Dallas) using 16 wires. With the help of temperature sensors and the corresponding software, you can monitor the temperature level. The use of 4 digital sensors significantly improves the performance of system management and increases the speed of data processing. The possibilities of configuring sensors for Arduino platforms, as well as a solar collector control scheme, were considered. This article scientifically analyzes the operation of a new controller for controlling the solar thermal system using 6 digital temperature sensors using the Arduino platform to determine the control of the entire solar thermal system. The most profitable use of solar collectors in the industry is to replace human intervention with wireless sensor networks. A standard solar collector system consumes an average of 30% of the heat due to poor management and configuration. Our monitoring and control system makes it possible to increase the efficiency of heating industrial and domestic premises using solar collectors for hot water supply.

Keywords Solar collector · Controller · Arduino · Monitoring

1 Introduction

Renewable energy sources usage promotes constant demand increase on the energy and ecologically friendly technologies outspread [9, 16]. Despite practically they extension applicable only large-scale very cost of PLC. Currently is interested in,

Y. Amirgaliyev · M. Kunelbayev (✉) · T. Sundetov
Institute of Information and Computing Technologies of CS MES RK, Almaty, Kazakhstan
e-mail: murat7508@yandex.kz

M. Lahby et al. (eds.), *Computational Intelligence Techniques for Green Smart Cities*, Green Energy and Technology, https://doi.org/10.1007/978-3-030-96429-0_15

completely allow several systems into a common operating system. Equally important for the management of control through [7] power source with photoelectric panels, [8, 10]. To the present day, implementation of such a system has been possible only with PLCs, the building's extension capacity being several controllers. Nevertheless, those devices' software is still closed. Along with the proliferation of cheap control controllers, it began to adapt with other controllers. As well, should be noted, that the controllers thereof can be updated supported. Regardless of substandard installations comply, specified. There is constantly grows the interest of professionals in using the common and adjustable equipment. In recent years there have appeared and advanced several joint projects and reasonable alternative technologies, which allows the end-users to operate electronics easily and promptly A «creator» an approach «do it yourself», approach, excluding was developed in [12]. Those movements outspread has permitted to distribution the device, always connected to a switched network, that is, the object networks are connected to the network and united [5]. The DIY era, relies on the functionality of reading/recording of the Internet and digital design/production, for the common people could produce commodities. Any person time can fulfill the principles of the philosophy «do it yourself» [2–4], using advanced technologies. In [14, 15], a monitoring system was developed that is accessible through a personal computer via a serial RS232 device that is. The article [13] introduced an intelligent energy management system at home based on ZigBee for monitoring the energy consumption of household appliances and lighting. The work [7, 6, 11, 17], developed are the cause of installation failures.

2 Method of Research

In 2019 there was developed the system of solar heating was, the solar energy E with temperature t0 is absorbed by solar collector 1, with temperature t1, heating solar energy flow, goes through semitransparent insulation glass 2. The heat, received from the solar flow, heats the liquid in coils 3, which is removed from the collector, and cold water occupies its place from the pipeline with cold water tap 8, and distributor tank siphon 7 there takes place constant circulation of thermal siphon by the usage of circulation pipe 10. Further, the liquid enters the thermal pump 11, which consists of condenser-evaporator 12 with temperature т2, in which a heat exchanger is fabricated in spiral form, absorbing the transfer medium heat, lowering its temperature below ambient temperature (Q2), using throttling valve 14, thereby promoting additional absorbing the heat from atmospheric air. The diagram also shows the solar irradiation, reflected from the semitransparent coating (Q0) and absorbing panel surface (Q1). Transfer medium, having a relatively low temperature of condenser transfer medium 15 in the spiral form with higher temperature t2, increases square and speed of heat exchange. To fulfill such a cycle there is used a compressor 13 with temperature т3 with electric drive 17. Further with the help of condenser heat exchanger 15 with temperature t4 heat from the thermal pump (Q5) is transferred to the tank from heat exchanger Q6 with temperature t6 of heating 18. As the installation has two circuits, it

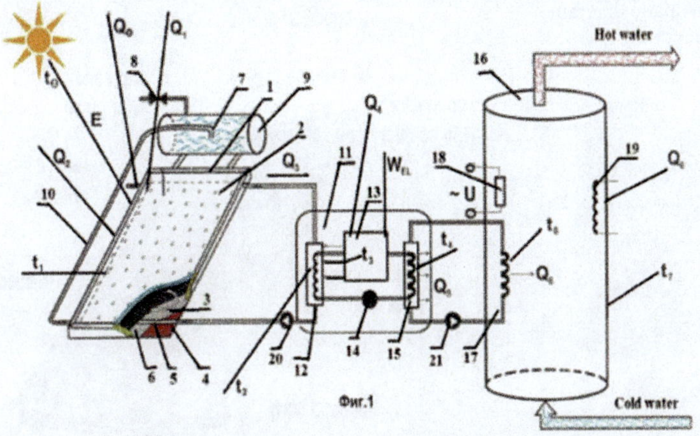

Fig. 1 A principal diagram of double circuit solar installation with thermosiphon circulation

is equipped with automatic circulation pumps 19 and 20 for liquid circulation between solar collector and evaporator, condenser, and storage reservoir. Water temperature is adjusted up to the required technological level and supplied to a customer for hot water supply and heating (Fig. 1).

The given research originality is the development of a double circuit solar system with thermosiphon circulation, which has a flat solar collector, representing heat-insulating transparent double-glazing unit with lower pressure, and transfer medium is made of the thin-walled corrugated stainless tube. Heat, obtained from solar flow, heats the liquid, which is removed from the collector, and its place is occupied with cold water from the siphon and there occurs constant thermal circulation, which upgrades heat transfer efficiency, at the expense of eliminating additional webs between a panel and heat insulation. There exists also a thermal pump, where the condenser and evaporator are made in the form of the heat exchanger of "spiral in spiral" type, heat exchanger pipelines are located one over another, increasing the square and heat exchange intensity (Fig. 2, 3 and 4).

3 The Counterpart of the Modular Controller for the Solar Thermal System

The controller on Fig. 5 controller prototype consists of a central module, controlling principle extension modules: system electric power.

Solar controller is based on platform (Fig. 5). Additionally, a block (for instance, accessible PTM the management). That amount input/ adjust the software for the analogous Arduino IDE software medium.

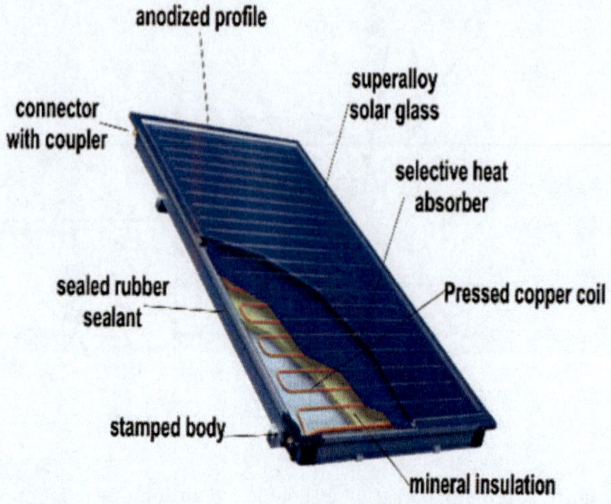

Fig. 2 Schematic diagram of a flat solar collector

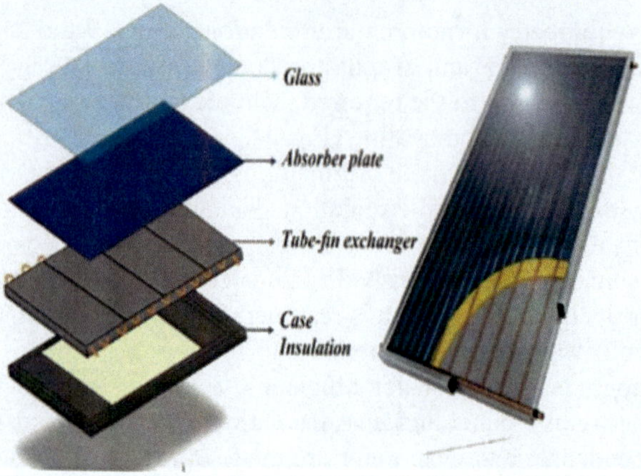

Fig. 3 Flat solar collector in parts

Where RJ45-Ethernet port; Reset KEY-Ethernet shield drop off and Arduino upon pressing; SD-card-support of Micro SD-card in FAT16 or FAT32; maximum memory volume is 2 Gb; IC W5200- hardware controller TCP/IP Ethernet. In the presented controller there anticipated several monitoring functions for the solar plant. Arduino logger, recording plant (Fig. 6) an example of plant recorded data. Measuring gives a possibility supplement at collector inlet/outlet, as well, a flowmeter. Using the sensors thereof allows implementing a heat meter, which might be used for the analysis of the efficiency. A chance for monitoring advanced via Internet (Fig. 7).

Language XML facilitates processing interpretation of electronic programs. States valves presented.

Monitoring managing plant, using Arduino Mega board, there has been described each element operation principle, based on which there will be executed a controller of the network controlling and monitoring.

Fig. 4 Flat solar collector

Fig. 5 Block managing

Built-in stress regulation, as well, microcontroller communication integrated compatible Arduino, allowing easy program and a microcontroller. Figure transformation functions alarm of over the temperature. Might prescribe nonvolatile memory of exchange microcontroller communication line, using the protocol of the interface 1-Wire. The sensor supplies, having been reserved capacity (Fig. 8).

Every micro scheme with a length discharges allows switched with one microcontroller port the regime quite to be used systems of ecological modules.

The feature of the DS1307 is its completely built-up delivery and programmed current time (it is needed only to fix). With the help packaged can operate power supply (Fig. 9).

Apart from the real-time clock micro scheme, the module has a micro scheme I2C EEPROM 24C32 and an interface for connecting the temperature sensor DS18B20.

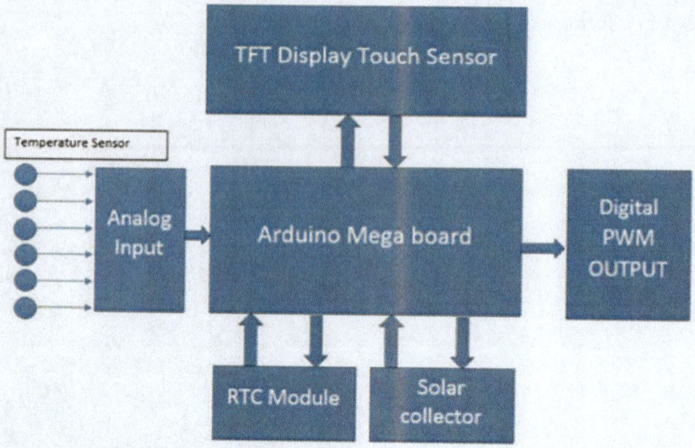

Fig. 6 Management controller block plant

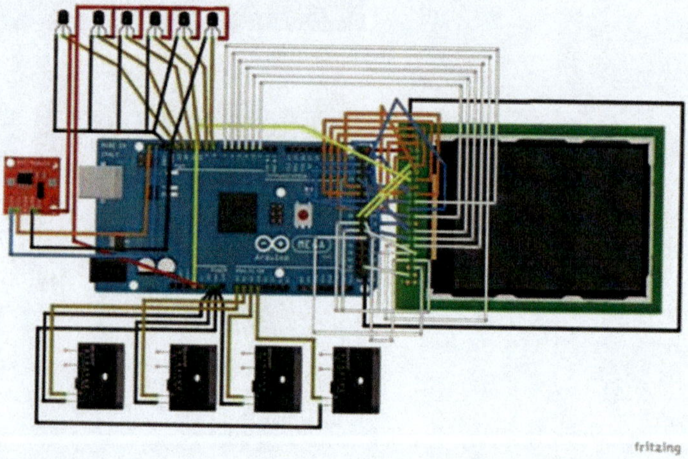

Fig. 7 Arduino Mega board to sensors of the controller management system

Fig. 8 Digital temperature meter DS18B20

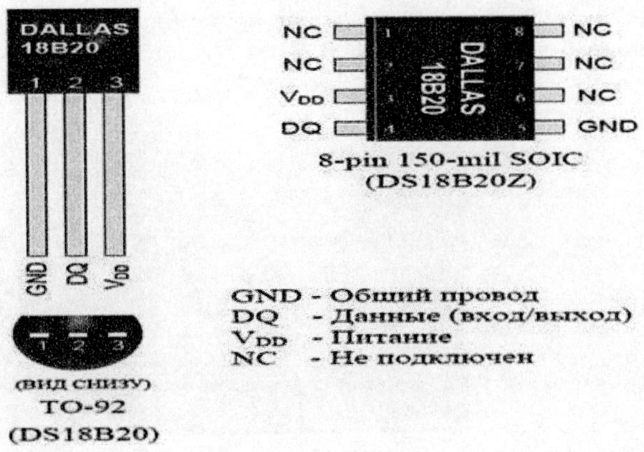

Fig. 9 Displays

Pulling the resistor gives a possibility. We have developed the controller temperatures. The figure presents an internal control unit: controller assemblage (Fig. 10).

The Control unit body has been designed as appropriate and implemented using has 3D box form and consists of 2 elements (Fig. 19): bottoms, covers (Figs. 11, 12, 13, 14, 15, 16, 17 and 18).

The solar collector monitoring controller has the processor ESP 32 1, which initializes and starts the solar collector 2 temperature data assembling and controls

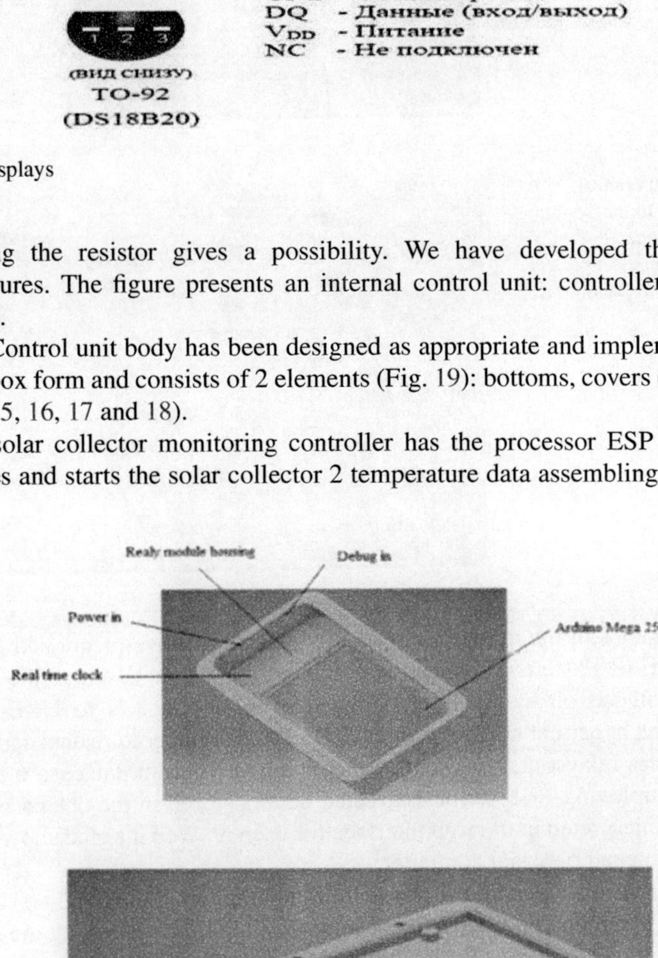

Fig. 10 Internal control unit: controller assemblage

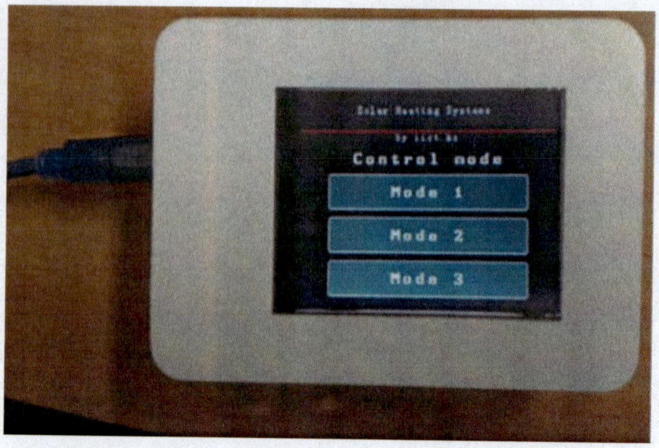

Fig. 11 The solar system management controller

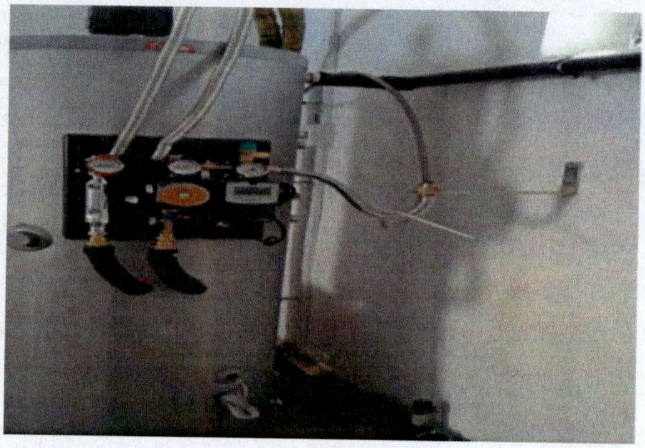

Fig. 12 Internal control unit: placement in the system Control unit is located inside, next to tubes, connecting a solar panel to a storage reservoir

valves relays states in the master controller 3. After the above-mentioned process, the processor ESP 32 starts a connection with Wi-Fi and switches on to the Internet 4. Having switched on to the Internet the ESP 32-1 connects to URL-domain address with the hypertext transfer protocol (HTTP) 5. Temperature data and master controller valves relay states are sent to the database 6. From database 6 the data is extracted employing PHP script. Extracted data is stored in the database 8. The stored data is interpreted in the web interface for users 9. Web interface operates for a mobile version and personal computer.

The link between a client and server is fulfilled employing the hypertext transfer protocol (HTTP). In that protocol a client initializes the link, requesting the definite web page with the help of HTTP. One of the main features of ESP 32 consists in the fact, that it can both be switched on to the Wi-Fi network and acts as a web-server, and can adjust the own network, permitting other devices to be connected directly to it and receive access to the web pages. It is possible because ESP32 can operate in free different regimes: station regime, soft access point regime, and both simultaneously. It provides the possibility to construct grid networks.

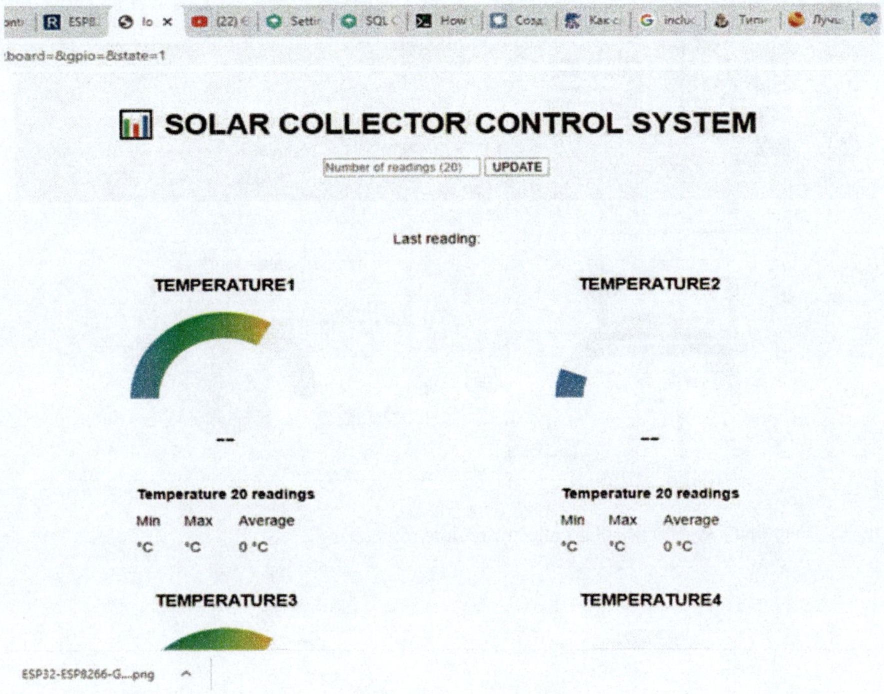

Fig. 13 System of net collecting, storing, and processing information from solar collectors

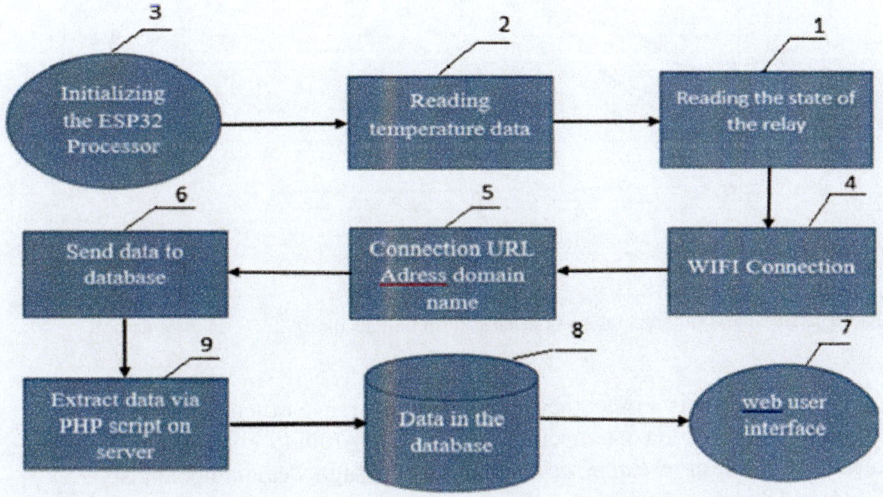

Fig. 14 Solar collector remote monitoring system

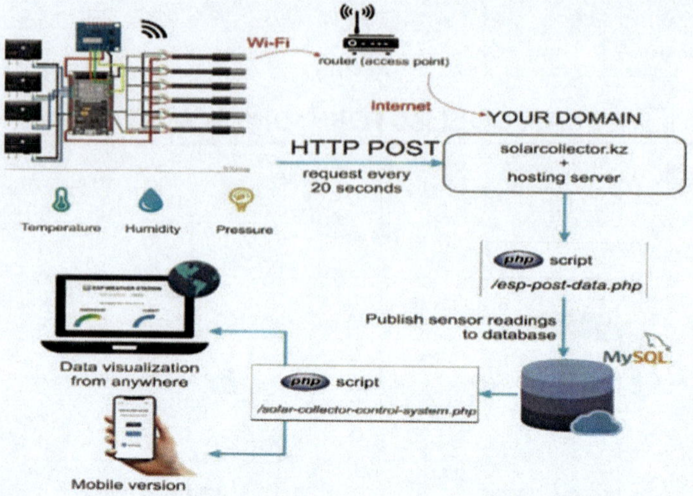

Fig. 15 Stationary system of solar collector remote monitoring

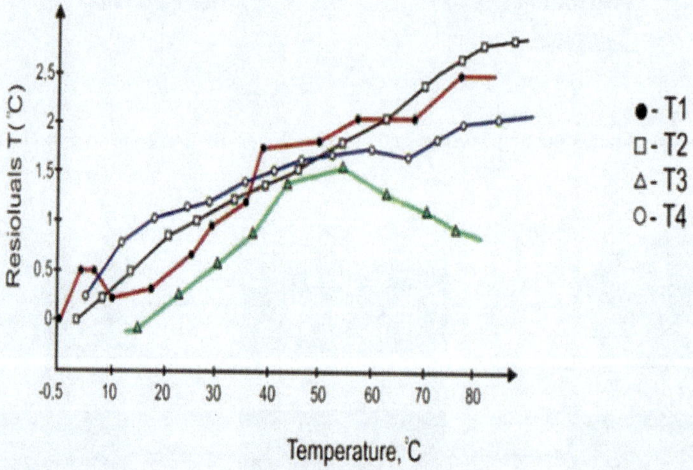

Fig. 16 Assessment of temperature external control unit accuracy

Figure 16 presents an accuracy assessment in a range of temperatures.

Figure 17, correction of temperatures various of Arduino, control, and monitoring unit. In the result of research, we can note, secure high, deviant upon raise.

Figure 18 shows different range temperature changes in the period from April 19 to June 9, 2021. As it is seen from Figure 14, indications, observed on April 19 from 07:30 to 09:00, are similar, though with less number of switching on/switching, comparing to June 9.

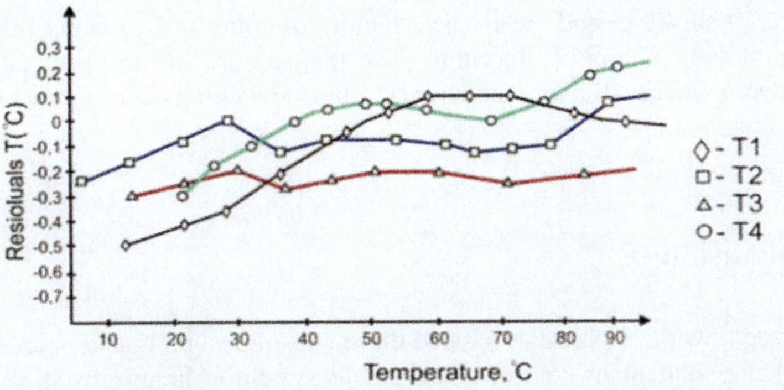

Fig. 17 Corrections of temperature

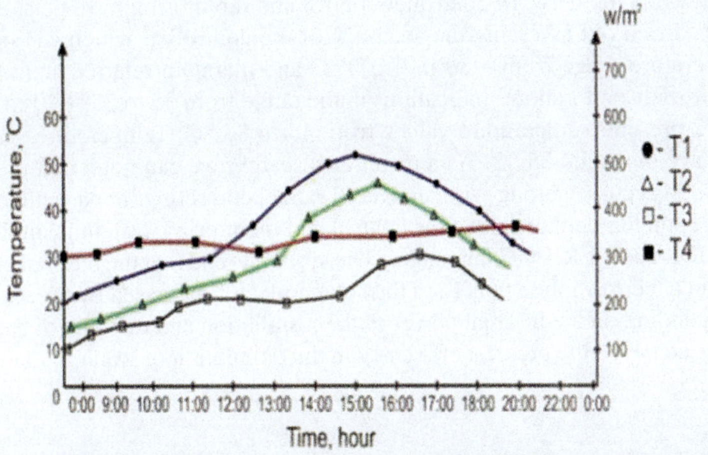

Fig. 18 A temperature change of sensors and heat pump according to the time

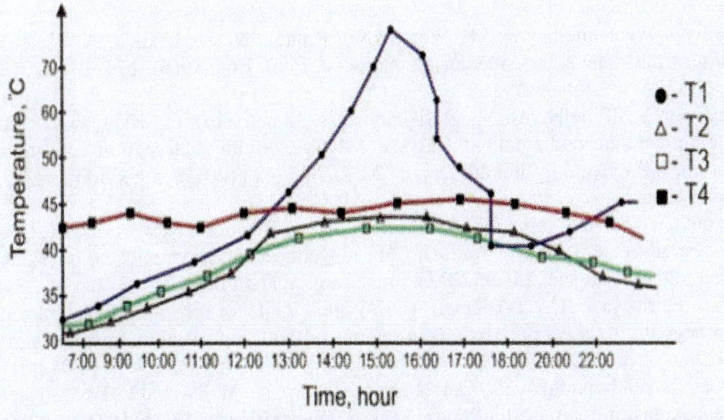

Fig.19 Periods of heat pump switching on / switching off

Figure 19 shows periods of P1 heat pump switching on/switching off, accordingly, employing a control algorithm. The temperature in the upper part of flat solar collector energy. Raising the internal, pump switching occur frequently, while exceeds switched.

4 Conclusion

In the present work, we have developed the stages of designing and practical application of the solar plant control management system (Almaty city, Kazakhstan). To create and research the solar plant thermal system monitoring platform, based on using the platform Arduino Mega, we have described every element operation, based on which the network controller control and monitoring have been executed. We have carried out assessing the accuracy of the controller, which can operate in the temperature range from −30 to +100 °C and maintain relative humidity from 10 to 90%. Sensor T1 shows indications in the range from 35 to 55 °C. Temperature sensor T2 presents temperature values from 45 to 85 °C. Temperature sensors T3 and T4 have the value 85 °C. As a result of research, we can note, that the sensors provide high accuracy along with an overall range, correcting, in particular, deviant behavior upon temperature rise. The control and monitoring system is implemented in the VHDL and VERILOG languages. The system found that the thermal efficiency of water in a thermosiphon tank for a flat solar collector increased by 5%. Solar radiation, depending on the thermal power of the installation and the time of heating the water, has achieved the greatest efficiency in the circulation of water in the metering tank.

References

1. Amirgaliyev, Y., Kunelbayev, M., Wójcik, W., Kataev, N., kozbakova, A.: Calculation and selection of flat-plate solar collector geometric. J. Ecol. Eng. **19**(6), 176–181 (2018). https://doi.org/10.12911/22998993/91882
2. Amirgaliyev, Y., Kunelbayev, M., Kalizhanova, A., Auelbekov, O., Katayev, N., Kozbakova, A.: Theoretical and mathematical analysis of double-circuit solar station with thermosiphon circulation. J. Polytech. Politeknik Dergisi **22**(2), 485–493 (2019). https://doi.org/10.2339/politeknik.491246
3. Anderson, C.: The New Industrial Revolution. Crown Business, New York, NY, USA (2010)
4. Fox, S.: Paradigm shift: Do-It-Yourself (DIY) invention and production of physical goods. J. Manuf. Technol. Manage. **24**(2) (2013). https://doi.org/10.1108/17410381311292313
5. Fox, S.: Third-wave Do-It-Yourself (DIY): potential for presumption, innovation, and entrepreneurship by local populations in regions without industrial manufacturing infrastructure. Technol. Soc. **39**, 18–30 (2014). https://doi.org/10.1016/j.techsoc.2014.07.001
6. Ghayvat, H., Mukhopadhyay, S., Gui, X., Suryadevara, N.: WSN- and IOT-based smart homes and their extension to smart buildings. Sens. J. (Switzerland), **15**(5), 10350–10379 (2015b). https://doi.org/10.3390/s150510350

7. Ghayvat, H., Liu, J., Babu, A., Alahi, E., Gui, X., Mukhopadhyay, C.: Internet of things for smart homes and buildings: opportunities and challenges. Published on Australian J. Telecommun. Digit. Econ. **3**(4), 33–47 (2015a). https://doi.org/10.18080/ajtde.v3n4.23
8. Grassie, T., MacGregor, K., Muneer, T., Kubie, J.: Design of a PV drove low flow solar domestic hot water system and modeling of the system collector outlet temperature. Energy Convers. Manage. **43**, 1063–1078 (2002). https://doi.org/10.1016/S0196-8904(01)00090-5
9. Kalogirou, S.: Solar thermal collectors and applications. Prog. Energy Combust. Sci. **30**, 231–295 (2004). https://doi.org/10.1016/j.pecs.2004.02.001
10. Porzuczek, J.: Zabezpieczenie ciągłości zasilania w małych systemach HVAC. Przegląd Naukowo-Metodyczny **24**, 665–670 (2014). https://doi.org/10.4467/2353737XCT.16.064.5413
11. Rehman, Z., Al-Bahadly, I., Mukhopadhyay, S.: Multi input DC-DC converters in renewable energy applications—an overview. Renew. Sustain. Energy Rev. **41**, 521–539 (2015).https://doi.org/10.1016/j.rser.2014.08.033
12. Salamone, F., Belussi, L., Danza, L., Ghellere, M., Meroni, I.: Design and development of nEMoS, an all-in-one, low-cost, web-connected, and 3D-printed device for environmental analysis. Sensors **15**, 13012–13027 (2015). https://doi.org/10.3390/s150613012
13. Sanz-Bobi, M.A.: Green energy and technology: experiences and future approaches. In: Use, Operation and Maintenance of Renewable Energy Systems. Springer Int. Publishing (2014)
14. Visconti, P., Lay-Ekuakille, A., Primiceri P., Cavalera, G.: Wireless energy monitoring system of photovoltaic plants with smart anti-theft solution integrated with household electrical consumption's control unit remotely controlled by the internet. Research Article Int. J. Smart Sens. Intell. Syst. **9**(2), 681–708 (2016). https://doi.org/10.21307/ijssis-2017-890
15. Viswanath, S., Belcastro M., Barton, J., O'Flynn, B., Holmes, N., Dixon, P.: Low-power wireless liquid monitoring system using ultrasonic sensors. Int. J. Smart Sens. Intell. Syst. **8**(1), 26–44 (2015). https://doi.org/10.21307/ijssis-2017-747
16. Wang, Z., Yang, W., Qiu F., Zhang, X., Zhao, X.: Solar water heating: from theory, application, marketing and research. Renew. Sustain. Energy Rev. **41**, 68–84 (2015). https://doi.org/10.1016/j.rser.2014.08.026
17. Zhenghua, X., Guolong, C., Li, H., Song, Q., Hu, L., Lei, C., Youwen, M., Yexiang, X.: The smart home system based on the IAP15F2K61S2 and GSM. Int. J. Smart Sens. Intell. Syst. 7(4), 1789–1806 (2014)

Case Studies and Smart Applications

Smart Human–Computer Interaction Interactive Virtual Control with Color-Marked Fingers for Smart City

Ching Yee Yong and Kelvin Uei Han Chia

Abstract This study proposed to use a camera to understand the movement of color-marked finger through the Microsoft Visual Studio platform to build a gesture-based wearable technology. The main purpose of this study is to detect and understand human gesture by using a camera to show the interaction of human capabilities that connect with both real and virtual world. The most important thing is to show a bunch of functionality which is very useful in our daily life. In detail, the proposed study combined two major components, a camera and a projector. The development of this study includes the integration of an electronic circuit with software. When a human gesture was detected by the camera, the signal or information was sent to a computer or an Android phone for further processing to analyze the meaning of the movement. The process from signal capturing until the analysis of the information is performed in a real-time condition. This study supports two types of platform; they are Windows and Android, which depends on user demand. Furthermore, the camera consists of HD lens to capture the picture precisely with a 720p resolution for a more accurate result. It can be viewed on a wide range of screen which is good at capturing lots of gesture information. In a nutshell, users can use this kind of technology to trace their movement status in order to make the future much more interactive instead of using a smartphone screen. It is a concept to let people understand that gesture technology enables motion in a 3D environment in a wonderful and amazing way for smart city purposes. It may change people's lives and yet, due to the acceptable cost through computational intelligence technique for green smart cities.

Keywords Gesture recognition · Vision · Image processing · Color detection · Motion · Smart technology

C. Y. Yong (✉) · K. U. H. Chia
School of Engineering and Technology, University of Technology Sarawak, 96000 Sibu, Sarawak, Malaysia
e-mail: chingyee@uts.edu.my

M. Lahby et al. (eds.), *Computational Intelligence Techniques for Green Smart Cities*, Green Energy and Technology, https://doi.org/10.1007/978-3-030-96429-0_16

1 Introduction

When it comes to sensing the world around them, humans have evolved over millions of years. When we come into contact with something, someone, or a place, we use our five natural senses to discern information, and it is the information that allows us to make a decision and take the appropriate action. However, we cannot argue that the most valuable evidence that can help us make the right decision comes not just from our five senses, but also from data, information, and wisdom that humanity has gathered over time. While computing system miniaturization is getting smaller, easier, and smarter, allowing us to hold a computer in our pockets and stay linked to the real world, there are some big problems, such as unrelated connections between digital devices and the physical world. Traditionally, information is contained on paper or on a tablet. This technology takes the digital world into our physical world and helps us to communicate with this knowledge using our hands because our hands are the most accurate and relaxed, allowing us to free ourselves from our imagination in a variety of ways. It also liberates knowledge from its confinement by effortlessly merging it with experience, effectively transforming the whole universe into your machine for smart cities environment.

The basic concept is based on the film "Ironman," in which J.A.R.V.I.S. incorporates all of the high-tech AI systems. The graphical user interface in the film is futuristic, and that would occur because we began to flourish as a human race solely by relying on a glass screen for our whole lives [1].

The ideal case is for a person to be able to obtain some information about something we want from anywhere in a matter of seconds. Not only will we be able to communicate with items in a whole different way, but we will still be able to interact with individuals. The great feature of this form of technology is its ability to scan items and individuals by projecting them depending on what they are looking at [2].

2 Research Review

This human–machine interaction technology combines a camera and a projector by connecting to a computer using a wireless module to transmit the data. Human–machine interaction recognizes hand gestures through a camera by projecting the image on a surface [3]. This technology can access the information with fingers, make a call in front of the projected image, get to know the time by drawing a circle in front of the camera, take a photo by making a square with the finger, zoom in or out, and many other features. This is a device that has a large number of functions with its portable and easy-to-wear characteristic [4].

The user can draw anything on the surface by moving the index fingers. A consumer will take a picture of a scene scoped in the rectangle by organizing the pairs of thumbs and index fingers in a rectangular shape. This technology supports a live function to allow users to read or view information. The device can tell the departure and arrival or even the delayed time of airplanes. The good news to book lovers is that they can search for additional information on texts, comments, features, and ratings of the books by picking any page. Steve Mann created the first version. This prototype's features include hands-free and headwear-free operation, which enhances the visual experience with graphics and text that directly display real-world objects. He dubbed this system "Synthetic Synesthesia of the Sixth Sense." It was later tested in telemedicine and has the ability to significantly change health care. According to Mann, it has a great capability of effective communication with a precise diagnosis to help in making a better decision. Since it has a large impact on real-life application, therefore, more improvement is needed to be positioned for better accuracy [5].

The computer is controlled by a reality user interface (RUI), which directs interaction with the physical world. Furthermore, the platform used direct user interface (DUI) to meet the need for augmented reality by acting as a pointing device. What distinguishes it from other gadgets with the same features if it is a pointing device? To obtain a response, it is necessary to depict the device's operating process in a larger context.

A bond between the subjects separated by a long distance can be created. For example, if a buyer in the United States of America needs to buy clothes from a Chinese online shop, the person in China will need to wear a telepointer device and project the video onto a screen to the buyer in the United States of America. To pick the object, the person does not need to make any sound or feedback; all he or she or she has to do is point the object on his or her or her projected screen with a laser pointer, and the laser will point at both points A and B at the same time, allowing all of them to see the same object. It can be a useful tool in medicine because it allows for good communication over long distances [6].

Seven MIT students begin to incorporate knowledge in their surroundings without pulling their tablets out of their pockets. They then created a wristband that reads radio frequency recognition to identify various items, such as the book the person is keeping in the shop. They also created another ring that communicated through infrared. The ring searched for facts about brands on store smart shelves. The ring will flash in various LED lights to distinguish whether the substance is organic or not [7].

Gest is a wearable that brings virtual interaction to a whole new level. It works with smooth hands motion and has more intuitive control. When it comes to interacting with a computer, the options are either a mouse or binary inputs like the keys on the keyboard. Besides that, it will let the user map his/her hand gestures to "key-in" the inputs more easily so that he/she can control every app with faster speed. The user just needs to flick the finger to the right for activating the input and increase the volume by twisting the hand. The best part is the user can decide which gesture map to which actions.

This study's description [8] is from Arduino, which employs the ESP method for simple gesture recognition. For example, the user may identify various tennis game motions such as forehand, backhand, and serve; elements for dance and weight-lifting gestures, and so on. The movements are detected by an accelerometer and submitted to the computer's ESP program. To record the motion, ESP will use a basic machine learning algorithm that will be matched with live accelerometer data. The downside of this system is that it only considers movements that are identical to those in the archive, which analyzes individual occurrences of distinct gestures. Furthermore, the computer has little detail about how the gesture is done [9]. The positive thing is that it can be applied to a wide variety of immersive apps [10].

Project Soli is a modern technology whose aim is to create a sensor-to-radar interaction for motion detection of the human hand. It is the most advanced device currently available on the market for accurately detecting rapid motion. The sensor is able to track up to submillimeter motion at a high speed with great accuracy. Besides that, it creates a unique gesture interaction that allows a user to control devices with a simple and global set of gestures which functions much better than the products in the market. It envisions the future in which human hands are universal input devices for interaction with technology [11].

The Myo armband is used for a gesture-based movement that is intuitive and easy to control. It is based on five distinct hand gestures which combine with motions. The user can feel haptic feedback through short, medium, and long vibrations. It functions through reading the electrical pulse of the user muscles and motion of the arm and let the user wirelessly control the technology by using gesture and motion [12]. Normally, it is used for presentation by controlling the pointer and zoom function. Besides, it is also used to connect the computer, such as browsing a website, playing videos, and switching between applications. Furthermore, people also used it to fly the drone and make the radio control experience escalate to the next level [13].

In a nutshell, the research review can be concluded with these seven products that are related to the study. Obviously, the elaboration of these products is just briefly described what the product is and its major functions. For this study, there are many types of functions that can be used for identification purposes such as virtual/augmented reality or gesture recognition using a camera, computer/image-based algorithm, radio frequency identification, and ubiquitous computing. The best one is using gesture recognition by the camera due to the ease of implementation, user-friendly, reliability, and efficiency [14].

In order to make the system work and perform its tasks properly, a program is required to be developed necessarily. The most suitable program in the study is Microsoft Visual Studio [15]. The program is described as a strong generation of programming language because it is a fully featured IDE and productivity for every application [16]. The code can be written faster and debug with ease. It can be tested often with confidence and extended by customized based on the user interest [17].

3 Methodology

3.1 Research Design

The system consists of four major components, they are:

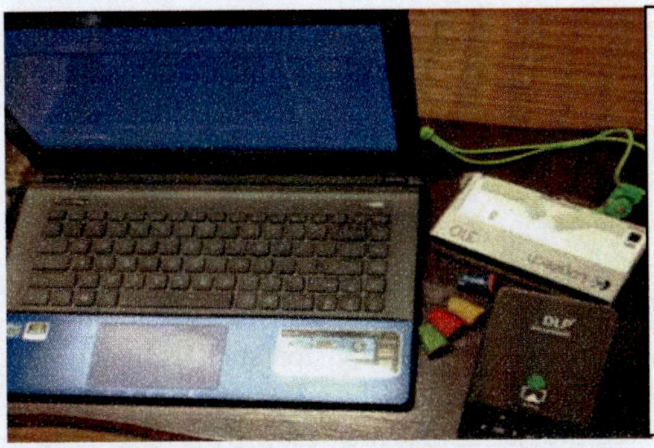

Components:
1. Computer/Laptop/Processor
2. 4 coloured finger caps
3. Portable projector
4. Hand-made hanger to carry the projector around the neck

Fig. 1 Major components of the study

- Camera—The camera is the major core hardware for this study that is used to understand the gesture movement in order to interact between the user and graphics by enriching the function to make it more beneficial.
- Projector—Projector is also another core hardware system that is used to project the output so that the user understands the function of the projected images.
- C Language Programming—The C programming language is a common, process-oriented computer programming language that is ideal for developing application software [18]. As a combination language, the program code can be read directly. This structured language is simple, clear, and easy to write. The C language's portability and cross-platform capabilities are exceptional [19].
- OpenCV—It is distributed under a BSD license and is open for academic and commercial use. It has interfaces in C++, C, Python, and Java that support Windows, Linux, Mac OS, iOS, and Android [20]. It is intended to be computationally effective, with a heavy emphasis on real-time applications. It can take advantage of multi-core computing by using C/C++. Furthermore, when using this platform, it will benefit from hardware acceleration [21] (Fig. 1).

3.2 Prototype Design

This research consists of a camera, a projector that is synced with a smartphone's Wi-Fi, and a pocket-sized projector. To begin, the camera will capture a stream of gesture movement of the user's fingers using color detection through the color marker fitted to the fingers. These movements were captured and sent to the processing unit for analysis and recognition. Programming codes were written to allow the system to understand the user command and then execute the next action requested by the user. The recorded gestures will be set as the instruction to building a certain useful function. An algorithm coded in the system will analyze the data in and out of the system in order to make an appropriate decision. At last, the system will project the output through a projector to the surface after processing the data. The output is based on the different gesture instructions that are processed by the processing unit.

In order to bring out the variations on the higher plane, the user may use the colored caps to fit on the fingers so that the system may easily differentiate the inputs and thus understand the variation in between the fingers movement for more efficient performance. The computer vision technique was used to detect and track the movement of the fingers. A user may customize his/her commands or applications as long as the movements are being captured and identified by the system through a camera. This study proposed a few functions related to vision such as the current time status by drawing a circle in front of the camera, having art drawing and painting on the screen by using fingers, shooting the current scenery view by arranging four fingers in a rectangle shape as a focus area, playing a video or image file from the gallery, and many others.

The main idea of the study is to interact with the computer for executing certain functions. The system was trained to recognize the identified gestures which related to the situation and try to make the best possible reaction based on the learning process. As you can know, the system is constantly trying to figure what is around the user, and what the user is trying to do. It will recognize the images the user sees, track the gestures, and measure the relevant data at the same time.

This technology normally recognizes three kinds of gestures; they are multi-touch gesture, freehand gesture, and iconic gesture. The multi-touch gesture is a technique in which when the user touches the screen, the projected outcome can be moved by pinching and dragging. Freehand gesture, for instance, by taking a photo shoot using an arrangement of fingers in a rectangle shape. While for the iconic gesture, the user can draw a star for the map. These are the techniques that are very useful in daily life and the interesting point is the proposed system covers all these techniques and the gestures can be customized by the user according to their will.

3.3 *Operational Procedures of the Prototype*

Figure 2 shows the concept of system operations of the developed prototype. First and foremost, the camera was used to analyze and recognize the gesture of the user's fingers. Since this analysis is largely based on C/C++ programming language, the processing unit was configured with Microsoft Direct X to handle multimedia tasks, such as Microsoft platforms related to video and game programming (Fig. 2).

Next, the system will be completed by setting up the demo mode, touchless mode, gestures mode, testing mode, menu control mode, and functions mode. Since the study covers a wide range of functions, thus, the process of every stage of the concept should strictly follow the flow.

Then, the touchless mode of the prototype that simulates the functions of the mouse and keyboard was designed. The environment code was also set up in order to exclude the background problem of the capture boundary. Furthermore, the system should ensure that the user draws the latest image from the active camera and draws

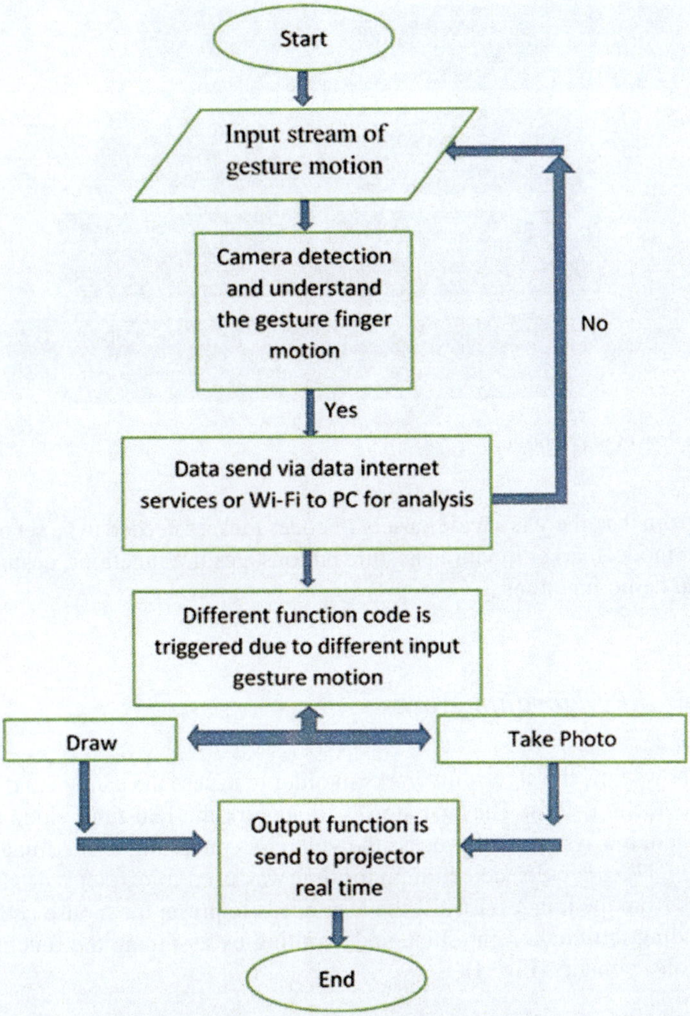

Fig. 2 Flowchart of the system process

the selection adornment which needs to set height, width, marker setting, and latest frame time segment. Besides, the system has to calculate the frame per second (FPS) in order to update the display once for every second and set it by saving the latest image for the drawing function. Meanwhile, for the camera issue, it is known that it needs to be refreshed the list of available input cameras. If the user selects a new input camera, the device will immediately delete the previous camera code in order to conform to the new situation. The camera code's primary role is to allow access to the marker mode until the camera is turned on.

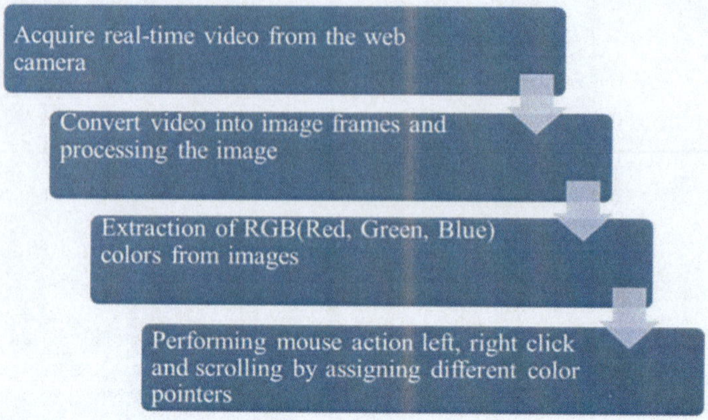

Fig. 3 The flow of the algorithm

Apart from that, there is a wide variety of codes that are needed to be set up such as the marker mode, marker function, gesture buttons, gesture functions, gesture mouse events, and demo functions.

3.4 How the Algorithm Works

Figure 3 shows how the algorithm works in order to access the camera and calibrate the color detection steps. The first step is to acquire the real-time video from the camera. Then, the system will process the video by converting it into image frames for analysis. Next, a color detection algorithm was used to extract red, green, and blue colors from the images. Lastly, the user needs to preset the mouse action functions including left-click, right-click, and scrolling by assigning the functions with different color pointers (Fig. 3).

4 Methodology

The result of the test function was shown in Fig. 4. The test function is a function to test whether the whole system is working properly or not. It is important due to the study is related to gesture recognition. A successful test function is by moving the mouse cursor in order to detect the color and start the interaction with the computer. The label on the top left corner shows the test word, which indicates that the system is activated on the test picture box by showing the test result. Four markers: red (R), blue (B), green (G), and yellow (Y) are recognized and displayed on the black background to ensure the colors do not interrupt each other. As the user is moving the four markers on the picture box, it will move around according to the user movement.

Fig. 4 The test function page of the system

This implies that the objective of the study which is to design a gesture recognition system is being achieved and run successfully.

Table 1 shows the list of proposed function in the system.

5 Conclusion

This study aimed to build an interactive control by using a Web camera based on a color detection technique for smart city. The camera needs to capture the movement of the color-marked fingers to activate and run certain functions and features. The first prototype was developed and relatively easy to operate. A camera can be replaced by a Web camera or even a smartphone as an image input for the processor. The processing unit will detect and analyze the movements of the fingers as input commands. The result will be displayed on a wall through a wired or wireless projector. The development of the prototype costs lower than the existing interactive products in the market. Overall, the system is able to cover up to eight functions for this study.

The applications of the prototype include:

- Drawing application
- Photo taking
- Browsing for a location using Google Map
- Photo viewing in the gallery
- Checking stock price in real-time
- Playing a video file
- Checking time
- Face recognition for security lock

The prototype is user-friendly by allowing the user to operate the system with simple movements. It integrates digital information into the physical world by

Table 1 Features and functions available in the proposed system

Feature	Function
	Drawing
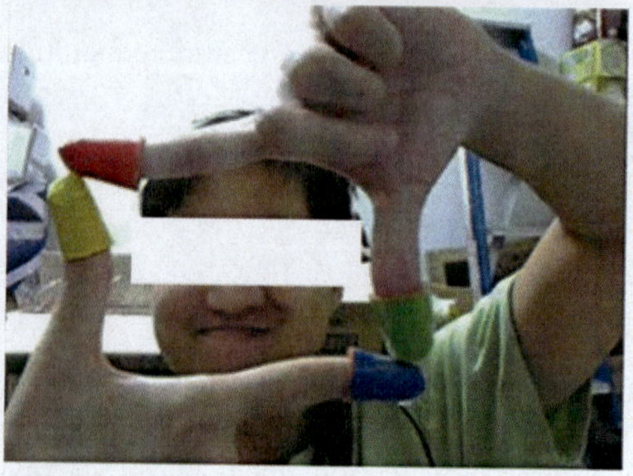	Photo taking
	Browsing for Google Map

(continued)

Table 1 (continued)

Feature	Function
	Photo viewing in gallery
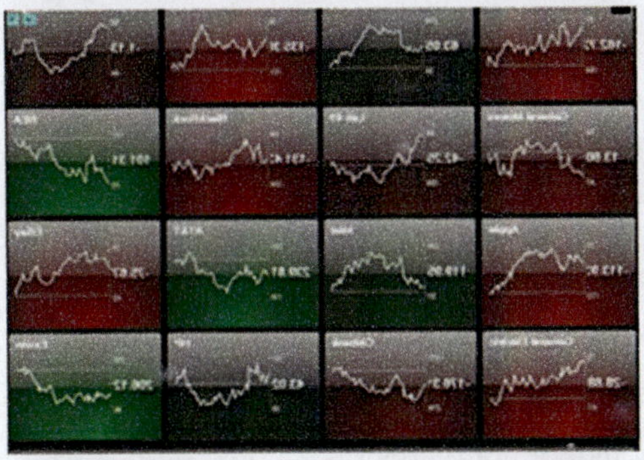	Stock checking
	Video playing
	Time checking

(continued)

Table 1 (continued)

Feature	Function
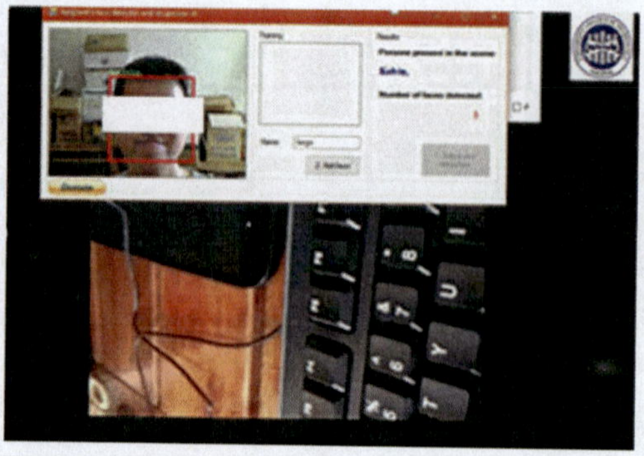	Face recognition for security lock

projecting the visual graphics to the surface. All the working mechanisms of the prototype are running in real-time. All the needed information can be accessed online. Besides, a user can carry this prototype anywhere since it is light and fulfills the portable characteristic. A user is not necessary to have knowledge on how the system is working because a user can customize every function by his/her own will (Fig. 5).

The prototype has high potential in interactive teaching and learning. It is beneficial and performs outstandingly neither indoor nor outdoor condition. Furthermore, it is a fantastic way for either business or academic presentation. The green way for the green cities.

However, there is still some room for improvement, such as increasing the efficiency of the detection and process algorithms by researching the written codes. Besides, more functions will be added in the future in order to furnish the prototype capabilities.

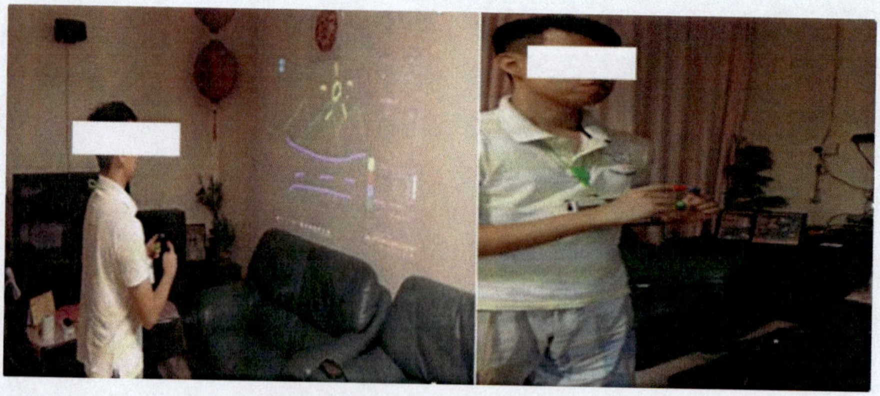

Fig. 5 A user was testing the drawing function

Acknowledgements Our gratitude to the University of Technology Sarawak for supporting and funding this study under University Research Grant (UCTS/RESEARCH/3/2019/07).

References

1. Shinde, A., Bagade, A., Thakur, R.: Computer Jarvis. Comput. Eng. **1**(2), 3 (2017)
2. Atzori, L., Lera, A., Morabito, G.: The internet of things: a survey. Comput. Netw. **54**(15), 2787–2805 (2010)
3. Yong, C.Y., Sudirman, R., Chew, K.M.: Motion detection and analysis with four different detectors. In: 2011 Third International Conference on Computational Intelligence, Modelling & Simulation, pp. 46–50. IEEE (2011)
4. Kumar, S.P., Pandithurai, O.: Sixth sense technology. In: 2013 International Conference on Information Communication and Embedded Systems (ICICES), pp. 947–953 (2013)
5. Mann, S.: Telepointer: Hands-free completely self-contained wearable visual augmented reality without headwear and without any infrastructural reliance. In: Digest of Papers. Fourth International Symposium on Wearable Computers, pp. 177–178 (2000)
6. Karim, R.A., Zakaria, N.F., Zulkifley, M.A., Mustafa, M.M., Sagap, I., Latar, N.H.: Telepointer technology in telemedicine: a review. Biomed. Eng. Online **12**(1), 21 (2013)
7. Konsynski, B., Smith, H.A.: Developments in practice x: Radio frequency identification (rfid) an internet for physical objects. Commun. Assoc. Inf. Syst. **12**(1), 19 (2003)
8. Shojima, H., Kuzunuki, S., Hirasawa, K.H.: On-line recognition method and apparatus for a handwritten pattern. U.S. Patent 4,653,107 (1987)
9. Wexelblat, A.: Research challenges in gesture: Open issues and un-solved problems. In: International Gesture Workshop, pp. 1–11. Springer, Berlin, Heidelberg (1997)
10. Chen, C., Jafari, R., Kehtarnavaz, N.: A survey of depth and inertial sensor fusion for human action recognition. Multimedia Tools Appl. **76**(3), 4405–4425 (2017)
11. Lien, J., Gillian, N., Karagozler, M.E., Amihood, P., Schwesig, C., Olson, E., Raja, H., Poupyrev, I.: Soli: Ubiquitous gesture sensing with millimeter wave radar. ACM Trans. Graph. (TOG) **35**(4), 142 (2016)
12. Murillo, P.U., Moreno, R.J.: Individual robotic arms manipulator control employing electromyographic signals acquired by myo armbands. Int. J. Appl. Eng. Res **11**, 1241–11249 (2016)
13. Sathiyanarayanan, M., Rajan, S.: MYO Armband for physiotherapy healthcare: A case study using gesture recognition application. In: 2016 8th International Conference on Communication Systems and Networks (COMSNETS), pp. 1–6 (2016)
14. Tannenbaum, A.R., Zetts, J.M., An, Y.L., Arbeitman, G.W., Greanias, E.C., Verrier, G.F.: International Business Machines Corp, Graphical user interface with gesture recognition in a multiapplication environment. U.S. Patent 5,252,951 (1993)
15. Randolph, N., Gardner, D., Anderso, C., Minutillo, M.: Professional Visual Studio 2010. John Wiley & Sons (2010)
16. Stubbs, P., Nordlund, P.N., Shepard, J.A., Quinn, T.E., Hodges, C.D.: Microsoft Corp, 2009. Managing application customization. U.S. Patent 7,530,079 (2009)
17. Laganière, R.: OpenCV Computer Vision Application Programming Cookbook Second Edition. Packt Publishing Ltd. (2014)
18. Carver, J.C., Kendall, R.P., Squires, S.E., Post, D.E.: Software development environments for scientific and engineering software: A series of case studies. In: Proceedings of the 29th international conference on Software Engineering, IEEE Computer Society, pp. 550–559 (2007)

19. Kernighan, B.W., Ritchie, D.M.: The C Programming Language (2006)
20. Deitel, P.J.: Java how to program. Pearson Education India (2007)
21. Morgan, S.: Programming Microsoft® Robotics Studio. Microsoft Press (2008)

Machine Learning for Green Smart Video Surveillance

Jose Filipe, Antonio Navarro, Luis Tavora, Sergio M. M. de Faria, and Pedro A. Amado Assuncao

Abstract The forthcoming smart environments will have advanced technology for capturing, processing, and analysing high-resolution and high-quality visual information. In future green smart cities, advanced video surveillance with dense deployment of video sensors is definitely necessary to build intelligent and secure urban environments. Since large-scale compression of high-quality video with low bit rate is computationally expensive, a significant impact on energy consumption is inevitable, and green computing approaches become increasingly relevant. The best candidate for video compression is the recently approved Versatile Video Coding (VVC) standard, but its huge computational power requirements have been under investigation to achieve low-energy consumption by means of reducing its algorithmic complexity. This chapter addresses the main sources of computational complexity in the VVC and presents a thorough review of machine learning approaches recently investigated for reducing this burden. Some recent research results, specifically targeted for omnidirectional video surveillance systems, are also presented, including future research directions towards green computing for video coding.

J. Filipe · A. Navarro · S. M. M. de Faria · P. A. A. Assuncao (✉)
Instituto de Telecomunicacoes, Aveiro, Portugal
e-mail: amado@co.it.pt

J. Filipe
e-mail: jose.filipe@ieee.org

A. Navarro
e-mail: navarro@av.it.pt

S. M. M. de Faria
e-mail: sergio.faria@ipleiria.pt

J. Filipe · A. Navarro
University of Aveiro, Aveiro, Portugal

L. Tavora · S. M. M. de Faria · P. A. A. Assuncao
Polytechnic of Leiria, Leiria, Portugal
e-mail: luis.tavora@ipleiria.pt

M. Lahby et al. (eds.), *Computational Intelligence Techniques for Green Smart Cities*, Green Energy and Technology, https://doi.org/10.1007/978-3-030-96429-0_17

Keywords Green video coding · Video surveillance · Video codec complexity · Machine learning · Versatile video coding

1 Introduction

Smart video surveillance is a fast evolving field of research and technological development, where machine learning has been undoubtedly contributing for many of the major recent advances. The challenges brought by massive expansion of surveillance video data will have significant impact on several elements of the near-future intelligent systems, including acquisition, processing, communication, analytics, and storage [1]. Smart surveillance is also being fueled by the most recent 5G communication infrastructures, which combine a great deal of video processing power at network edges along with increased transmission capacity to deliver video data with very low delay [2, 3]. In these intelligent environments, green smart cities are contributing to push forward the requirements of video systems with increasingly higher complexity and dealing with huge amounts of data in energy-constrained deployments. Consequently, efficient video coding plays a major role in achieving the best possible operational conditions, which maximise the quality of visual information using the minimum amount of compressed data and energy consumption.

However, while the progress of video encoding techniques has been enormous in the last decades, the most recent standard encoder, the Versatile Video Coding (VVC) [4], is much more efficient than its predecessors, namely the high efficiency video coding (HEVC) [5], but the required computational power is about one order of magnitude higher. Different studies have demonstrated that, depending on the codec configuration, the computational complexity of VVC can be about 5 to 8 times more complex than its predecessors [6, 7]. The contribution of different coding tools and configurations modes for the overall computational complexity was addressed in [8] for HEVC and in more recent studies for the VVC [9, 10]. For the HEVC standard, the relationship between coding efficiency, energy consumption, and algorithmic complexity was investigated in [11, 12]. For the VVC, relevant evidence about the relationship between energy consumption and computational complexity of software-based video decoders has been recently presented in [13]. The dependency between energy consumption and standards coding tools and configurations was also demonstrated by the same authors, along with modelling and optimisation, in [14] and [15], respectively.

These recent research outcomes demonstrate that computational complexity, including both algorithmic and hardware optimisation, has a direct impact on the energy consumption of modern video codecs. Therefore, the efficiency of future video surveillance systems in green smart cities depends, to a great extent, on successful development and implementation of low-complexity codecs. Furthermore, smart video surveillance may rely on ultra-high definition (UHD) images and video for monitoring and capturing high-quality visual information, which is meant to be delivered and processed through intelligent machine vision algorithms. In massive deployments of smart video surveillance architectures [16], the number of VVC encoders is equal to the number of video sensors, i.e. cameras, scaling up the energy consumption, which becomes a non-negligible factor in the design of green computing environments.

Therefore, since green computing in video surveillance systems is tightly tied to the algorithmic complexity of video encoding, this subject is driving significant research activity worldwide to devise efficient methods for reducing and limiting the computational complexity of standard video encoders. To this aim, machine learning approaches have demonstrated superior performance in reducing the computational complexity of the most recent highly efficient standard video encoders.

In the next sections, this chapter addresses machine learning approaches to implicitly reduce the energy consumption of standard video encoders by means of reducing and controlling the computational complexity. Particularly focusing the most recent VVC standard, the main sources of complexity are firstly identified along with supporting evidence about the contribution of different coding tools. Then a review of the most relevant research works is presented and discussed. A case study presenting a machine learning method for complexity reduction of block split decisions in the complex data structures known as quaternary tree and multi-type trees (QTMTs) is also described. Final conclusions and future research topics are discussed in the last section.

2 Computational Complexity of Video Encoders

The main sources of computational complexity in video encoders, which are implicitly responsible for most of the energy consumption, are the full-search algorithms required for finding optimal coding decisions, i.e. those which achieve the best possible trade-off between compressed rate and image distortion. These decisions are taken after analysing a huge number of possibilities generated by nested video data structures comprising blocks of pixels with different sizes and shapes. Thus, in the design and implementation of energy-aware video encoding systems, these are the crucial aspects that must be analysed and evaluated. This is the aim of Sect. 2.1, where the characteristics of VVC algorithm and the coding tools that can be tackled to reduce the complexity of VVC are described.

2.1 The VVC—Data Structures and Coding Tools

In VVC, a video signal is organised into pictures (or frames), which are divided into several square blocks called coding tree units (CTUs). When compared to the previous standard (HEVC), the maximum size of the CTUs has been increased, from 64×64 pixels to 128×128 pixels. A typical CTU comprises three coding tree blocks (CTBs): one containing the luma information and the other two the corresponding chroma components. During the coding process, CTUs are recursively partitioned into smaller blocks, called coding units (CUs). While in HEVC, this partition is done using a quaternary tree (QT) structure, the VVC uses a quaternary tree multi-type tree (QTMT), which is a process considerably more complex and intricate, as explained below.

Firstly, CTUs (128 × 128 pixels blocks) are always divided, using QT, into four CUs of 64 × 64 pixels. Then, each of these CUs can be further divided into four CUs of 32 × 32 pixels, or not divided at all. For CUs of 32 × 32 pixels or smaller, new partition types are allowed. After using QT partitioning, the block can be divided again by half, either vertically or horizontally, i.e. vertical binary tree (BTV) and horizontal binary tree (BTH), or divided into three parts, also either vertically or horizontally, i.e. vertical ternary tree (TTV) and horizontal ternary tree (TTH). This group of non-square partitions is known as multi-type tree (MT) partitions, as shown in Fig. 1.

An example of partitioning a 32 × 32 pixels CU is illustrated in Fig. 2. The QTMT block partitioning scheme used in the standard allows all partitions represented in Fig. 1. It is important to notice that a CU originated from any MT partition cannot be further split with QT, but it can be split using MT. According to [18], the QTMT partition process is responsible for the majority of VVC's complexity. This is because the optimal partition structure for each CTB can only be found through exhaustive search of the best rate-distortion trade-off, across all possible block combinations.

Another source of complexity is the decision process of intra coding prediction modes, where each frame is encoded on its own without using information from any

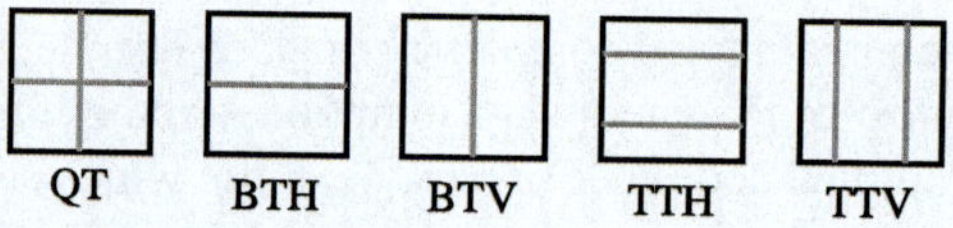

Fig. 1 Multi-type partition schemes used in VVC [17]

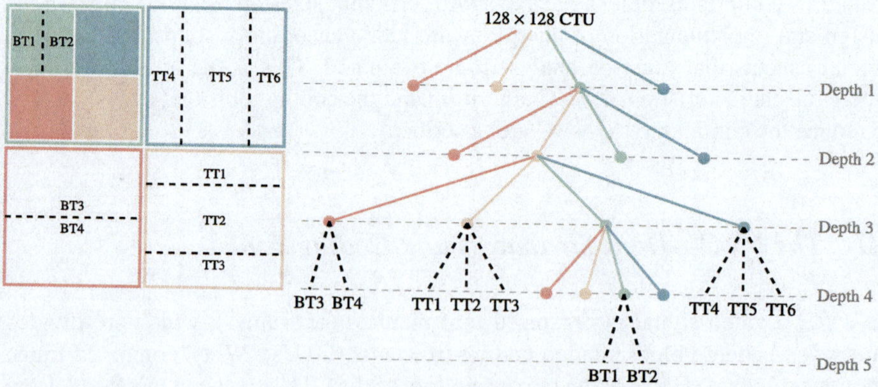

Fig. 2 Example of a 32 × 32 pixels CU partitioned using QTMT. Tree on the right shows partitions that were used

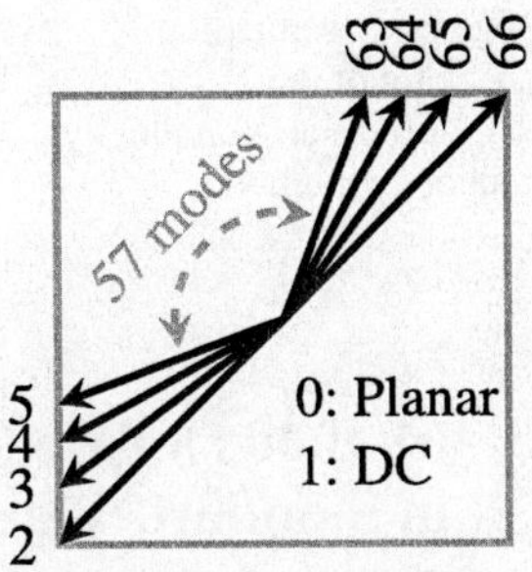

Fig. 3 Directional intra prediction modes in VVC [17]

previously coded frames. The VVC steps up the number of possible intra prediction modes from the previous 35 (planar mode, DC mode, and 33 directional modes) in the HEVC, to 67 (planar, DC, and 65 directional modes) in the VVC. These are schematically shown in Fig. 3. According to the study presented in [18], the increased number of intra prediction modes is the second biggest contributor for the computational complexity of the VVC. This is due to the full-search methods that are commonly used to find the optimum intra mode for each CU, which requires computing the coding rate and distortion across all possible options.

The third major contributor to VVC's complexity is affine motion compensation (AMC). While previous coding standards always assume translational motion for the purpose of motion estimation (ME), the VVC further supports AMC, which includes modelling more complex motion types, such as scaling, shearing, and rotations. In practice, these transformations are handled by computing two or three motion vectors (MV) for each block. Once more, finding the optimum decision requires significant computational complexity due to the large size of the searching space.

Overall, in order to achieve the maximum possible compression efficiency, it is necessary to find the best rate-distortion trade-off for each coding tool from a huge number of possible options, e.g. the best partition structure, the best intra mode or MVs, among others. This is how the various models are adapted to the non-stationary characteristics of each video sequence with the aim of reducing both the spatial and temporal redundancies. In practice, to find the best configuration for each coding tool, the so-called rate-distortion optimisation (RDO) is used. This approach consists in coding each CTU using every possible combination of tools and configurations and also calculating its rate-distortion cost (RD-Cost). The configuration that achieves the lowest RD-Cost is the optimum and thus used to encode the corresponding block.

In general, the research works described in the literature reveal that low complexity and energy reduction approaches usually tackle one or more coding tools, by constraining the optimisation procedures, i.e. not all possible configurations are tested, avoiding exhaustive RDO evaluation for all coding tools and all data structures. As previously discussed [18], the most energy-intensive tools of VVC are QTMT partitioning, intra mode prediction, and AMC. As such, these are the focus of most complexity reduction approaches described in the literature and also the motivation for further research, as the standard algorithms evolve towards higher efficiency.

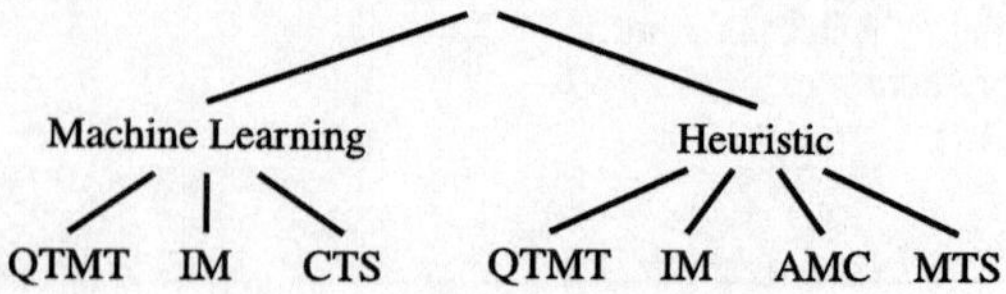

Fig. 4 Classification structure of the state-of-the-art complexity reduction methods

3 Methods for Reduction of Computational Complexity in Standard Video Encoders: A Review

To reduce the computational complexity of the coding tools and associated data structures presented in Sect. 2, which are the most prominent contributors for the complexity burden of video coding standards, several methods have been investigated in the recent past. These can be classified according to the nature of the corresponding algorithms, as based on either machine learning or heuristic approaches. Then, as shown in Fig. 4, a finer classification further splits each approach according to the main coding tool that is targeted to reduce the complexity of the encoding process. To this aim, the most relevant coding tools of the VVC standard are the following: quaternary tree and multi-type tree (QTMT), intra mode (IM), affine motion compensation (AMC), coding tools selection (CTS), and multiple transform selection (MTS). Nevertheless, it is worthwhile to notice that the majority of the works available in the literature are mainly concerned with the QTMT partitioning process. This is due to the fact that this process is considerably more complex than others due to binary tree (BT) and ternary tree (TT) partition types. It is also relevant to point out that, if the partition structures were known a priori for all CTU, then it would be possible to reduce the computational complexity of video encoders by up to 97% [18]. Therefore, there is plenty of room for reducing the computational resources required by advanced video coding algorithms.

3.1 Machine Learning Methods

In general, the methods based on machine learning approaches are developed in two stages, where the first one builds one or more data sets with real data extracted from the process to be modelled. In the second stage, such data is used for learning the model parameters through a training procedure. These models are then used to provide estimates of the optimal rate-distortion decisions, replacing the conventional RDO process that is commonly used to find the best possible coding options for each specific coding tool, such as QTMT, IM, and CTS. Furthermore, there are two main approaches for data extraction and training, according to their relationship with the coding process: in-loop, where the data is extracted and models are trained during the encoding process itself, and off-loop, where the data and models are trained beforehand, i.e. outside the encoding process. Both approaches have advantages and disadvantages. While off-loop approaches tend to achieve higher complexity reduction than in-loop approaches, given that there is no complexity overhead caused by the model training process, these models do not adapt to the time-varying characteristics of video sequences as good as the in-loop methods, resulting in higher coding efficiency loss.

Quaternary Tree and Multi-Type Tree (QTMT) Many of the machine learning approaches rely on features capable of providing discriminative information for the various coding mode decision processes. For instance, the method proposed in [19] uses Bayesian inference to identify situations when multi-type tree (MTT) (i.e. BT and TT partitions) is not used, in order to avoid testing these partition types in the RDO process, thus reducing the coding complexity. In the first stage, the a priori probability of MTT partitions being tested during RDO is estimated. Then, features that are indicative of when MTT is not used are learned by finding those features that decrease the *a posteriori* probability (when compared with the base a priori probability), according to Bayes' theorem. For instance, it was found that, in most cases, if the RD-Cost of a BTV partition is smaller than that of a BTH partition, then the probability of choosing a BTH partition is less than 5%. Therefore, it is possible to discard MTT partitions with low probability of being chosen, to reduce the complexity of the encoding process. Namely it is able to achieve a complexity reduction of 37.00%, with a Bjøntgaard delta rate (BD-Rate) loss of 1.84%.

Other approaches, such as support vector machine (SVM), have been proposed to speed up previous encoders HEVC [20, 21] and motivated further research for the recent VVC. In [22], the authors propose two SVM models to speed up the RDO optimisation process of VVC. The first model acts as a binary classifier to predict whether a CU should be further split, such that early termination of the partitioning process is triggered when such prediction indicates a no-split decision. Otherwise, in case of a split decision, another binary classifier is used to predict whether the vertical or horizontal partitions should be skipped, i.e. testing only the QT and horizontal BT and TT, or the QT and vertical BT and TT partitions, respectively. The features used by both models include, for instance, the quantisation parameter (QP), variance of the CU, difference between vertical and horizontal gradients of sub-blocks of the CU, and maximum magnitude of the gradient vector of the CU. It is worthwhile to notice that, for both classifiers, different models are trained for different CU sizes, in order to increase the accuracy of the models. This partition pruning and early stop method greatly reduces the number of partition configurations tested during RDO, achieving a complexity reduction of 63.16%, with a BD-Rate loss of 2.71%.

Other types of approaches are based on convolutional neural networks (CNNs), which learn how to compute the relevant features from the training data rather than using handcrafted ones. Inspired by ResNet [23], a CNN architecture was devised in [24] to process each 64×64 pixel CUs as the input data unit of the CNN. The proposed CNN produces an output vector, mapping the probability of horizontal and vertical splits. For instance, if a 64×64 block can be split at each pixel row and column, then the CNN outputs a 128-length vector, where the first 64 values map the probability of vertical splitting the CU at the horizontal position given the index of the probability vector. The last 64 values refer to the probability of splitting horizontally, following the same logic. In practice, more possibilities are considered in this work, such as sub-pixel splitting resulting in a 480-length vector. These probabilities are then used to compute the split likelihood of the block using QT, BTH, BTV, TTH, or TTV, and the most probable is then used to partition the 64×64 CU. By determining the full partitioning structure of each 64 CU, without resorting to RDO, this method is able to reduce the computational complexity of the encoder by 51.50%, with a BD-Rate loss of 1.45%.

The split or non-split decision at CU level can also be estimated through a shape-adaptive CNN, as proposed in [25]. This is achieved by using a pooling layer before the CNN itself, whose size varies depending on the shape of the residuals of each CU. Through a combination of pooling and convolutional layers, the input residuals are reduced to a 4×4 matrix that is fed into a fully connected layer. At this point, the information of height and width of the CU, as well as the QP, is added to the fully connected layer, which finally provides the binary output, either split or no-split of the input CU. However, since a CNN is a reasonably complex algorithm, a pre-processing step is used to decide whether the CNN should be used. The decision of split/no-split is immediately taken if the average sum of the gradients squares of the residuals is either above or below two corresponding thresholds (based on the QP). Otherwise, if such value falls between the two thresholds, the residuals of the CU are fed into the CNN that makes the split/no-split decision. This early termination method avoids testing non-optimal partition configurations through the RDO process, reducing the overall complexity by 33.41% and resulting in a BD-Rate loss of 0.99%.

A different approach is to design a CNN to predict the final CU configuration for each CTU. This is proposed in [26], where the input of a six-stage CNN is the luminance of a 128×128 CTU and each stage is designed to apply one partition, thus any given block can only be the result of six partitions at most. The output of each stage is fed to a sub-net to determine whether the CTU block needs to be further split or the process can be early terminated. Finally, the CNN outputs the boundaries of all CUs of a given CTU. This allows the a priori determination of the partitioning structure of a whole CTU, avoiding the RDO process, thus attaining a complexity reduction of 44.65%, with a BD-Rate loss of 1.32%.

Nevertheless, traditional CNNs give rise to complex models, which may penalise the objective of reducing the overall computational complexity either in split/no-split or early termination decisions. Moreover, among all partition types, TT is the one that requires most of the computational effort needed by the QTMT partitioning process. To take advantage of this characteristic, the authors in [27] propose a method to skip redundant TT partitions during the RDO optimisation process, leveraging deep learning (DL) models and features derived from the encoding process such as block size and shape, QT depth, MTT depth, block shape ratio, and intra mode. In essence, only features that can be readily obtained from the running process or required very little processing (as it is the case of the block shape ratio, which only entails dividing the block's height by its width) are used. These features are then fed to a light neural network model, which is composed by a fully connected neural network (NN) with five input nodes, 30 hidden nodes, and one output node. The binary output indicates whether the TT partition should be skipped or not for a given CU. By skipping the test of MTT partitions that prove unnecessary, this method is able to attain an average complexity reduction of 32.00%, with a BD-Rate increase of 0.49%. The complexity of CNNs models is significantly higher than traditional machine learning models, during the inference process, which poses extra research challenges because it is necessary to compensate for their own intrinsic complexity. To cope with such drawback, the CNN and DL-based models are discussed above, either run outside the RDO loop (and therefore are only called once for each CTU) [24–26], or used lightweight versions of CNN/DL [27]. Thus, it is very unlikely that in-loop full fledge CNN/DL models are viable for complexity reduction.

However, while the methods discussed so far in this section tackle complexity reduction for intra coding, others have been investigated for inter coding in VVC. For instance, in [28], a fast method for inter prediction CTUs is proposed. The authors begin by observing the distribution of the RD-Cost for split and no-split CUs. The parameters of both distributions, that is both mean and standard deviation, are learned using maximum likelihood estimation (MLE), by training data obtained from encoding the first frames with unaltered VVC. Then, these parameters are updated every few frames, using the Bayesian rule. Therefore, depending on whether the RD-Cost of a given CU belongs to the split or no-split distribution, an early termination algorithm estimates the best decision based on the confidence interval, which is also used as a tuning parameter to adjust the coding efficiency/complexity trade-off. The RDO process is resumed only in the case where the decision is for split; otherwise, the splitting process is terminated. This method achieves an average complexity reduction of 56.08%, with a BD-Rate loss of 1.30%.

Lastly, the methods proposed in [29, 30] tackle low-complexity QTMT decisions in both intra and inter frames. In [29], random forests classifiers were proposed as binary classifiers to estimate the optimal CU split/no-split, using either QT or BT. The binary classification of random forests models is based upon a predefined set of features, such as QP, pixel variance, vertical and horizontal gradients, variance of four subdivisions of the CU, and the difference between them, among others. In this method, four different models are trained, according with the CU size, i.e. 128×128 pixels, 64×64 pixels, 32×32 pixels, and 16×16 pixels. Nonetheless, no distinction between CTUs that use either inter or intra prediction is made. However, the trained models achieve low accuracy, when compared with similar models previously used in HEVC. In order to improve such limitation, a QP-dependent threshold is introduced so that the decision made by the classifiers is taken into account only when the percentage of decision trees in the random forest that voted in the final classification is higher than this threshold. Otherwise, if it falls under the threshold, the partition type is computed as usual, using RDO full search optimisation. Overall, an accuracy of 98% is achieved. A more flexible implementation of this method was proposed later in [30], by allowing the users to adjust the classifier threshold, and consequently the coding efficiency/complexity trade-off, according to their own criteria. The first method is able to achieve an average complexity reduction of 30.00%

(with respective BD-Rate loss of 0.57%), while the second is able to achieve an average complexity reduction of 30.08% (with a respective BD-Rate loss of 0.70%).

Intra Mode (IM) Machine learning methods have also been explored to find an accurate approximation of the optimal intra prediction mode, in rate-distortion (RD) sense, using fast decision methods to avoid searching overall the 67 possibilities (see Sect. 2.1). These are not completely independent from the QTMT split/no-split decision at CU level because the best intra prediction mode also depends on the CU size. A good example of this approach is described in [31], by first using decision trees (DTs) to decide about split/no-split of a CU and then to estimate the best intra prediction mode to be used in the resulting CUs. In the first step, DTs models are trained for each partition type, converting a multi-classification into multiple binary classification (split/no-split) problems. These models are supported by spatial features such as the spatial gradients of the CUs, the difference of gradients between different regions of the CUs, and information of neighbour CUs, such as total partition depth. In the second step, the best intra mode for a given CU is estimated using gradient descent and Hadamard cost as the objective function. It is worthwhile to notice that the initial search point is very important; otherwise, the convergence time may increase and/or it may converge to a local *minima* that is not the ideal solution regarding coding efficiency. In this method, the initial search point is determined by choosing the most probable mode from the list of modes with minimum Hadamard cost.

A different approach with the same objective of simplifying both decision processes of QTMT partition and intra mode decision is proposed in [32]. For the QTMT partitioning, a binary random forest classifier is used to predict whether or not each CU should be further split. The random forest leverages the four directions of the luminance co-occurrence matrix, the entropy of the CU, the contrast of the CU, as well as the inverse different moments of the CU. Then, intra modes are estimated by first computing the gradients of the CU through Sobel filters and then computing the pixel-wise angle of the gradient vector. The orientation of the texture is, by definition, perpendicular to the gradient vector. Therefore, it is possible to project the pixel-wise luminance orientations in one of four groups (0°, 45°, 90°, and 135°). Since the probability of CU intra mode belonging to that group is given by the integration of each projection over the area of the CU, it is considered that the luminance orientation of the CU belongs to the group that maximises this probability. Finally, the best intra mode is determined by only testing through RDO the modes that belong to the group of a given CU, thus greatly reducing the number of tested modes.

Analysing the results of these methods, Yang et al. [31] achieved a complexity reduction of 63.40%, with a BD-Rate loss of 2.19%, while [32] attained an average complexity reduction of 48.58%, with a BD-Rate loss of 0.91%. Despite these methods cannot be directly comparable, since the first was tested in VVC test model (VTM) 2.0 and the second in VTM 7.0, it is possible to draw some conclusions from the wide gap in performance between them. Regarding the QTMT partition stage, both methods rely on decision trees to learn the best CU structure, by completely avoiding the RDO process. Given that the complexity of both tree-based methods is not usually significant, a difference in the accuracy of the models could explain the significantly worse performance of the methods proposed by Yang et al. [31], regarding BD-Rate loss. However, the biggest difference between them is the methodology used to estimate the intra mode. In fact, the methods proposed by [32] estimate the general direction of the CU and then perform RDO around that direction to find the best intra mode, while [31] leverages gradient descent to directly estimate the best intra mode, avoiding RDO at all, which could explain the better performance complexity-wise of this method, as well as the worse performance in regard to RD-Cost.

Another recent method structured in three processing modules is described in [33]. This first module is specifically tailored for the two new tools introduced in VVC, namely intra sub-partition (ISP) and intra block copy (IBC) modes. This is based on two decision trees (DTs) to decide whether a given CU should use ISP and/or IBC modes. These modes are not tested, i.e. skipped if the DT output is a "no". The decision process is learned from features such as the consistency of the luminance histogram, the number of high and zero gradient pixels in the CU, the context information, the distance of the pixel values, and the maximum gradient magnitude. The second module aims to prune the classical modes and reorder the testing sequence of the most probable modes. To make this decision, an ensemble decision strategy is designed by using a set of indicators of the best intra mode. Then, based on the number of indicators that chose a given mode, MLE is used to compute the probability of the best mode. This way, it is possible to construct a small list of the most probable modes that will be tested through RDO. Finally, the third module introduces an early termination model, using again DTs to decide if a given mode already achieves the optimal RD-Cost. If this is the case, then no other intra modes are tested; otherwise, the next most probable intra mode is tested. These models leverage features in common with the ones used for the first module, with added information about the current RD-Cost, as well as spatial gradients of the CU. It is worthwhile to mention that, in the third module, different models are trained for different CU sizes, in order to improve accuracy. Overall, this method is able to achieve an average complexity reduction of 51.07%, with a BD-Rate loss of 1.21%.

Coding Tools Selection (CTS) Since many of the coding tools introduced by VVC were designed for very high resolutions, such as 4k and 8k, it is plausible to assume that such tools play little to no part to encode low-resolution video and can therefore be disabled, allowing the reduction of complexity of the overall coding process. This is the aim of the research described in [34], which was specifically designed for low-complexity encoding of low resolution and low bitrate video. In order to characterise such tools, a set of video sequences was encoded enabling a single tool at a time, for different video resolutions. This allowed to individually assess the impact of each tool in terms of both computational complexity and bitrate, according to the resolution. Finally, the problem is formulated as a constrained optimisation using a branch-and-prune approach to identify the ideal set of tools that should be disabled in each resolution in order optimise the coding efficiency/complexity trade-off. This dynamic allocation of coding tools results in a complexity reduction of 38.44%, with a BD-Rate loss of 1.45%.

For other types of applications, low-complexity screen content encoding has also been under investigation. This type of visual information has quite different characteristics from natural video. For instance, the high-frequency content is much more important because it is necessary to preserve readability of text, fidelity of graphics, and other computer generated content. For this reason, a couple of coding tools specifically targeting screen content were specifically introduced in the VVC standard. In these types of applications, a possible approach to reduce the overall encoding complexity is to start by classifying the image content in regions of either natural or synthetic content. After such classification, it is possible to adjust the best set of coding tools, avoiding the need to extensively test all coding tools and parameters and therefore achieving complexity reduction of the encoding process. This type of approach was followed in [35], where a CNN is proposed to classify CUs into either natural or computer generated content. An extensive study was carried out to determine which set of tools is more appropriate to each type of CU. If the content is deemed natural, intra prediction and pallet tools are used, while for synthetic content intra block copy and pallet are the primary tools should be used. A confidence interval is also introduced, so that the classification of the CNN is only taken into account when the certainty falls outside the confidence interval. If it falls within, all three tools (intra prediction, intra block copy, and pallet) are used. This method achieves a complexity reduction of 29.88%, with a BD-Rate loss of 2.42%.

3.2 *Heuristic Methods*

In this section, complexity reduction methods based on heuristic algorithms are presented and discussed. Unlike machine learning algorithms, that are trained based on example data, heuristic methods are handcraft based on assumptions or hypotheses that follow from some kind of logic reasoning about the encoding process and a priori statistical analysis.

Quaternary Tree and Multi-Type Tree(QTMT) Heuristic methods to estimate QTMT partitions are based on statistical models derived from sets of representative video sequences. An example of these kinds of method can be found in the work described in [36], where the authors propose to classify the CUs into three categories (T_{CTU}), based on its variance, according to Eq. 1.

$$T_{CTU} = \begin{cases} \text{Simple}, & \sigma \leq \tau_1 \\ \text{Common}, & \tau_2 < \sigma \leq \tau_1 \\ \text{Complex}, & \sigma > \tau_2 \end{cases}, \text{ were } \tau_1 < \tau_2 \qquad (1)$$

If the variance is below a threshold τ_1, then the CU is classified as simple, and an early termination is triggered. If the variance is above a second threshold τ_2, then the CU is classified as complex, and the CU is partitioned (i.e. the skip mode is not used). Otherwise, if the variance of the CU falls between both thresholds, the

CU is classified as common. In this case, no partition type or no-split is discarded a priori. Both τ_1 and τ_2 are dynamically set, based on the probability distributions of splitting and not splitting the CU as a function of its variance. It is worthwhile to notice that the first frame must be encoded with unaltered VVC, in order obtain the first estimation of these distributions.

Furthermore, for common and complex CUs, splitting modes are further pruned, based on Eq. 2, where σ_V is the vertical variance and σ_H is the horizontal variance. If the horizontal variance is greater than the vertical variance, the BTH and TTH partitions are skipped. Conversely, if the horizontal variance is smaller than the vertical variance, the BTV and TTV partitions are skipped.

$$\begin{cases} \text{Skip Horizontal}, & \sigma_H \leq \sigma_V \\ \text{Skip Vertical}, & \text{else} \end{cases} \tag{2}$$

However, this pruning mechanism is only accepted if the absolute difference between the vertical and horizontal variance is smaller than the overall variance, or, in other words, when the horizontal and vertical variances are significantly different. By pruning the number of CU partitions that are tested during the RDO process, it is possible to greatly reduce the computational complexity of the encoding process and implicitly the energy consumption. This method, in particular, achieves an average complexity reduction of about 61.72%, with a BD-Rate loss of 2.53%.

In [37], the authors propose a simplified process, designed for estimating the best intra mode of CUs partitioned using MTT, by leveraging a metric called sum of the Hadamard transform coefficients of the residuals (SATD) and testing only seven main intra modes. The authors refer to this process as SATD-based mode decision. The best intra mode (out of the seven tested) is determined for a CU partitioned with QT (i.e. before any MTT partition mode is tested), by finding the mode that minimises the SATD cost. Then, all possible MTT partitions are tested, and the cost of each partition using the predetermined best intra mode is assessed using once more the SATD cost. Finally, RDO is utilised to refine the intra modes around the previously best mode found, using SATD for each of the spawned MTT CUs. This way, this method is able to strongly prune the number of intra modes to be tested. Only seven cases are tested, followed by a refinement around the best one, and by averaging SATD for computing the MTT partition structure, which is a much less complex metric to compute than RD-Cost. Overall, this method is able to achieve a complexity reduction of about 36.10%, with a BD-Rate loss of 0.69%.

As previously explained, the CTUs are defined as blocks of 128 × 128 pixels, but this size is not directly used for coding, since the VVC standard does not use CUs of this size. The first QT partitioning in four CUs of 64 × 64 pixels is mandatory, and next one must also be QT, but it is decided through RDO optimisation whether the blocks are kept with 64 × 64 pixels or split into four sub-blocks of 32 × 32 pixels. The resulting 32 × 32 pixels block can however be split (or not) using one out of five available options: QT, BT (horizontal and vertical), or TT (horizontal and vertical). Conversely, determining the optimal partition structure for small CUs is relatively fast, and for this reason speed up the splitting decision of blocks with 32 × 32 pixels has been addressed in [38]. This decision process comprises three steps. The first one, inspired by [39], assesses the variance of the 32 × 32 CU. If it is lower than a given threshold, it is deemed that the CU is located in a smooth image region, and therefore, no further splitting is done, avoiding all further test cases. If this is not the case, then the sum of the absolute gradients of the CU, in both horizontal and vertical directions, is determined by using the Sobel operator. If the ratio between highest and smallest gradient is around one (in practice, a threshold is defined) and the individual gradients are not small, then it means that vertical or horizontal features are not very strong within the CU, and therefore, the block is split using QT. If this is not the case, then the CU decision process advances to step three, where the five possible partitions (QT, BTV, BTH, TTV, and TTH) are tested. Then, for each sub-block, the variance is computed, e.g. four variances are obtained for QT, while three are obtained for TT and two for BT. Finally, the block is partitioned with the partition type that maximises this variance. The underlying idea is that partitions with different textures are likely to be split into different sub-blocks, in order to achieve a better prediction performance. This translates into sub-blocks with very different spatial variances, which in turn means that they have very high variance between themselves. The complexity improvements brought by this method are twofold: first, the introduction of an early stop mechanism that avoids unnecessary testing of sub-optimal partition configurations, and second, the introduction of a mechanism to prune the testing of partition schemes when the early termination is not triggered. This method is able to achieve a complexity reduction of 49.27%, with a loss of BD-Rate of 1.63%. The authors in [40] also propose a very similar approach, with a complexity reduction of 53.17%, with a loss of BD-Rate of 1.62%.

Another method based on two adaptive thresholds was proposed in [41], using the angular second moment (ASM) as a measure of texture complexity. If the ASM of a CU is below the lower threshold, then the texture structures are considered simple, and no further divisions of the CU are needed. If the ASM is higher than the highest threshold, then the CU is still texture-wise complex, implying that the CU needs to be split. If the ASM of the CU falls between these two values, the split decision is handled by the regular VVC RDO optimisation procedure. Nonetheless, even when ASM is higher than the highest threshold, the decision to split the CU is made, but at this point there is no information on which type partition to use. To solve this ambiguity, the ration between horizontal and vertical texture complexities is computed to indicate whether the CU texture has a vertical, horizontal, or neutral direction. Therefore, if this ratio indicates that the CU texture has a vertical direction, only BT and TT vertical modes are tested, while in the case of horizontal direction, BT and TT horizontal partitions are tested. Finally, if the direction is neutral, QT partition is directly used to partition the CU. This method also introduces an early termination algorithm and a pruning mechanism, to greatly reduce the number of block partitions tested in the RDO process. It is able to achieve an average complexity reduction of 55.78%, with a BD-Rate loss of 0.96%.

The authors in [42] focus on reducing the complexity of MTT partioning, i.e. only BTH / BTV and TTH / TTV. It is based on a couple of very simple conditions, reducing the number of possible splits modes from four to only two. First, if the horizontal variance of the CU is smaller than the vertical and the intra prediction mode is vertical, or the intra sub-partition is vertical, then horizontal BT and TT should be tested. Otherwise, if the horizontal variance of the CU is bigger than the vertical and the intra prediction mode is horizontal, or the intra sub-partition is horizontal, then vertical BT and TT are tested. If none of these conditions verify, the block is not further split. This partition pruning and early stop method greatly reduces the number of partition configurations tested during RDO, resulting in a complexity reduction of 31.41%, with a BD-Rate loss of 0.98%. Other methods follow similar approaches, such as [43], which is based on early termination for both BT and TT partition types. The main novelty of this work is the use of the difference between the original CU and the its intra mode with the lowest RD-Cost as a basis to extract features, rather than using the luminance of the CTU itself. Then, heuristic conditions are built around the mean and standard deviation of this distortion, and if such conditions are met, the current partition type is early terminated; otherwise, the RDO proceeds in default mode. The heuristic conditions are dependent upon a set of thresholds that can be adjusted to tune the coding complexity/efficiency trade-off. This method achieves an average complexity reduction of 22.60%, with a BD-Rate loss of 0.56%.

Statistical modelling has also been investigated to estimate the bitrate and the RD-Cost of a given CU [44]. This type of approach allows emulation of the RDO process, using statistical models to avoid the much more complex computation of the actual information. For instance, this is much more efficient to calculate than the Hadamard transform that is usually required in the estimation of the true RD-Cost. Therefore, the partition scheme that minimises the estimated RD-Cost is selected, as usually done under traditional RDO. Thus, this method is able to achieve a complexity reduction of 20.02%, with a BD-Rate loss of 0.44%.

While the heuristic methods discussed so far in this section were designed to tackle intra frames specifically, the method proposed in [45] is aimed at intervening in both intra and inter frames. Considering the QTMT partitioning process for both intra and inter blocks, two approaches are proposed in [45] to speed up sequential split/no-split decisions. For intra CTUs, an early termination based on the smoothness of the current CU is first applied. If the variance of the current CU is below a given threshold, early termination is triggered, and the block is not further split. Otherwise, the Canny edge detector is used to extract edges. If either the vertical or horizontal edges of the CTU are dominant, then optimal partitioning, either BT or TT, is more likely to be found in the dominant direction. This reduces the number of possible partitions to test from four to two, since non-square CUs cannot be further split using QT. For inter prediction CTUs, the authors argue that, although it may not be the optimal choice in terms of BD-Rate, if a given inter CU is mostly similar to a block located in either of the reference frames, then the loss of BD-Rate is small, because such block can be easily predicted, and the search for the optimal partition can stop early, avoiding exhaustive testing. This is implemented by first computing binary maps from the pixel-wise absolute difference between the current frame and each of its two references, where the zeros represent similar regions between the current frame and its corresponding reference. After merging the two binary maps through a logical "AND", the binary values of the merged map are inverted, such that the normalised integration over a region indicates the percentage of pixels that are similar to at least one of the reference frames. Then, if this map is integrated for the coordinates and shape of the CU being partitioned, and at least 95% of the pixels are equal to the references, then the partitioning process is terminated, because the block can be easily predicted. Otherwise, the Canny edge method proposed for intra modes is used to split this CU, and the similarity between these new blocks and the references is checked again. By severely limiting the number of hypothesis tested by the RDO approach, both in intra and inter prediction CTUs, this method is able to reach an average complexity reduction of 33.41%, with a BD-Rate loss of 0.99%.

Intra Mode (IM) Heuristic methods based on statistical analysis and modelling are the most commonly used in the current conventional approaches that are being progressively replaced by more efficient models based on machine learning. In the case of VVC, there is an exploratory work using an early test model for VVC using only quaternary ree and binary tree (QTBT) partitioning, which was developed for JEM. This model, described in [46], presents a statistical study on the most frequently used modes at each BT depth as a function of the QP. Then, a list of the six most probable modes to be tested during RDO is obtained. Since the probability of finding the optimal mode within this set of modes is high, the need to exhaustively test all modes is thus avoided. Thus, by reducing the number of intra modes tested by RDO, this method is able to achieve a complexity reduction of 56.32%; however, the BD-Rate loss is not available.

Affine Motion Compensation (AMC) As mentioned before, affine motion compensation (AMC) is also a major contributor for the overall complexity of the VVC standard, which motivates research efforts towards low-complexity implementations with minimum efficiency loss. A relevant work is presented in [47], proposing a method based on Bayesian inference. It is demonstrated that the *a posteriori* probability of using AMC greatly decreases when the parent CU has chosen the skip mode, and the best motion estimation mode is unilateral prediction. Therefore, when both of these cases are verified, the AMC can also be skipped with no efficiency loss. Then, a further improvement was designed to allow sub-CUs in recursive block partitioning to use the information already gathered from previous sub-CUs. By adopting this strategy, redundant testing of AMC vectors is avoided when trying out the various types of block partitioning (QT, BT, TT) in subsequent nodes. Another recent research work that is worthwhile to mention is presented in [48], which consists in a experimental study on the efficiency of several coding tools such as AMC, integer motion vector, and generalised bi-prediction. This is done by determining and comparing the coding complexity and efficiency of frames encoded with these tools enabled and disabled, one at a time and per CTU. This study concludes that AMC and integer motion vector tools bring negligible coding efficiency gains in borderline CTUs (CTUs that are near the limits of the frames), at the cost of significant complexity. This is because both tools are designed to accurately describe complex motions between frames, but very often objects near the frame boundaries are not present in neighbouring frames. Therefore, the encoder takes time trying to match a block between frames, where it might not exist, making these tools increase the amount of time it takes doing so. Therefore, the authors propose to exclude these tools for the borderline CTUs.

Both these methods achieve significant complexity reduction by disabling AMC under some specific conditions where the improvement of coding efficiency is consistently low. Furthermore, the method proposed in [47] achieves even lower complexities by using the MV of parent blocks as the starting search point for the MV of the current CU, promoting a potentially faster convergence to the best MV. On the other hand, the method proposed in [48] tackles not only AMC, but also some other tools that prove to be unhelpful in improving the coding efficiency specifically at frame boundaries, such as triangular partition mode (TPM), bidirectional optical flow (BIO), and generalised bi-prediction (GBI). Thus, the first method attains a complexity reduction of 5.00% (and a BD-Rate loss of 0.10%), while the method proposed in [48] attains a complexity reduction of 28.31% (and a BD-Rate loss of 0.64%). It is worthwhile to notice that these two methods are somewhat complementary and can be implemented together.

Multiple Transform Selection (MTS) As previously stated, VVC includes the option of using different transforms, such as DST7 and DCT8, to compact the energy of the residue signal of a given CU. This allows the encoder to chose the transform that optimises the RD cost for a given CU. However, this means that all possible transforms must be evaluated for each different configuration of the remaining coding modes MV and partition structures, which requires a significant number of tests in the RDO process. Thus, if the transform to be used in each CU is known a priori, then a significant complexity reduction will be achieved, as described next.

A first statistical study conducted for JEM (the early version of VVC) in [49] provided evidence to support the claim that in CUs partitioned with BT, the DCT5 and DST1 transforms are rarely selected by the RDO process. Therefore, the authors propose their removal, simplifying the transform selection process, which yields an average complexity reduction of 41.05%. Nevertheless, the authors do not present the BD-Rate loss. Another method to simplify the decision process between DST7 and DCT8 transforms was proposed in [50], based on a two-stage algorithm. The first stage verifies if the sum of the RD-Cost of the sub-CUs, computed only with the primary transform (DCT-II), is higher than the RD-Costs of the non-partitioned CUs. If this is the case, there is no need to test another transform, since there is a low probability that the transform alone would change the RD-Cost. The second stage targets the best intra mode, which should be determined using the RDO procedure, but instead of testing all possible transforms for each mode, as usual, this is done by only using the primary transform. Then, after determining the best intra mode, a list of the most probable transforms is built by checking the transforms used in neighbouring CTUs and sorting them by relative frequency. As the most frequent transform in the neighbourhood is the most likely to be chosen, this is the first one to be tested. If its RD-Cost is lower than the DCT-II transform, then the second most frequent transform is tested. If the RD-Cost is lower than the RD-Cost of the first transform of the list, the third transform is tested; otherwise, the process is stopped and the first transform of the list is used, and so on. This simplification of the number of tests performed by the RDO process, during the evaluation of the intra mode, yields an average complexity reduction of 23.00% (resulting in a BD-Rate loss of 0.16%). This meaningful reduction in complexity, allied with the almost negligible loss in BD-Rate, results in the best method analysed, when it comes to complexity reduction by percentage of BD-Rate loss.

3.3 Comparison of Different Approaches for Reduction of the Computational Complexity of VVC

Since different approaches have been used with the aim of reducing the computational complexity of video coding algorithms, their performance must be evaluated by weighting the combination of two adversarial metrics: the actual computational complexity reduction, defined as the average reduction of the encoding time in comparison with a reference implementation, and its counter-effect in coding efficiency, usually measured in BD-Rate loss. Since, in general, there is a trade-off between reduction of computational time and coding efficiency loss, it is useful to merge both into a single metric in order ease performance comparison. The ratio between the percentage of computational complexity reduction and BD-Rate has been used for this purpose, which represents the amount of relative complexity reduction per 1% of BD-Rate loss. These metrics are shown in Table 1 for all methods cited and discussed in previous sections, sorted from the highest complexity percentage gain per 1% of BD-Rate loss, to the lowest.

In previous sections, the methods have been classified according to the algorithmic approach, either heuristic methods or based on machine learning. Then, these are further divided according to the coding tools, either intra coding, inter coding, inter and intra, or transform selection. The first two classes (intra coding and inter coding) have been further divided according to the specific coding modes, such as QTMT partitioning, intra mode, and AMC. Naturally, since the QTMT partition is mainly responsible for the VVC's complexity, as discussed in [18], this is also the focus of most publications, as can be seen in Table 1. In fact, regarding QTMT in intra coding, eight heuristic methods were analysed, with an average complexity reduction of 41.26% and a BD-Rate loss of 1.18%. Regarding machine learning methods, six works were analysed, with an average complexity reduction of 43.62% and a BD-Rate loss of 1.47%. When it comes to reduce the complexity of optimal intra prediction using machine learning, three methods were discussed, with an average complexity reduction of 54.35% and a BD-Rate loss of 1.44%. Regarding heuristic methods for reducing complexity of the AMC estimation in inter frame coding, two works were analysed, with an average complexity reduction of 16.66% and a BD-Rate loss of 0.37%. Concerning both intra and inter frame prediction simultaneously, two machine learning methods were discussed, with an average complexity reduction of

Table 1 Summary of the results achieved by the studied methods

Method	Tool	Type	Coding	BD-rate increase (%)	Complexity reduction (%)	Ratio
[50]	IM	H	Intra	0.16	23.00	143.75
[27]	QTMT	ML	Intra	0.49	32.00	65.31
[41]	QTMT	H	Intra	0.91	48.58	53.38
[32]	IM	ML	Intra	0.96	55.78	58.10
[29]	QTMT	ML	Both	0.57	30.00	52.63
[37]	QTMT	H	Intra	0.69	36.10	52.32
[47]	AMC	H	Inter	0.1	5.00	50.00
[44]	QTMT	H	Intra	0.44	20.02	45.50
[48]	AMC	H	Inter	0.64	28.31	44.23
[28]	QTMT	ML	Inter	1.3	56.08	43.14
[30]	QTMT	ML	Both	0.7	30.08	42.97
[33]	IM	ML	Intra	1.21	51.07	42.21
[51]	QTMT	ML	Intra	1.37	56.25	41.06
[43]	QTMT	H	Intra	0.56	22.60	40.36
[24]	QTMT	ML	Intra	1.45	51.50	35.52
[26]	QTMT	ML	Intra	1.32	44.65	33.83
[25]	QTMT	ML	Intra	0.99	33.41	33.75
[40]	QTMT	H	Intra	1.62	53.17	32.82
[42]	QTMT	H	Intra	0.98	31.41	32.05
[38]	QTMT	H	Intra	1.63	49.27	30.23
[31]	IM	ML	Intra	2.19	63.40	28.95
[34]	TS	ML	Both	1.45	38.44	26.51
[45]	QTMT	H	Both	1.34	31.43	23.46
[22]	QTMT	ML	Intra	2.71	63.16	23.31
[36]	QTMT	H	Intra	2.53	51.72	20.44
[19]	QTMT	ML	Intra	1.84	37.00	20.11
[35]	TS	ML	Both	2.42	29.88	12.50
[46]	IM	H	Intra	–	56.32	–
[49]	QTMT	ML	Intra	–	41.05	–

30.04% and a BD-Rate loss of 0.64%. In the category of coding tools selection methods, two methods were analysed, with an average complexity reduction of 34.16% and a BD-Rate loss of 1.95%. The remaining categories only had one work each, thus there were no collective performance to analyse. However, from this collective analysis, as well as from Table 1, it is possible to conclude that, although QTMT partitioning is the main contributor to VVC's complexity [18], the methods that target low-complexity intra mode estimation are able to achieve greater reductions than others. In fact, regarding the relative complexity reduction per 1% of BD-Rate loss, the methods targeting intra mode estimation outperform those that target intra QTMT partitioning, no matter whether heuristic or machine learning methods are considered.

Furthermore, it is also interesting to analyse the comparative performance of heuristic and machine learning methods aiming to reduce the computational complexity of intra QTMT partitioning. Although the latter are able to achieve marginally higher reductions of complexity, the former cause a smaller impact in the BD-Rate, coming out ahead in performance measured as relative complexity reduction per 1% of BD-Rate loss. In fact, the best performing methods are those based on machine learning targeting both intra and inter coding simultaneously.

4 Complexity-Aware Omnidirectional Video Coding: Case Study

The specific case of omnidirectional video coding has been addressed on its own due to the particular characteristics of planar projections, such as equirectangular projection (ERP), as well as the UHD resolutions. In regard to complexity-aware encoding of omnidirectional video, most research carried out so far deals with the HEVC, thus to a great extent this is yet to be explored for the VVC standard. The approach described in [52] was specifically designed to reduce intra coding complexity of omnidirectional video in ERP format using HEVC. It is shown that in ERP images, the increased horizontal redundancy in the regions near the poles leads to less than 18 significant intra modes. Then, based on such characteristics, two methods are proposed to speed up the coding process: one for fast intra mode decision and an another for early termination of split/no-split decisions. The results show that complexity can be reduced by 24.50% at the cost a negligible loss of 0.20% in coding efficiency. A similar research problem was addressed in [53], also exploiting the fact that the sampling density is higher in regions near the poles in order to avoid motion estimation with quarter pixel resolution. It is proposed to gradually decrease the resolution of the MVs with respect to the increase of the latitude. Furthermore, it is shown that, due to the oversampling near the poles, CUs tend to be larger near the poles and smaller near the equator. Based on this observation, an algorithm is devised to dynamically change the minimum CU size, according to the latitude. The method proposed in this work is able to achieve a complexity reduction of about 15.00%, with a BD-Rate increase of only about 0.50%.

Other methods, proposed in [54], also target reduction of computational complexity for ERP video using HEVC. A scoring system is devised based on a reduced list of intra modes that most likely include the best mode for each CU, thus reducing the number of modes evaluated during the RDO process. To exploit the ERP characteristics, frames are divided into three and five horizontal bands. The scoring system is based on three types of information: (i) frequency of occurrence of a given mode in the same band and prediction unit (PU) size, (ii) correlation between the mode decision of the rough mode decision (RMD) and the final decision of the RDO process, and (iii) whether the intra mode was also derived as a most probable mode. Overall, this work was able to achieve a complexity reduction of about 15.20%, with a BD-Rate increase of 0.40%

Finally, to the best of the authors knowledge, this is the only research work dealing with the VVC and ERP video sequences. In, [51], three machine learning models [either RF or extremely randomised trees (ERT)] were proposed to predict the maximum partition depth for each partition type, at the CTU level. The features leveraged by these models are specially designed to take advantage of the geometric nature of ERP projection. If the maximum depth is reached in a given CTU for a given partition type, no further partitions of that same type are further tested by the RDO process. This method is able to achieve an average complexity reduction of 56.25%, with a BD-Rate loss of 1.37%. Although the method was designed and tested for VVC, it is still possible to conclude that, when compared with the remaining works discussed in this section, which were designed for HEVC, this method provides significantly higher complexity reductions, at a reasonable increase of the BD-Rate loss.

As pointed out before, so far there are not many research works focused on complexity reduction of omnidirectional video and even less leveraging the new specific tools introduced in the VVC standard. Furthermore, the potential for exploring other types of projections, such as Cubemap Projection (CMP), is still open for research. Additionally, while in complexity reduction for regular video, machine learning approaches are quite common, and in omnidirectional video, these approaches have not been fully investigated, which also opens room for further research.

4.1 *Performance Evaluation Study: HEVC Versus VVC*

The actual computational complexity of standard video coding obviously depends on each particular implementation and hardware platform. Nevertheless, it is quite useful for researchers and engineers to know the evolution of standard coding algorithms and to establish common references for comparison and performance evaluation. To this aim, this section presents a performance evaluation study between HEVC and VVC, for omnidirectional video, including quality, bitrate, and computational complexity. Then a method to reduce the computational complexity of omnidirectional VVC based on machine learning is presented and discussed.

In this study, all omidirectional video sequences from the common test conditions [55] were encoded, using both HEVC and VVC using the main 10 and next profiles, respectively. Four different QPs (22, 27, 32, and 37) were used with random access configuration. The quality was measured using PSNR, S-PSNR, WSPSNR [55], and nearest neighbour interpolation for the luma channel. Since spherical images are always represented in some planar projection, which may influence the coding performance, in order to eliminate any possible bias, the source images should be reprojected back to the spherical domain, downsampled, and then converted to the target projection to be encoded [56]. After decoding, all images can be converted again to the sphere, upsampled, and finally converted to the target projection and resolution. All the performance metrics used in this study were calculated between the downsampled reference sequence (before coding) and the decoded sequence (before upsampling). However, the E2E-WSPSNR (end-to-end WSPSNR) is measured between the reference and decoded images in their original resolutions. Then, the Bjøntgaard delta rate (BD-Rate) is calculated for each quality metric [57] using the HEVC as reference. Furthermore, the computational complexity of both encoders is also assessed, by measuring the time that each encoder spent to encode the same sequence with the same QP, running in the same hardware platform. The average encoding time per sequence for the four QPs is normalised to the average processing the time of HEVC. This is expressed by (3), where TN represents the normalised time, $\mu(\cdot)$ denotes the average, and t_i^m denotes the processing time spent by encoder m (VVC or HEVC) to encode a video sequence with QP i.

$$TN = \frac{\mu([t_{22}^{VVC}, t_{27}^{VVC}, t_{32}^{VVC}, t_{37}^{VVC}])}{\mu([t_{22}^{HEVC}, t_{27}^{HEVC}, t_{32}^{HEVC}, t_{37}^{HEVC}])} \tag{3}$$

The results of this experimental study are shown in Table 2. For every metric, the VVC saves on average at least 41% in bitrate when compared with HEVC. Furthermore, higher savings are obtained when evaluating the quality of the image via PSNR. This is due to the fact that PSNR treats all pixels of the video sequences equality, disregarding the spherical nature of the ERP representation, resulting in regions near the equator being regarded as equally important as those near the poles. *Ergo* distortions near the equator are not as important for the overall quality, resulting in a inflated BD-Rate. Moreover, both WSPSNR and S-PSNR present very identical results, while E2E-WSPSNR presents a slightly worse BD-Rate. This was expected, since this metric includes not only the distortions introduced by the encoding process, but also the upsampling distortions.

However, as expected, the superior coding efficiency of VVC is obtained at the cost of significantly higher computational complexity. On average, the computational time of VVC is 4.96 times longer than HEVC, but the maximum is 9.10 times in this simulation study (BranCastle sequence). Nevertheless, this increase in complexity is highly content dependent, as can be seen in column TN of Table 2. This is explained by the fact that many of the coding tools that most contribute for increasing the coding efficiency depend on the CU partitioning stage applied at the CU level. Consequently, scenes with higher complexity have smaller CUs, resulting in more CUs, while simpler scenes are encoded with bigger CUs, resulting in a smaller number of CUs. Therefore, since the coding tools are applied at the CU level, sequences with more CUs contribute more to the overall computational complexity.

Table 2 Comparison of HEVC and VVC—BD-rate (%) and normalised time

Sequence	PSNR	WSPSNR	S-PSNR	E2E-WSPSNR	*TN*
Landing2	44.70	43.34	43.36	43.16	6.64
Trolley	27.77	25.92	25.92	24.99	3.50
ChairliftRide	63.53	59.05	58.98	58.88	4.52
Balboa	66.56	65.15	65.12	62.53	4.67
Harbour	30.23	28.11	27.98	27.56	3.23
Broadway	61.30	59.85	59.79	57.28	5.16
Gaslamp	33.18	32.16	32.22	31.49	2.08
SkateboardInLot	47.92	46.83	46.81	46.36	6.00
BranCastle	35.77	35.13	35.13	38.58	9.10
KiteFlite	27.51	26.20	26.32	25.04	4.73
Average	43.85	42.17	42.16	41.75	4.96

4.2 A Machine Learning Approach for Low-Complexity Coding of Omnidirectional Video

This section presents a low-complexity method to encode omnidirectional video, addressing the case of QTMT partitioning in intra frames. This is based on multiple random forests (RF) which learn, from different types of features, to distinguish block partitions that can be skipped from those that are likely to be the best coding option. The split/no-split decision is modelled as a binary classification problem for each type of CU by using different RF classifiers for each type of partitioning.

In this case study, two different machine learning approaches are proposed. Given that we are targeting omnidirectional video, it is known that the most common projection (ERP) suffers from distortions near the poles [58]. Therefore, regions near the poles tend to be less spatially complex due to oversampling than regions near the equator. It is worthwhile to take into consideration that, in this work, the regions near the poles are defined as the regions with latitude comprehended in the intervals [45°, 90°[and [−45°, −90°[, while the region in]45°, −45°[is considered the equator region. This means that complexity reduction methods can be more aggressive in the equator regions than in the regions near the poles. For these reasons, we have used the method proposed in [51] to early terminate the QTMT process in the equator region. This method consists in using three RF models, one for each partition type (i.e. QT, BT, and TT), in order to predict the maximum depth achieved by each partition type in a given CTU, based on a set of spatial and coding features extracted from that block. This means that the prediction is made only once per CTU, which results in a very low complexity overhead. Once the predicted maximum depths are reached during the QTMT partition process, no further partitions of that type are used in that CU, greatly reducing the number of QTMT configurations that are tested during RDO.

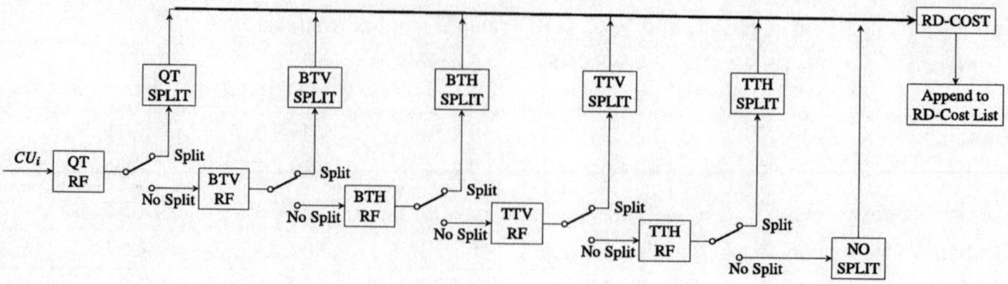

Fig. 5 Early split/no-split decision process to avoid exhaustive calculation of all RD-Costs

To tackle the pole regions, five other RF models (for QT, BTV, BTH, TTV, and TTH partitions) are used to decide, at CU level, if that block should or not be partitioned using the respective partition type. The classifiers are inserted in the RD-Cost calculation loop, as shown in Fig. 5. For a given CU i, all partition methods are tested sequentially, as shown in the image. Traditionally, an RD-Cost is calculated for each partition type, which is a complex operation. The proposed split/no-split decision methods intervene in the process right before each of the partitions types is tested in the conventional RD loop. The RF models are used to determine whether each CU is worth split using a specific partition type or not. If the models deem it is better to be partitioned, then the CU is split, and an RD-Cost is computed. However, if the RF modes classify a partition type for the given CU as no-split, then that CU is skipped altogether, and no RD-Costis computed, avoiding wasting resources with sub-optimal partitions.

This implementation approach introduces some extra complexity overhead, since the prediction models are used much more often. However, instead of reducing the number of possible partitioning configurations, this approach fully replaces the RDO process by a much simpler function during QTMT partition, resulting in a great deal of complexity reduction. These classification models leverage a set of coding and spatial features, that are detailed in Table 3, to make the split/no-split classification. These features are first extracted from the training data to build (training) the RF classification model, and then they are computed from the actual data to be encoded while running the low-complexity encoder. Note that the computational resources required to calculate these features are quite low, thus not contributing to significant extra overhead. The models of the case study presented in this section are trained and implemented following the methodology detailed in [59].

It is also worthwhile to point out that RF models are composed by several individual decision trees (in this work, 50 trees were used) that make the decision of whether a CU should be split. The decision of the whole RF model is then obtained by majority voting of the individual decision trees, as shown in Fig. 6. In order to limit the impact on the bitrate, in this case the no-split decision is only taken when a majority of 85% or above is reached.

Results In Fig. 4, the results obtained by the method described above are shown, in regard to complexity reduction and bitrate overhead. On average, a complexity

Table 3 List of features used by all five models

Name	Description	Models
f_1	Estimated number of bits needed to encode CU	QT, BTH, TTH
f_2	Width of CU	QT, QT, BTV, TTV
f_3	Depth of QT partitioning so far	QT
f_4	Total partitioning depth so far	QT
f_5	RD-Cost of parent CU	QT
f_6	Height of CU	QT, BTV, TTV
f_7	Distortion based on Hadamard transform	QT
f_8	Ratio between CU height and width	BTV, TTV
f_9	Std. Dev. of horizontal component of CU gradient	BTV, TTV
f_{10}	Std. Dev. of horizontal component of parent CU gradient	BTV, TTV
f_{11}	Std Dev. of magnitude of CU gradient	BTV, TTV, TTH
f_{12}	Current RD-Cost of CU	BTH
f_{13}	Estimated number of bits needed to encode parent CU	BTH
f_{14}	Std. Dev. of vertical component of CU gradient	BTH, TTH
f_{15}	Std. Dev. of magnitude of parent CU gradient	TTV
f_{16}	Std. Dev. of vertical component of parent CU gradient	TTH

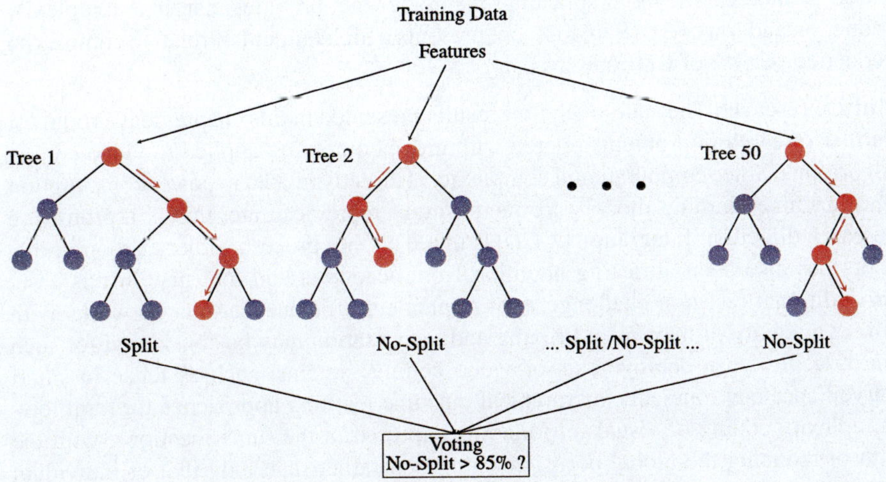

Fig. 6 Illustration of the voting mechanism used by random forests

Table 4 Summary of the results achieved by the studied methods

Sequences	BD-Rate increase (%)	Complexity reduction (%)	Ratio
Harbour	−6.25	71.39	11.42
Landing2	−6.68	80.44	12.04
Skateboard	−3.40	68.34	18.47
Gaslamp	−6.93	72.21	10.42
Average	−5.89	73.10	13.09

reduction of about 73.10% can be attained, with a coding efficiency loss of about 5.89%. It is worthwhile to notice that the average complexity reduction is higher than all the methods shown in Table 1, but so is the BD-Rate loss, which results in lower amount of relative complexity reduction per 1% of BD-Rate loss. This also highlights the downside of measuring performance as the relative complexity reduction, because methods with low-complexity reduction and BD-Rate losses much smaller than 1 yield high scores, while methods with high complexity reductions and BD-Rate losses greater than 1 achieve a low score. In fact, a BD-Rate loss of 6% still gives the VVC room to outperform HEVC by more than 30%, according to Table 2, while benefiting from a complexity reduction of over 70%. Depending on the application, this RF-based method can be more appealing than other that provides marginal complexity reduction and lower BD-Rate loss, but presents a high amount of relative complexity reduction per 1% of BD-Rate loss.

Future research The state-of-the-art results presented in this chapter leave room for further research in problems related with green computing applied to video coding algorithms. Since computational complexity is directly related to power consumption and machine learning models are increasingly more accurate, this is a promising research direction. Integration of CNNs in the coding process, either in-loop or off-loop, has also been attracting attention from academia and industry. In this case, an additional research challenge is to implement efficient learning networks with low complexity of their own. Pruning and quantisation may be considered for such purpose. In dense deployments of video coding systems, such as those in smart surveillance environments, hierarchical machine learning approaches for joint low-complexity coding of visual information, captured in the same locations, with the aim of reducing the global power consumption rather than only that of individual encoders are also seen as a promising research direction towards green computing in video coding standards.

5 Conclusions

Over the years, video encoders have become increasingly more power hungry, in the pursuit of higher and higher coding efficiencies [60]. As shown in many studies [61–63], this increased demand for complexity is intrinsically tied with the exponential increase in the number of possible configurations and coding tools, that are extensively and recursively exploited, in order to find the optimal coding mode for blocks of different size. Thus, efficient methods for pruning a priori some of the possible coding modes and block sizes were presented in this chapter to reduce the computational complexity of standard video encoders, such as the VVC. For this purpose, a machine learning approach based on RF was proposed to significantly reduce the complexity (and power consumption) of the VVC standard, i.e. up to 73%.

The contribution of energy-efficient video coding systems for green smart cities greatly depends on their computational complexity. Therefore, optimised methods for finding the best performance, in terms of rate-distortion complexity, are of utmost importance to achieve overall objectives of energy efficiency and sustainable technological development. Some remarks and possible routes for future research work in this direction were drawn in the last subsection of this chapter.

Acknowledgements The authors would like to acknowledge the financial support of Programa Operacional Regional do Centro, Project ARoundVision CENTRO-01-0145-FEDER-030652, FCT/MCTES Grant 2020.06341.BD and Project UIDB/EEA/50008/2020 through national funds and when applicable co-funded EU funds, Portugal.

References

1. Shao, Z., Cai, J., Wang, Z.: Smart monitoring cameras driven intelligent processing to big surveillance video data. IEEE Trans. Big Data **4**(1), 105–116 (2018)
2. Rao, S.K., Prasad, R.: Impact of 5G technologies on smart city implementation. Wirel. Pers. Commun. **100**(1), 161–176 (2018)
3. Bellavista, P., Chatzimisios, P., Foschini, L., Paradisioti, M., Scotece, D.: A support infrastructure for machine learning at the edge in smart city surveillance. In: 2019 IEEE Symposium on Computers and Communications (ISCC), pp. 1189–1194 (2019)
4. Bross, B., Chen, J., Liu, S.: JVET-M1001: versatile video coding (draft 4). In: Joint Video Experts Team (JVET), 13th Meeting: Marrakech, MA, Tech Rep, p. 1 (2019)
5. Sullivan, G.J., Ohm, J.-R., Han, W.-J., Wiegand, T.: Overview of the high efficiency video coding (hevc) standard. IEEE Trans. Circ. Syst. Video Technol. **22**(12), 1649–1668 (2012)
6. Filipe, J., Carreira, J.F.M., Tavora, L., Faria, S.M.M., Navarro, A., Assuncao, P.A.: Omnidirectional video coding: new coding tools and performance evaluation of vvc. In: Conference on Telecommunications—ConfTele (2019)
7. Zhang, F., Katsenou, A.V., Afonso, M., Dimitrov, G., Bull, D.R.: Comparing VVC, HEVC and AV1 using objective and subjective assessments (2020)
8. Correa, G., Assuncao, P., Agostini, L., da Silva Cruz, L.A.: Performance and computational complexity assessment of high-efficiency video encoders. IEEE Trans. Circ. Syst. Video Technol. **22**(12), 1899–1909 (2012)
9. Bossen, F., Sühring, K., Wieckowski, A., Liu, S.: VVC complexity and software implementation analysis. IEEE Trans. Circ. Syst. Video Technol., pp. 1–1 (2021)

10. Pakdaman, F., Adelimanesh, M.A., Gabbouj, M., Hashemi, M.R.: Complexity analysis of next-generation VVC encoding and decoding. In: 2020 IEEE International Conference on Image Processing (ICIP), pp. 3134–3138 (2020)
11. Monteiro, E., Grellert, M., Zatt, B., Bampi, S.: Rate-distortion and energy performance of HEVC video encoders. In: 24th International Workshop on Power and Timing Modeling. Optimization and Simulation (PATMOS), vol. 2014, pp. 1–8 (2014)
12. Shafique, M., Henkel, J.: Low power design of the next-generation high efficiency video coding. In: 2014 19th Asia and South Pacific Design Automation Conference (ASP-DAC), pp. 274–281 (2014)
13. Kränzler, M., Herglotz, C., Kaup, A.: A comparative analysis of the time and energy demand of versatile video coding and high efficiency video coding reference decoders. In: 2020 IEEE 22nd International Workshop on Multimedia Signal Processing (MMSP), pp. 1–6 (2020)
14. Kränzler, M., Herglotz, C., Kaup, A.: Decoding energy modeling for versatile video coding. In: 2020 IEEE International Conference on Image Processing (ICIP), pp. 3144–3148 (2020)
15. Herglotz, C., Bader, M., Fischer, K., Kaup, A.: Decoding-energy optimal video encoding for x265. In: 2020 IEEE 22nd International Workshop on Multimedia Signal Processing (MMSP), pp. 1–6 (2020)
16. Shorfuzzaman, M., Hossain, M.S., Alhamid, M.F.: Towards the sustainable development of smart cities through mass video surveillance: A response to the COVID-19 pandemic. Sustain. Cities Soc. **64**, 102582 (2021). [Online]. Available: https://www.sciencedirect.com/science/article/pii/S2210670720308003
17. Chen, J., Ye, Y., Kim, S.H.: JVET-M1002: algorithm description for versatile video coding and test model 4 (VTM 4). In: Joint video experts team (JVET), 13th Meeting: Marrakech, MA, Tech. Rep., 1 (2019)
18. Tissier, A., Mercat, A., Amestoy, T., Hamidouche, W., Vanne, J., Menard, D.: Complexity reduction opportunities in the future VVC intra encoder. In: IEEE 21st International Workshop on Multimedia Signal Processing, MMSP 2019 (2019)
19. Park, S.H., Kang, J.W.: Context-based ternary tree decision method in versatile video coding for fast intra coding. IEEE Access **7**, 172597–172605 (2019)
20. Liu, Y.-C., Chen, Z.-Y., Fang, J.-T., Chang, P.-C.: Svm-based fast intra cu depth decision for hevc. In: 2015 Data Compression Conference, pp. 458–458 (2015)
21. Grellert, M., Zatt, B., Bampi, S., da Silva Cruz, L.A.: Fast coding unit partition decision for hevc using support vector machines. IEEE Trans. Circ. Syst. Video Technol. **29**(6), 1741–1753 (2019)
22. Wu, G., Huang, Y., Zhu, C., Song, L., Zhang, W.: SVM based fast CU partitioning algorithm for VVC intra coding. In: IEEE International Symposium on Circuits and Systems (ISCAS), no. Mic, vol. **2021**, pp. 1–5 (2021)
23. He, K., Zhang, X., Ren, S., Sun, J.: Deep Residual Learning for Image Recognition (2015)
24. Tissier, A., Hamidouche, W., Vanne, J., Galpin, F., Menard, D.: CNN oriented complexity reduction of VVC intra encoder. In: Proceedings—International Conference on Image Processing, ICIP, pp. 3139–3143 (2020)
25. Tang, G., Jing, M., Zeng, X., Fan, Y.: Adaptive CU split decision with pooling-variable CNN for VVC intra encoding. In: 2019 IEEE International Conference on Visual Communications and Image Processing, VCIP 2019, pp. 2019–2022 (2019)
26. Li, T., Xu, M., Tang, R., Chen, Y., Xing, Q.: DeepQTMT: a deep learning approach for fast QTMT-based CU partition of intra-mode VVC. IEEE Trans. Image Process. **7149**(c), 1–14 (2020). [Online]. Available: http://arxiv.org/abs/2006.13125
27. Park, S.H., Kang, J.: Fast multi-type tree partitioning for versatile video coding using a lightweight neural network. IEEE Trans. Multimedia **41566**(c) (2020)
28. Zhang, Q., Zhao, Y., Jiang, B., Wu, Q.: Fast CU partition decision method based on Bayes and improved de-blocking filter for H.266/VVC. IEEE Access **9**(Cclm), 70382–70391 (2021)
29. Amestoy, T., Mercat, A., Hamidouche, W., Bergeron, C., Menard, D.: Random forest oriented fast QTBT frame partitioning. In: ICASSP 2019—2019 IEEE International Conference on Acoustics, Speech and Signal Processing (ICASSP), pp. 1837–1841 (2019). iSSN: 2379-190X

30. Amestoy, T., Mercat, A., Hamidouche, W., Menard, D., Bergeron, C.: Tunable VVC frame partitioning based on lightweight machine learning. IEEE Trans. Image Process. **29**, 1313–1328 (2020)
31. Yang, H., Shen, L., Dong, X., Ding, Q., An, P., Jiang, G.: Low-complexity CTU partition structure decision and fast intra mode decision for versatile video coding. IEEE Trans. Circ. Syst. Video Technol. **30**(6), 1668–1682 (2020)
32. Zhang, Q., Wang, Y., Huang, L., Jiang, B.: Fast CU partition and intra mode decision method for H.266/VVC. IEEE Access **8**, 117539–117550 (2020)
33. Dong, X., Shen, L., Yu, M., Yang, H.: Fast intra mode decision algorithm for versatile video coding. IEEE Trans. Multimedia **9210**(c), 1–15 (2021)
34. Aklouf, M., Leny, M., Dufaux, F., Kieffer, M.: Low complexity versatile video coding (VVC) for low bitrate applications. In: Proceedings—European Workshop on Visual Information Processing, EUVIP, vol. 2019-October, no. Vvc, pp. 22–27 (2019)
35. Tsang, S.H., Kwong, N.W., Chan, Y.L.: FastSCCNet: fast mode decision in VVC screen content coding via fully convolutional network. In: 2020 IEEE International Conference on Visual Communications and Image Processing, VCIP 2020, no. December, pp. 177–180 (2020)
36. Peng, S., Peng, Z., Ren, Y., Chen, F.: Fast intra-frame coding algorithm for versatile video coding based on texture feature. In: 2019 IEEE International Conference on Real-Time Computing and Robotics, RCAR 2019, pp. 65–68 (2019)
37. Lei, M., Luo, F., Zhang, X., Wang, S., Ma, S.: Look-ahead prediction based coding unit size pruning for VVC intra coding. In: Proceedings—International Conference on Image Processing, ICIP, vol. 2019-September, pp. 4120–4124 (2019)
38. Fan, Y., Chen, J., Sun, H., Katto, J., Jing, M.: A fast QTMT partition decision strategy for VVC intra prediction. IEEE Access **8**, 107900–107911 (2020)
39. Mallikarachchi, T., Fernando, W., Kodikara Arachchi, H.: Efficient coding unit size selection based on texture analysis for HEVC intra prediction. In: 2014 IEEE International Conference on Multimedia and Expo (ICME), vol. 2014, pp. 1–6 (2014)
40. Chen, J., Sun, H., Katto, J., Zeng, X., Fan, Y.: Fast QTMT partition decision algorithm in VVC intra coding based on variance and gradient. In: 2019 IEEE International Conference on Visual Communications and Image Processing, VCIP 2019, no. c (2019)
41. Zhang, Q., Zhao, Y., Jiang, B., Huang, L., Wei, T.: Fast CU partition decision method based on texture characteristics for H.266/VVC. IEEE Access **8**, 203516–203524 (2020)
42. Saldanha, M., Sanchez, G., Marcon, C., Agostini, L.: Fast partitioning decision scheme for versatile video coding intra-frame prediction. In: 2020 IEEE International Symposium on Circuits and Systems (ISCAS), pp. 1–5 (2020)
43. Li, Y., Yang, G., Song, Y., Zhang, H., Ding, X., Zhang, D.: Early intra CU size decision for versatile video coding based on a tunable decision model. IEEE Trans. Broadcasting, pp. 1–11 (2021)
44. Li, W., Fan, C., Ren, P.: Fast intra-picture partitioning for versatile video coding. In: 2020 IEEE 5th International Conference on Signal and Image Processing, ICSIP 2020, pp. 108–111 (2020)
45. Tang, N., Cao, J., Liang, F., Wang, J., Liu, H., Wang, X., Du, X.: Fast CTU partition decision algorithm for VVC intra and inter coding. In: 2019 IEEE Asia Pacific Conference on Circuits and Systems (APCCAS), pp. 361–364 (2019)
46. Zouidi, N., Belghith, F., Kessentini, A., Masmoudi, N.: Fast intra prediction decision algorithm for the QTBT structure. In: IEEE International Conference on Design and Test of Integrated Micro and Nano-Systems, DTS 2019 (2019)
47. Park, S.H., Kang, J.W.: Fast affine motion estimation for versatile video coding (VVC) encoding. IEEE Access **7**(Vvc), 158075–158084 (2019)
48. Rezaeieh, A., Roodaki, H.: A method for rate-distortion-complexity optimization in versatile video coding standard. In: 26th International Computer Conference. Computer Society of Iran (CSICC), vol. **2021**, pp. 1–5 (2021)
49. Abdallah, B., Belghith, F., Masmoud, N.: Low-complexity transform algorithm for versatile video coding. IEEE International Conference on Design and Test of Integrated Micro and Nano-Systems, DTS, vol. 2019, pp. 2019–2021 (2019)

50. Fu, T., Zhang, H., Mu, F., Chen, H.: Two-stage fast multiple transform selection algorithm for VVC intra coding. In: Proceedings—IEEE International Conference on Multimedia and Expo, vol. 2019-July, pp. 61–66 (2019)
51. Filipe, J.N., Carreira, J., Tavora, L.M.N., de Faria, S.D., Navarro, A., Assunção, P.: Tree-based ensemble methods for complexity reduction of vvc intra coding. In: 2021 Telecoms Conference (ConfTELE), pp. 1–6 (2021)
52. Wang, Y., Li, Y., Yang, D., Chen, Z.: A fast intra prediction algorithm for 360-degree equirectangular panoramic video. In: IEEE Visual Communications and Image Processing (VCIP), vol. 2017, pp. 1–4 (2017)
53. Ray, B., Jung, J., Larabi, M.-C.: A low-complexity video encoder for equirectangular projected 360 video content. In: 2018 IEEE International Conference on Acoustics, Speech and Signal Processing (ICASSP), pp. 1723–1727 (2018)
54. I. Storch, B. Zatt, L. Agostini, G. Correa, L. A. da Silva Cruz, and D. Palomino, "Spatially adaptive intra mode pre-selection for erp 360 video coding," in *ICASSP 2020 - 2020 IEEE International Conference on Acoustics, Speech and Signal Processing (ICASSP)*, 2020, pp. 2178–2182
55. Hanhart, P., Boyce, J., Choi, K.: JVET-L1012: JVET common test conditions and evaluation procedures for 360° video. In: Joint video experts team (JVET), 11th meeting: Ljubljana, SI, Tech. Rep., 7 (2018)
56. Yu, M., Lakshman, H., Girod, B.: A framework to evaluate omnidirectional video coding schemes. In: 2015 IEEE International Symposium on Mixed and Augmented Reality, pp. 31–36 (2015)
57. Bjøntegaard, G.: VCEG-M33: calculation of average PSNR differences between RD-curves. In: Telecommunications standardization sector (ITU), thirteenth meeting: Austin, Texas, USA, Tech. Rep., 4 (2001)
58. Filipe, J.N., Carreira, J., Tavora, L.M.N., Faria, S.M.M., Navarro, A., Assuncao, P.A.A.: Complexity estimation for load balancing of 360-degree intra versatile video coding. In: IEEE Workshop on Signal Processing Systems (SiPS), vol. 2020, pp. 1–5 (2020)
59. Abuzaid, F., Bradley, J., Liang, F., Feng, A., Yang, L., Zaharia, M., Talwalkar, A.: Yggdrasil: An optimized system for training deep decision trees at scale. In: Proceedings of the 30th International Conference on Neural Information Processing Systems, ser. NIPS'16. Curran Associates Inc., Red Hook, NY, USA, pp. 3817–3825 (2016)
60. Uitto, M.: Energy consumption evaluation of h.264 and hevc video encoders in high-resolution live streaming. In: 2016 IEEE 12th International Conference on Wireless and Mobile Computing, Networking and Communications (WiMob), pp. 1–7 (2016)
61. Sidaty, N., Heulot, J., Hamidouche, W., Nogues, E., Pelcat, M.: Reducing computational complexity in HEVC decoder for mobile energy saving. In: 2017 25th European Signal Processing Conference (EUSIPCO), pp. 1026–1030 (2017)
62. Khan, M.U.K., Shafique, M., Grellert, M., Henkel, J.: Hardware-software collaborative complexity reduction scheme for the emerging hevc intra encoder. In: Proceedings of the Conference on Design, Automation and Test in Europe, ser. DATE'13. EDA Consortium, San Jose, CA, USA, pp. 125–128
63. Chen, F., Wen, P., Peng, Z., Jiang, G., Yu, M., Chen, H.: Hierarchical complexity control algorithm for hevc based on coding unit depth decision. EURASIP J. Image Video Process. **2018**(1), 96 (2018)

ArkiCity: Analysing the Object Detection Performance of Cloud-Based Image Processing Services Using Crowdsourced Data

Mehrdad Amirghasemi, Ekin Arin, Rasmus Frisk, and Pascal Perez

Abstract ArkiCity is an innovative smartphone application that enables citizens to share local knowledge and ideas about their city by taking a photo, augmenting that photo with a set of 2D and/or 3D objects and submitting both images to a backend storage service. As an effective crowdsourcing tool for smart cities, ArkiCity advocates for a more inclusive approach to urban design and bridging the gap between citizens, planners and decision-makers. ArkiCity has been trialled in ten local city councils in Australia and Denmark and has received near 1,000 end-user submissions. In this chapter, the utilisation of such crowdsourced data for training object detection models is proposed, and the performance of a cloud-based computer vision service, in terms of its accuracy in identifying the cutout objects used in these submissions, is analysed.

1 Introduction

Technology, citizens and communities are considered to be three interconnected components of the green smart cities paradigm [16, 25]. Crowdsourcing could be seen as a tool that effectively integrates and utilises these three components to make cities sustainable, safe and inclusive. While crowdsourcing has traditionally been used to collect and analyse structured data from citizens through online surveys and questionnaires, gathering and analysing unstructured data, such as digital photos and images, have been a challenging endeavour.

M. Amirghasemi (✉) · P. Perez
SMART Infrastructure Facility, University of Wollongong, Wollongong, Australia
e-mail: mehrdad@uow.edu.au

P. Perez
e-mail: pascal@uow.edu.au

E. Arin · R. Frisk
Arki_Lab ApS, Copenhagen, Denmark
e-mail: ea@arkilab.dk

R. Frisk
e-mail: rf@arkilab.dk

M. Lahby et al. (eds.), *Computational Intelligence Techniques for Green Smart Cities*, Green Energy and Technology, https://doi.org/10.1007/978-3-030-96429-0_18

This chapter presents ArkiCity, a cloud-based digital platform for green smart cities, which enables a democratic approach to urban planning. Images containing multiple cutout objects are collected via the ArkiCity mobile application and are utilised to train a cloud-based computer vision model to effectively classify users—submissions. ArkiCity is supported by an efficient cloud-based serverless backend service which aims to streamline data acquisition, data labelling and model training into an effective computer vision pipeline. Overall, nearly 160 submitted collages are labelled and utilised to train and validate computer vision models using Amazon Rekognition [2], a high-performing cloud-based computer vision service. Multiple train and test data sets are constructed, and the performance of these models is analysed.

It should be noted that the term green in green smart cities could refer to several interrelated concepts such as sustainability, green economy or greener, natural landscapes [14, 27]. ArkiCity contributes to a greener city in two ways. Firstly, in terms of green technology [14], through employing a cloud-based, serverless architecture, ArkiCity aims to reduce the amount of IT resources required to help to minimise the environmental impacts. This has recently been emphasised as one of the best practices in cloud computing, as part of a new sustainability pillar in Amazon Web Services (AWS) well-architected framework [1]. Secondly, ArkiCity incorporates several, highly used cutout objects containing plants and animals, promoting urban greening through addition of a variety of natural features to city areas.

The rest of this chapter is organised as follows. The next section provides an overview of ArkiCity and its serverless architecture. Section 3 briefly outlines related work, in terms of similar tools and related cloud-based computer vision concepts. Section 4 presents the experiment results and Sect. 5 presents some concluding remarks.

2 ArkiCity

2.1 A Democratic Approach to Urban Planning

Traditional methods in urban planning are slowly becoming obsolete, giving way to innovative and alternative approaches. Towards this, ArkiCity provides a digital space, where citizens share local knowledge and ideas on how to improve spaces around their city. Such a digital platform has the potential to transform the way we approach citizen involvement, through generating a dialogue between citizens and decision-makers, as illustrated in Fig. 1.

The main idea behind the app is to help to identify the critical problems or improvements needed within the city through giving citizens the spotlight, being the ones who use the city the most. Through giving citizens the power to address the issues cities face today and tomorrow, ArkiCity shifts the focus from the decision-makers—table to the streets, ensuring a more direct and democratic insight into urban life. This

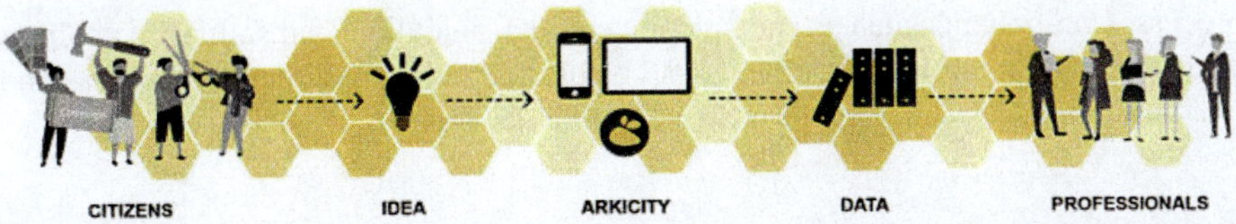

Fig. 1 ArkiCity connects ideas from citizens to data, providing new insights for decision-makers

process creates a whole new scene for the future development of our urban context; a future where bottom-up and top-down planning, pixels and participants work hand-in-hand in co-creating the urban landscape.

The ArkiCity app is designed to be user-friendly and accessible to everyone. Users simply take a photo of an area of interest and use cutouts and drawing tools to illustrate their vision for the area. The app database has a wide range of built-in cutouts, but also has the potential to integrate site-specific cutouts, tailored towards particular projects. For instance, if the project is focussed more on the streetscape, site-specific cutouts such as landmarks, trash cans, street signs and so on, can be added. This allows users to better connect with a familiar setting, enabling them to incorporate local knowledge of the area.

As shown in Fig. 2, ArkiCity can be used both in mobile phones and tablets, providing a variety of platforms and screen sizes to work on. When redesigning the area, decision-makers, designers and planners are now able to access the extensive collection of citizen inputs, communicating their suggestions and wishes for the area.

ArkiCity has several potential applications. Even though the initial idea was solely concentrated on connecting citizens to municipalities and public projects, as we developed ArkiCity further, we quickly realised the wide range of potential uses that were not only limited to urban contexts.

For instance, ArkiCity could be a great tool for use in schools or other learning environments, offices or residential spaces. In school settings, the app can be used to extend teaching outcomes by giving students more agency over their learning. Due to the variety of applications for ArkiCity, different versions and packages were developed. They are composed of various components that can be curated for each

Fig. 2 ArkiCity is accessible on multiple mobile devices

process. Components such as site-specific cutouts, analytical and statistical analysis of the data and many more elements can be used to help strengthen the process and overall findings.

2.2 Technical Architecture

The ArkiCity platform consists of a mobile app and a cloud-based backend service. The mobile app has been developed using the open-source React Native framework [20] and released in both Google Play and Apple App stores. The app is supported by a backend architecture, shown in Fig. 3, which consists of multiple interacting micro-services deployed on the AWS cloud.

In effect, the backend architecture integrates authentication, compute, storage and database microservices. While the compute module supports the backend API and interacts with the database, a storage service supports storing user submissions. For a low-latency user experience, the submissions are directly transmitted from the mobile app to the storage service. This loosely coupled architecture is designed to ensure modularity, scalability and security in a cloud environment. It should be noted that the entire architecture is serverless and is deployed mainly on open-source frameworks such as the serverless framework [23]. While the compute service is deployed on the Amazon API Gateway and AWS Lambda, the storage and database services are deployed using Amazon S3 and AWS DynamoDB, respectively. Lastly,

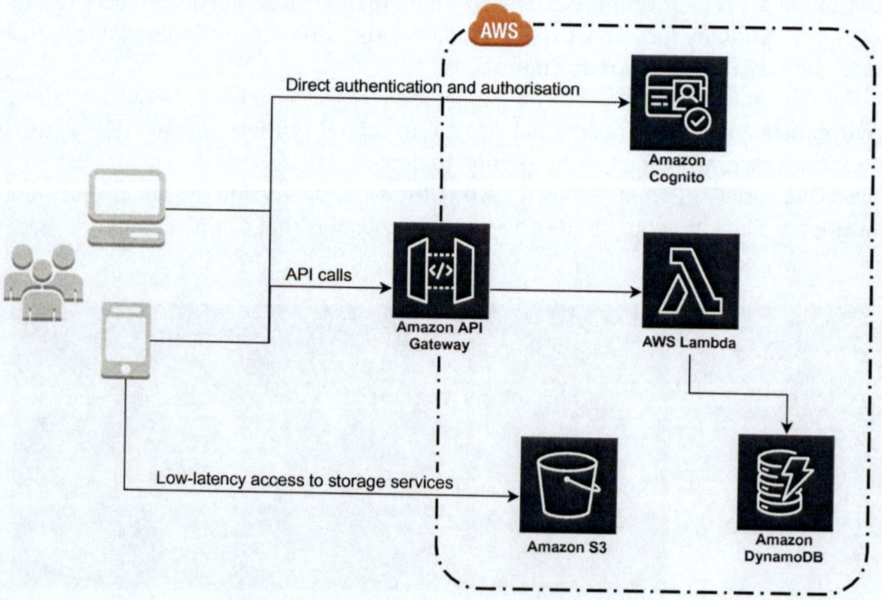

Fig. 3 ArkiCity serverless architecture diagram

Amazon Cognito secures the access to all other services through performing relevant authentication and authorisation tasks.

The mobile app user interface follows a simple, easy-to-navigate design as well, with two main navigation tabs, as depicted in Fig. 4. While the home tab shows all the other users' submissions within the same project, the camera tab is used for taking a photo and augmenting that photo using the set of cutouts and drawing tools provided. The user also has the capacity to import a photo from the their device photo gallery. Figure 5 shows how a user can create two before/after collages using the available cutouts.

3 Related Work

In this section, a brief outline of similar technological platforms is firstly presented, and then, we briefly outline the related cloud-based computer vision services.

Fig. 4 The ArkiCity user interface

3.1 Similar Digital Platforms

Several digital technologies have delivered their underlying promise of creating more democratic, inclusive and user-centred processes in different fields. Architecture and urban planning have also experienced a paradigm shift through their increasing integration of different digital tools, from open-source data collections to crowdfunding possibilities. ArkiCity takes its place amongst various digital platforms that have been produced around the world to bridge the gap between design, planning and citizens.

Both public and private groups have been interested in utilising the potential of digital platforms. While initiatives such as FluxMetro [6] target developers, smartphone apps like Civitist [19] and CitySwipe [28] enable citizens to be involved in conversations about urban development.

However, most of these initiatives have either been abandoned or stuck in the proof-of-concept stage. This could be due to various factors, most of which relate to funding and lack of publicity and demand. Nevertheless, those platforms that are integrated into the public sector, municipalities, and so on, tend to have a longer lifespan, as their range of impacts grow.

3.2 Computer Vision as a Service

In this subsection, the intersection between computer vision and cloud computing is briefly outlined. We start by providing a brief introduction to computer vision and then go on to define how cloud computing and software as a service concept are utilised for computer vision problems. Finally, we review some industry-leading cloud-based computer vision services.

Computer vision, which can be seen as a sub-field of computational intelligence [15], has recently become increasingly popular amongst researchers. One of the main nature-inspired techniques used in computer vision is artificial neural networks (ANNs). The modern expansion of traditional ANNs into new paradigms such as

Fig. 5 Creating a collage using the cutouts available in ArkiCity

(deep) convolutional neural networks has led to a new era for computer vision in the past decade. In effect, the unprecedented progress of such techniques for computer vision, amongst other machine learning problems, has created an entire new field of deep learning, with a large number of books and academic journals devoted to the subject [11, 13, 18]. Computer vision has a wide array of real-life applications, ranging from medical diagnoses [8] to self-driving vehicles [5] and to smart cities [26].

The main objective of computer vision algorithms is to extract meaningful, (numerical) information from digital images and videos. This can be achieved in a variety of ways. For instance, in an image classification problem, the algorithm aims to detect whether some objects (or object classes) exist in a given image. In a semantic segmentation problem, however, the focus is on how an image can be divided into *regions* belonging to various object classes [29]. It is worth mentioning that computer vision is not limited to object recognition tasks, and it also includes the task of intelligently generating or augmenting images, using techniques such as style transfers [10].

Having briefly explained computer vision, we now outline cloud computing and Software as a Service (SaaS) concepts and explain how computer vision can be provided as a service, on the cloud.

Cloud computing refers to the on-demand provision of various IT resources such as computing, storage, databases and networking services over the Internet, in order to improve reliability, scalibiliy and fault tolerances, while reducing the total cost of ownership. As one of the key enabling technologies for Industry 4.0 and smart cities paradigms, it enables small-medium enterprises (SMEs) as well as local councils and other government entities to access the IT capabilities that have traditionally only been available to large technological corporations. The worldwide cloud computing market is estimated to reach USD 832.1 billion by 2025, from USD 371.4 billion in 2020 [21].

SaaS can simply be defined as migrating existing software applications to the cloud, and hosting those applications as a service. The "as a service" concept has been expanded into computational intelligence by introducing concepts such as AI as a Service (AIaaS) [17] and Machine Learning as service (MLaaS) [22]. A related but more general term is that of Everything as a Service (XaaS), presented in [7]. The main benefits of running computational intelligence techniques as a service are scalibility, cost-effectiveness and significantly reducing time-to-insight for big data and machine learning pipelines.

Computer vision services are offered by major cloud computing service providers, namely Amazon Web Services (AWS), Google Cloud Platform (GCP) and Microsoft Azure (MA). The standard computer vision offerings are Amazon Rekognition [2], Google Cloud Vision API [12] and Azure Computer vision [3], respectively. Whereas, all the mentioned products include pre-trained models capable of image labelling and text detection, they also offer the ability to train a custom model via services such as Amazon Rekgnition Custom Labels, Google AutoML and Azure Custom vision. In this study, we used Amazon Rekognition Custom Labels to train computer vision models. It is worth noting that a benchmark experiment has been performed on all three aforementioned services in [24], using the Pascal Visual Object Classes (VOC) data set [9], with Amazon Rekgnition Custom Labels outperforming the other two services in terms of mean average precision (mAP) [4] metric.

4 Computational Analysis

As of now, ArkiCity has been used in 10 projects with over 1000 registered users contributing submissions. These projects are listed in Table 1. The projects include four council workshops in New South Wales, Australia, one project for the initial launch of the project, as well as five projects in different cities and municipalities in Denmark. It is worth mentioning that several of the projects in Denmark have been conducted in an educational setting, with school students actively contributing submissions. All these projects have been a source of valuable feedback for iteratively improving ArkiCity, both in terms of user experience and backend services.

The scope of our experiments constituted a total of 162 image submissions received from the workshops conducted in local councils in NSW, Australia, namely WCC01, LCC01, PENRITH01 and CBCC01. The main reasons for these being selected were: (i) the identical format and set of cutouts provided for these projects and (ii) the fact that some projects in Denmark have not yet been completed and are still open for user submissions.

Table 1 ArkiCity projects in Australia and Denmark

Project code	Location/event	Country	Year
WCC01	Wollongong City Council	Australia	2019
LCC01	Liverpool City Council	Australia	2019
PENRITH01	Penrith City Council	Australia	2019
CBCC01	Canterbury Bankstown City Council	Australia	2019
LAUNCH	Initial Launch event of ArkiCity	Australia	2019
AALBORG	City of Aalborg	Denmark	2021
STENGAARD	Stengård neigbourhood, Copenhagen	Denmark	2019
VIG	City of Vig	Denmark	2021
AARS	City of Aars	Denmark	2020
SUNDBY	Sundby neigbourhood, Copenhagen	Denmark	2021

Table 2 Labels associated with each cutout category

Label ID	N	Included cutout objects
01 People	18	One or more persons standing walking or sitting
02 Activities	20	Different activities such as bike riding, playing sports and playing music
03 Transportation	9	Cars, buses and other transport-related cutouts
04 Objects	24	Outdoor objects such as chairs, tables, signs etc.
05 Nature	25	Trees, planys, pets, domestic animals, etc.
06 Other	9	Surface cutouts such as tiles, roads and sand
07 Indoor	13	Objects mostly used for indoor decorations
Total	118	

4.1 Data Preparation and Labelling

For training the cloud-based computer vision model, all 162 images from four council workshops conducted in Australia were considered. The images were manually verified in order to remove any invalid submissions, i.e. those without any cutouts. In total, 22 invalid submissions were excluded and 140 images were kept in the data set. Next, all the images were labelled according to the seven object classes presented in Table 2. These classes indicate the category of cutouts a user can choose from in the app. Figure 6 shows some example cutouts from the Objects category (04).

It is worth noting that all 118 cutout images were also included in the data set. This makes a total of 258 labelled images that can be used for training and validation purposes. In the next subsection, we explain how the entire data set is split into separate training and test subsets to train and validate different models.

4.2 Experiments Setting

All experiments have been performed on the cloud using Amazon Rekognition [2]. The custom label feature of the Rekognition service was utilised to train a custom model with the generated data set. In order to find the best-performing model, multiple train–test split settings were used. The models all used the 118 cutout images in the training data set, yet different percentages of user submissions were selected for each scenario. The train–test splits are shown in Table 3. The columns in Table 3 show the

Fig. 6 Sample cutouts in Objects category

Table 3 Different models based on various train–test split of the data set

Model #	N_{total}	N_{train} (%)	N_{test} (%)	SUB%
1	258	118 (46%)	140 (54%)	0
2	258	188 (73%)	70 (27%)	50
3	258	203 (79%)	55 (21%)	60
4	258	230 (89%)	28 (11%)	80

number of total images in data set, N_{total}, the number of images used for training the model, N_{train}, the number of images used for testing and validation of the model, N_{test} and the percentage of user submissions included in the training data set, SUB%.

4.3 Results

Full validation results of all models are presented in Tables 4, 5, 6 and 7. Precision, recall and F1 scores, three key metrics, are reported for each label class. While precision indicates the fraction of correct predictions over all predictions, recall shows the sensitivity of the model to detecting all existing (ground-truth) labels of a given test image. In other words, precision and recall are defined as TP/(TP+FP)

Table 4 The detailed performance of model #1

Label ID	F1 score	Test images	Precision	Recall	Assumed threshold
01 People	0.618	42	0.500	0.810	0.6
02 Activities	0.763	76	0.680	0.868	0.6
03 Transportation	0.303	19	0.188	0.789	0.6
04 Objects	0.800	79	0.767	0.835	0.6
05 Nature	0.847	97	0.752	0.969	0.6
06 Other	0.586	52	0.481	0.750	0.6
07 Indoor	0.500	9	0.571	0.444	0.6
Overall	0.631		0.563	0.781	

Table 5 The detailed performance of model #2

Label ID	F1 score	Test images	Precision	Recall	Assumed threshold
01 People	0.667	26	0.643	0.692	0.6
02 Activities	0.775	36	0.705	0.861	0.6
03 Transportation	0.667	10	0.750	0.600	0.6
04 Objects	0.784	39	0.655	0.974	0.6
05 Nature	0.839	48	0.734	0.979	0.6
06 Other	0.769	18	0.714	0.833	0.6
07 Indoor	0.500	9	0.368	0.778	0.6
Overall	0.714		0.653	0.817	

and TP/(TP+FN), where TP, FP and FN are true positive, false positive and false negative instances, respectively. The reported F1 score is simply the harmonic mean of precision and recall for each label class. Finally, an assumed prediction threshold of 0.6 is internally set for all label classes by Amazon Rekognition.

Figure 7 shows the summarised performance of all models, with respect to average precision, recall and F1 score. As is seen, model 4 has the best F1 score, which could be attributed to several factors including that of a larger training set and smaller test set compared to other models.

5 Concluding Remarks

ArkiCity provides a digital space that bridges the gap between citizens and decision-makers by allowing citizens to contribute to the (re)design of their neighbourhood in a playful and exciting way. Citizens are given a digital platform to create and share their vision of a more liveable public space, street or precinct.

Table 6 The detailed performance of model #3

Label ID	F1 score	Test images	Precision	Recall	Assumed threshold
01 People	0.788	16	0.765	0.813	0.6
02 Activities	0.792	23	0.700	0.913	0.6
03 Transportation	0.727	6	0.800	0.667	0.6
04 Objects	0.755	22	0.645	0.909	0.6
05 Nature	0.920	26	0.958	0.885	0.6
06 Other	0.846	13	0.846	0.846	0.6
07 Indoor	0.750	5	1.000	0.600	0.6
Overall	0.797		0.816	0.805	

Table 7 The detailed performance of model #4

Label ID	F1 score	Test images	Precision	Recall	Assumed threshold
01 People	0.744	16	0.593	1.000	0.6
02 Activities	0.789	16	0.682	0.938	0.6
03 Transportation	0.750	7	0.667	0.857	0.6
04 Objects	0.800	15	0.800	0.800	0.6
05 Nature	0.875	21	0.778	1.000	0.6
06 Other	0.947	10	1.000	0.900	0.6
07 Indoor	0.857	8	1.000	0.750	0.6
Overall	0.823		0.789	0.892	

We have leveraged the valuable, crowdsourced, unstructured data collected through ArkiCity to train high-performing, cloud-based computer vision models to classify the users' submissions in an effective and efficient way. Computational analyses showed that a total F1 score of 82.3% could be achieved through having an optimal split between training and testing data sets.

This paves the way for new and innovative approaches to future research. First, the entire pipeline could be streamlined by using automation to train a new model once a certain number of submissions have been received. As the data will be labelled automatically, this could supply a constant stream of high-quality training data for deep learning computer vision models. Second, the entire architecture could be extended to include short videos featuring augmented reality 3D objects. This would ultimately result in the production of highly effective, serverless deep learning pipelines for training computer vision models on the cloud.

Acknowledgements ArkiCity, is the recipient of the *Shape Your Future Award* at the 2020 Amazon Web Services (AWS) *City on a Cloud competition*. The ArkiCity development has been supported by a grant awarded by *Urban Development Institute of Australia (UDIA) NSW* through their City Life Labs program.

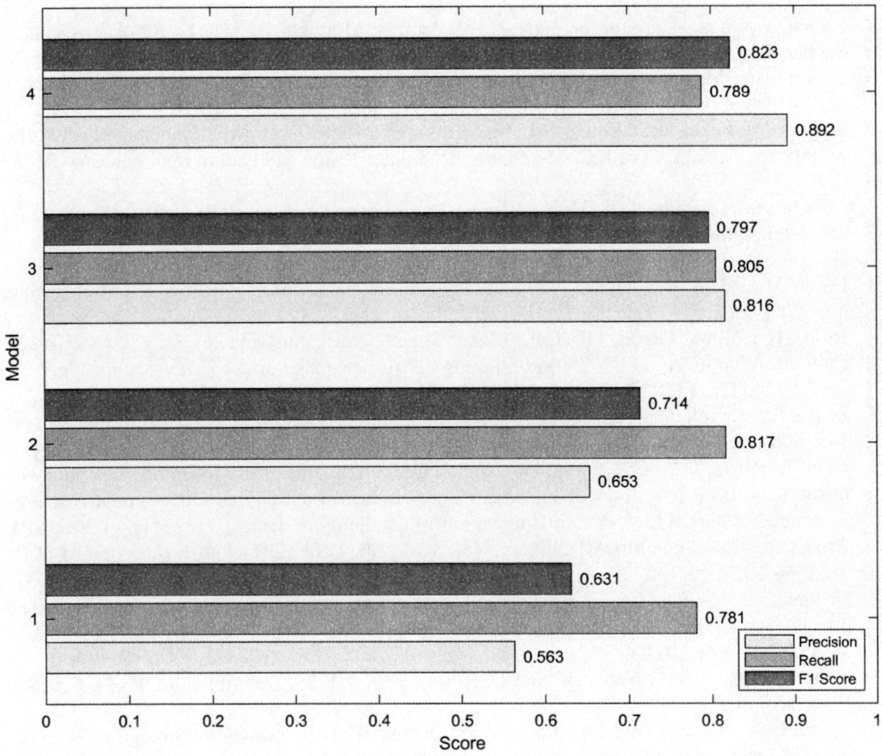

Fig. 7 The performance summary of all models

References

1. Sustainability Pillar—AWS Well-Architected Framework. https://docs.aws.amazon.com/wellarchitected/latest/sustainability-pillar/sustainability-pillar.html (2021). Accessed 7 Dec 2021
2. Amazon Rekognition—automate your image and video analysis with machine learning. https://aws.amazon.com/rekognition/. Accessed 12 Nov 2021
3. Computer Vision—An AI service that analyzes content in images and video. https://azure.microsoft.com/en-us/services/cognitive-services/computer-vision/. Accessed 12 Nov 2021
4. Cartucho, J., Ventura, R., Veloso, M.: Robust object recognition through symbiotic deep learning in mobile robots. In: 2018 IEEE/RSJ International Conference on Intelligent Robots and Systems (IROS), pp. 2336–2341 (2018)
5. Daily, M., Medasani, S., Behringer, R., Trivedi, M.: Self-driving cars. Computer **50**(12), 18–23 (2017)
6. Didech, K.: Flux Metro: A Better Way to Visualize Development Code. https://law.stanford.edu/2015/03/05/flux-metro-a-better-way-to-visualize-development-code/ (2015). Accessed 12 Nov 2021
7. Duan, Y., Fu, G., Zhou, N., Sun, X., Narendra, N.C., Hu, B.: Everything as a service (xaas) on the cloud: Origins, current and future trends. In: 2015 IEEE 8th International Conference on Cloud Computing, pp. 621–628 (2015). https://doi.org/10.1109/CLOUD.2015.88

8. Esteva, A., Chou, K., Yeung, S., Naik, N., Madani, A., Mottaghi, A., Liu, Y., Topol, E., Dean, J., Socher, R.: Deep learning-enabled medical computer vision. NPJ Digit. Med. **4**(1), 1–9 (2021)
9. Everingham, M., Van Gool, L., Williams, C.K.I., Winn, J., Zisserman, A.: The pascal visual object classes (voc) challenge. Int. J. Comput. Vis. **88**(2), 303–338 (2010)
10. Gatys, L.A., Ecker, A.S., Bethge, M.: Image style transfer using convolutional neural networks. In: Proceedings of the IEEE Conference on Computer Vision and Pattern Recognition (CVPR) (2016)
11. Goodfellow, I., Bengio, Y., Courville, A.: Deep Learning. MIT Press (2016). http://www.deeplearningbook.org
12. Google Cloud Vision API. https://cloud.google.com/vision. Accessed 12 Nov 2021
13. Hardt, M., Recht, B.: Patterns, Predictions, and Actions: A Story about Machine Learning. https://mlstory.org (2021)
14. Joshi, N.: How 'Green IT' Can Make Smart Cities Sustainable. https://www.forbes.com/sites/cognitiveworld/2020/06/05/how-green-it-can-make-smart-cities-sustainable/?sh=7f875bd476f3 (2020). Accessed 7 Dec 2021
15. Kruse, R., Borgelt, C., Klawonn, F., Moewes, C., Steinbrecher, M., Held, P.: Introduction, pp. 1–5. Springer London (2013). https://doi.org/10.1007/978-1-4471-5013-8_1
16. Nam, T., Pardo, T.A.: Conceptualizing smart city with dimensions of technology, people, and institutions. In: Proceedings of the 12th Annual International Digital Government Research Conference: Digital Government Innovation in Challenging Times, dg.o '11, pp. 282–291. Association for Computing Machinery, New York, NY, USA (2011). https://doi.org/10.1145/2037556.2037602
17. Newman, D.: Why AI as a service will take off in 2020. https://www.forbes.com/sites/danielnewman/2020/01/07/why-ai-as-a-service-will-take-off-in-2020/?sh=14a3548e3366. Accessed 12 Nov 2021
18. Nielsen, M.A.: Neural Networks and Deep Learning, vol. 25. Determination Press, CA, San Francisco (2015)
19. Ojgaard, A.: Civitist—design din by med en app (2019). https://www.magasinetkbh.dk/indhold/design-din-med-en-app. Accessed 12 Nov 2021
20. React Native—Learn Once, Write Anywhere (2021). https://reactnative.dev/. Accessed 12 Nov 2021
21. Cloud computing market by service, deployment model, organization size, vertical and region—global forecast to 2026. https://www.reportlinker.com/p05749258/Cloud-Computing-Market-by-Service-Deployment-Model-Organization-Size-Workload-Vertical-And-Region-Global-Forecast-to.html. Accessed 12 Nov 2021
22. Ribeiro, M., Grolinger, K., Capretz, M.A.: Mlaas: machine learning as a service. In: 2015 IEEE 14th International Conference on Machine Learning and Applications (ICMLA), pp. 896–902 (2015). https://doi.org/10.1109/ICMLA.2015.152
23. Do more with less. Serverless (2021). https://www.serverless.com/. Accessed 12 Nov 2021
24. Solawetz, J.: Benchmarking the Major Cloud Vision AutoML Tools (2020). https://blog.roboflow.com/automl-vs-rekognition-vs-custom-vision/. Accessed 12 Nov 2021
25. Srivastava, P., Mostafavi, A.: Challenges and opportunities of crowdsourcing and participatory planning in developing infrastructure systems of smart cities. Infrastructures **3**(4) (2018). https://doi.org/10.3390/infrastructures3040051. https://www.mdpi.com/2412-3811/3/4/51
26. Szeliski, R.: Computer Vision: Algorithms and Applications. Springer Science & Business Media (2010)
27. Toli, A.M., Murtagh, N.: The concept of sustainability in smart city definitions. Front. Built Environ. **6**, 77 (2020). https://doi.org/10.3389/fbuil.2020.00077. https://www.frontiersin.org/article/10.3389/fbuil.2020.00077
28. Wainwright, O.: Tinder for Cities: How Tech is Making Urban Planning More Inclusive (2017). https://www.theguardian.com/cities/2017/jan/24/tinder-cities-technology-making-urban-planning-interactive. Accessed 12 Nov 2021
29. Zhang, A., Lipton, Z.C., Li, M., Smola, A.J.: Dive into Deep Learning (2021). arXiv:2106.11342

Relevance of Green Manufacturing and IoT in Industrial Transformation and Marketing Management

Arshi Naim, Anandhavalli Muniasamy, Arockiasamy Clementking, and R. Rajkumar

Abstract In the current scenario technology has impacted all the sectors where materials and manufacturing sectors are also included. Green Manufacturing (GM) and Internet of Things (IoT) are two important applications applied in major business domains for positive results. This study shows the role of IoT as an example of emerging technology in industrial transformation (ITN) in general and in three areas of Marketing Management (M.Mgt) explained in five research proposals. This extended research article shows the relevance of IoT for achieving optimum growth, development and safe working for ITN and M.M.gt. This research paper is an instance-based qualitative analysis that explains how IoT contributed in the important fields of M.Mgt. These fields are Customer Relationship Management (CRM), Building advanced Business Process Models (BPM) and Product Life Cycle (PLC). The paper also covers a general impact and advantages of Green Manufacturing in BPM and PLC. The results show that GM and IoT are well integrated for ITN and M.Mgnt. Also the working of ITN and M.Mgnt became effective and contributed to the social benefits with the applications of GM and IoT.

Keywords Internet of Things · Industrial Transformation · Marketing Management · Green Manufacturing

A. Naim (✉) · A. Muniasamy
Department of Information Systems, College of Computer Science, King Khalid University, Abha, Kingdom of Saudi Arabia
e-mail: arshi@kku.edu.sa

A. Muniasamy
e-mail: anandhavalli@kku.edu.sa

A. Clementking
Human Resource Developments, Mount Carmel College, Autonomous, Bengaluru 560052, India
e-mail: clementking1975@gmail.com

R. Rajkumar
Department of Business Studies, Mount Carmel College, Autonomous, Bengaluru 560052, India
e-mail: srajmcc@gmail.com

M. Lahby et al. (eds.), *Computational Intelligence Techniques for Green Smart Cities*, Green Energy and Technology, https://doi.org/10.1007/978-3-030-96429-0_19

1 Introduction

Since Past one decade, the climatic concerns have aggravated, and researchers have been trying hard to bring solutions to reduce the pollution and wastes that are increasing due to industrialization and technological advancements. Green manufacturing (GM) can be seen as a major and important solution to this issue which aids in reducing wastes and pollution in the production processes [1, 2]. This change covers many domains of GM that result in better production design, introducing new BPM and using eco-friendly devices and materials as inputs in production processes. Climatic change across the world has brought new measures of GM and imposed competitive notion at all the business firms to adopt GM in their BPM. However not all sectors could utilize the GM at all level and its application is still at developing stage due to the unavailability of good replacements. Many researchers have explained the concepts and experiential learning related to green productivity (GP) and its advantages in profit maximization and achieving customer satisfaction. As mentioned earlier, many firms are still unaware of valid application of GM, therefore the factors related to successful implementation of GM have to be known to improve production processes and increase competitiveness. GM has significant advantages in M.Mgnt and ITN as this has changed the entire concept of BPM and contributed in environment friendly production and distribution [2, 3].

World's industrial companies are stressed by the competitive and divided markets, continuing economic uncertainties and growing operational costs (OC) [4]. All these issues have to be solved through reducing the OC, increasing productivity along with the achievement of competitive advantage. The solution is not easy to implement because that also requires some investment like increasing cost in technology.

Cost in Operation Technology (OT) may help in reducing OC but will cause increase in other technological related expenditures. Traditional industries have focused on OT for competitive advantages but with the introduction of Information Systems (IS) better decision making options came into existence. IS gave rise to new digital technological developments in ITN and offered solutions that could integrate OT and Information Technology (IT). This was the time of application of IoT in ITN. This combination could mean exploring more options for business processes and adding values to the products [4]. IoT creates a liaison between business and all the human resources and other tangible assets and resources that give rise to obtaining valuable business vision for industrial competitors [5].

ITN is a practical and synchronized principle to facilitate IoT to develop chronological changes and developments in operations. ITN is assessed by changes in the context of industrial structure, defined as the ratio of the variety of imitation—to innovation-based intermediate goods. IoT creates integration of digital technology into all areas of an ITN and aids in the methods of operations, CRM, BPM and other areas needed for transformations in industries. Apart from these changes IoT also brings changes in organizational culture [5].

The industrial Internet of things (IIoT) suggests interrelating sensors, instruments, and other devices networked together with computers' industrial applications, including manufacturing and energy management. This connectivity allocates for data collection, trade and investigation, beneficially directing to improvements in productivity and efficiency and economies of scale [6]. The IIoT is an evolution of a distributed control system (DCS) that allows for a higher degree of automation by using cloud computing to refine and optimize the process controls. There are three essential components of a digital transformation, namely, the repair of processes, the repair of operations and the rebuilding of relationships with customers. IoT provides three types of IT, for example Process Transformation (PTN), Business Model Transformation (BMT) Domain Transformation (DT) and Cultural/Organizational Transformation (COT) [7].

This era is termed as digital era and Internet of things (IoT) has helped in creating new and innovative products that can know, sense and share data as well as information to all the users such as companies, suppliers and consumers [7]. There are many retail products' examples, such as Sunsilk shampoo, Reebok sneakers or even beverages such as Pepsico or fruit juices like Tropicana, all these products have extended the fundamental applications of customary goods by offering the ability to collect and share data on the Internet [7]. This new category of data collecting and sharing products are called as IoT-ready products or simply IoT products (IoTP) [8].

Such IoTP may be seen as a provisional stage in the development of smart products (SP), which are able to analyze and, potentially, "interpret" usage data in a goal-oriented way. SP can make decisions that would require man's cognitive understanding and critical thinking. For example, DSS or machine learning methods can analyze the stage of SP and use data to evaluate customer behavior, preferences and complaints too that may contribute in the process of new product development considering the achievement of customer satisfaction and eventually depending on the product type and application purpose, decisions made by a SP can be used to provide users and the company with recommendations or rating and reviews for further research and development. Figure 1, shows the growth junctures from traditional products (TP) to (IoTP) to (SP), where TP has only basic functionalities, IoTP includes function of autonomous data collection and product analysis through virtual interfaces and finally SP encompasses data analysis, artificial applications, machine learning DSS applications for decision making processes [9].

There are many products in the current IoT environment marketed as SP but as a researcher we clearly comprehend that it is just a transition phase and developing SP involves many features and applications such as IoT with Customer Relationship Management (CRM) and Business Process Management (BPM). IoT has not left any field without its relevance in the process of growth such as in the medical sector, education, automobile, retailing, traffic control automation, smart cities and home automation, etc. In some trades, the contemporary step of growth is by now stirring

Fig. 1 Transformation from TP to SP [8, 9]

toward SP at a swift pace, such as cooking in the smart kitchen, self-driving cars in the automobile industry, etc. but we can still illustrate that IoTP constitutes a relatively new phenomenon for many sectors. This chapter discusses the positive impact of IoT for various M.Mgnt such as CRM, PLC and BPM. We have proposed five research ideas and analyzed qualitatively the impact of IoT and referred to a few real examples of firms showing the success with the application of IoT. This chapter is divided into five sub-topics; first topic will cover the historical aspects of the concepts used in this chapter in literature review, second topic gives a brief outline on the research methodology applied in this chapter, thirdly, all five research proposals are discussed for IoT for three M.Mgnt concepts; CRM, PLC and BPM. Discussion part also covers the description of GM for ITN and general advantage as well as challenges of applying GM in ITN. Fourth topic gives the results and the fifth topic illustrates the final outline as a conclusion.

2 Literature Review

Firstly in this part we have explained the definition of GM, advantages and its applications, and we have given a brief account on GM's role in PLC and BPM. In the second part of the same section, we highlight some of the related concepts in the applications of IoT techniques in Industrial Transformation (ITN) and M.Mgnt.

GM means as green and sustainable manufacturing process that classifies mechanization as the BPM for producing outputs from inputs for achieving customer satisfaction and earn good return on investments. In this process the focus should be of creating less pollution and reducing wastes from both materials and finished goods. Researchers have studied the past work to know the importance and results of sustainable manufacturing and significantly concluded many several benefits of GM at all levels of BPM and phases of PLC. The analytical results suggest a positive correlation between environmentally sustainable manufacturing practices and competitive outcomes and establish a relationship between GM and BPM as well as for PLC [10].

GM covers all factors which are important for conserving and integrating green inputs and outputs in the process of ITN, BPM and achieving CRM. GM has many advantages in applying it in BPM such as it can moderate and reduce the pollution level in air, water and land. This advantage is for all living things, flora and fauna because GM can reduce waste at the source. When GM is used in materials and processes of production are focuses to use the eco-friendly inputs, reuse of materials, recycling, eliminating toxic release in BPM and in all the phases of PLC. In past one decade many researchers have confirmed that GM is the most efficient method of improving BPM and reducing toxicities.

Firms who have adopted GM in their BPM have minimum adverse impact on environment and have optimum consumption during the manufacturing process, not only this also GM has resulted in increasing competitive advantages [11].

Firms which applied GM are able to achieve their organizational and strategic advantages through process design (PD), where the PD in GM aids in reducing the wastes, pollutions and achieve CRM and profit maximization. Studies have also shown that firms using GM have consideration in measuring performance dealing with two notions, GM dimension and GM factors. Dimension explains various qualitative aspects, view points for achieving optimum GM and factors elaborate the application part that may include product design, synthesis, processing, packaging, transportation and recycling in the BPM and all the phases of PLC [12].

Over the last decade, the term (IoT) has fascinated concentration by applying the prophetic vision to the global infrastructure of networked physical objects, allowing all time connectivity from any location. IoT is an open and comprehensive network of intelligent objects that have the capacity to auto-organize, share information, data and resources, reacting and acting in face of situations and changes in the environment [13].

This process of ITN started in the UK in the eighteenth century and from there it moved to other parts of the world. French writers mentioned the term ITN in 1852 during the Industrial Revolution but Britishers were the pioneers in making it known to the world. Economic historian from the UK in 1852 defined ITN as a way to UK's economic development [14]. One of the most famous examples of ITN is a Textile industry which was the first to use modern production methods. Connectivity and data collection as well as analysis are the major roles of IIoT for achieving competitive advantages and explain the fundamental principles of growth and development in ITN. IoT has brought major ITN by decreasing unpredicted economic depressions and increasing productivity, which is subjected to over scheduled repairs, reduce complete preservation costs and eradicate crashes [15]. It is assumed that IoT devices are incorporated into various forms of energy consuming devices that can communicate with the different requirements of firms to balance power generation and energy usage which will help in ITN eventually [16].

In general, IoT system has three parts namely, sensors, network connectivity and data storage applications as shown in Fig. 2. The figure explains the background material of IoT system which is important to learn when we are explaining the role of IoT in M.Mgnt. Also Fig. 2 provides listing of IoT services and its applications.

IoT is a universal network which offers the communication between human-to-human, human-to-things and things-to-things, which is anything in the world by providing unique identity to each and every object and anything can be connected and communicated in an intelligent fashion than ever before. It supports coding and networking of everyday objects and things to render them individually machine-readable and traceable on the Internet, like the content created through coded RFID tags and IP addresses linked into an electronic product code (EPC) network [17]. IoT services can accumulate requested data from various sources and transmit it to the other devices and systems automatically. This process aids in solving problems for routine-based scenario and provides solutions to BPM [18].

CRM [19] refers to all strategies, techniques, tools and technologies used by enterprises for developing retaining and acquiring customers. It is a comprehensive strategy and process of acquiring, retaining and partnering with selective customers

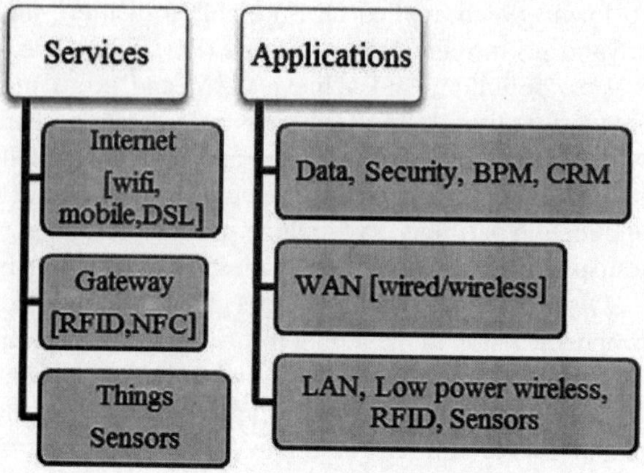

Fig. 2 IoT system [17]

Fig. 3 Scope of CRM [20]

to create superior value for the company and the customer. The scope of CRM is shown in Fig. 3.

CRM software available in the market ensures that every step of the interaction with consumers goes smoothly and efficiently in order to increase the overall profits [21]. The software gathers customer data from multiple channels. Hence, CRM stores detailed information on overall purchase history, personal information and even purchasing behavior patterns [22]. It entails the amalgamation of marketing, sales, customer service and the supply chain practices of the organization to realize more efficiencies and effectiveness in building the customer value [23, 24].

PLC is the historical study of sales of the product [25] covers four phases for its development. These phases are given below.

2.1 Introduction [26]

- Product is commenced.

- Slow growth of Sales because unawareness of goods.
- No gain in this situation; Advertising model is suggested to be applied.

2.2 *Growth [26]*

- Steep increment in sales
- Reminders or other strategies of advertising is suggested
- Low price for goods because more competitors enter in the market.

2.3 *Maturity [26]*

- Sluggish movement of sales increment
- More competition in the market.
- Sales discount or promotional methods are suggested because profit may start to reduce.

2.4 *Decline [26]*

- Decrease in sales and goods lose its value.
- No expenditure in advertising
- Suspend or stop completely the production

The value of the product can be measured by its competence to meet market's potentials. It lasts or exists as long as it satisfies its users [28]. Figure 4 presents the phases of product from initial stage to decline in terms of revenue and time factors.

BPM is a study involving any blend of modeling, automation, execution, control, measurement and optimization of business activity flows, in support of enterprise goals, spanning systems, employees, customers and partners within and beyond the

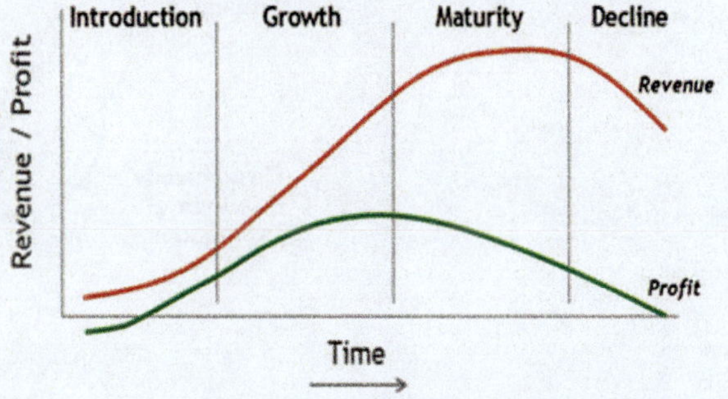

Fig. 4 PLC phases [27]

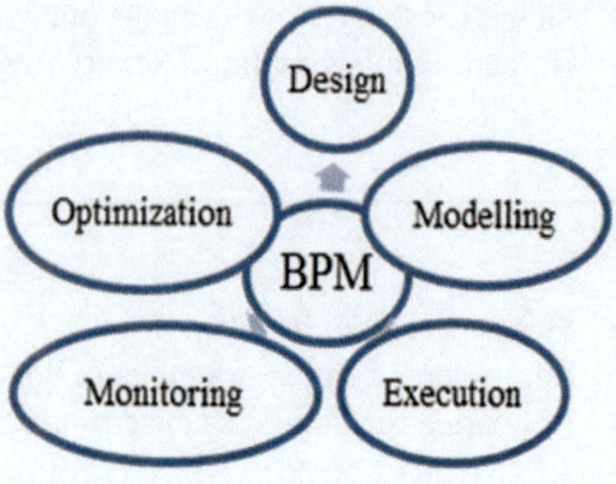

Fig. 5 Business process management (BPM) [30]

enterprise boundaries [29]. Figure 5 shows the activities of BPM to illustrate the relevance of this discipline in the business environment.

It involves 5 steps namely, design, modeling, execution, monitoring and optimization. Design is to identify current processes and design of proposed change, modeling used by What-if analysis to model change proposed on multiple variables, execution is to use, track and manage the software; monitoring keeps the track of competition level of changes and effectiveness and finally optimization identifies problems opportunities and apply changes for greater cost savings and efficiency [31].

CRM is the most growing branch and it applies technological related tools for its effective working and relevance. CRM systems are increasingly applying IoT technology to improve front-end processes [32]. With advancements in IoT, CRM is all set to change the BPM. By connecting devices, products and equipment to the Internet, IoT drives insights throughout sectors, including sales, marketing and customer service. The collective powers of IoT and CRM enhance efficiency and visibility to help BPM to provide swift benefits to consumers. Figure 6 shows the impact of IoT on one of the important application of M.Mktg [33].

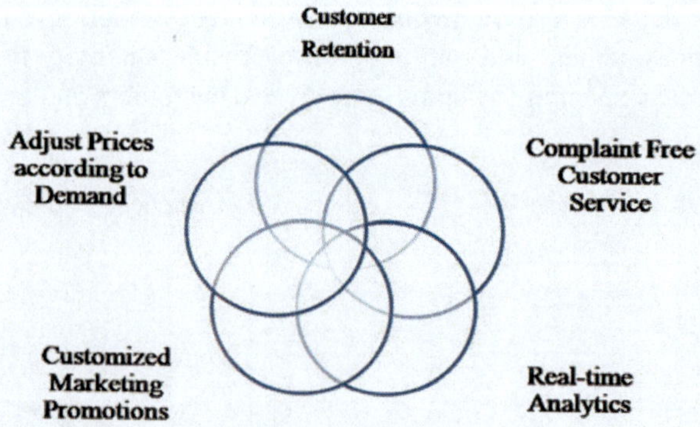

Fig. 6 Impact of IoT and CRM on enterprise performance [34]

2.5 *Improve Customer Retention [35, 36]*

This is the method to keep customers' for the firm's products and services, therefore IoT help in providing solutions to CRM so that customers do not change to another options and stay connected with the firm. Many suggestions can be illustrated here to retain the loyalty such as introducing the promotion, analyzing the customer's problem and finding solutions promptly by IoT.

2.6 *Complaint Free Customer Service [36]*

This is the process, where IoT is used to apply in the scenario in identifying the customer's issues and solve them promptly. This is a machine-based benefit to the CRM where before the identification of problems at consumer's end the solution is offered.

2.7 *Real-time Analytics [36]*

This is a proactive approach of IoT for CRM where the strategies are applied to analyze data for providing solutions in the real scenario. IoT is able to give real results to firms for developing campaigns for marketing, illustrating their promotional messages, work on dynamic pricing and eventually help in CRM.

2.8 *Customized Marketing Promotions [36]*

This is one of the approaches of IoT for CRM, where it is identifying the real customers and potential customers for the firm's products and services. Based on the characteristics of these customers the firm can design their effective marketing and advertising strategies for profit maximization and good communication.

2.9 *Adjust Prices according to Demand [36]*

IoT's applications can get to know the real data of customers, products acceptance, completion which can help in developing pricing policies, and create good options for other pricing strategies. IoT also help in knowing the real demand that can also aid in deciding the pricing.

Businesses need to overcome several business and service challenges to be able to realize smooth operation and manageability of disparate systems. Embedding intelligence for the real-time data gathering from gateways and devices and consuming them through business processes helps businesses achieve not just cost savings and efficiency but also generate more revenue patterns [36].

The IoT is changing the way we live our lives and that all businesses need to adapt to. Benefits of adopting IoT in business are as follow [37]:

- Data: Users can apply the automated systems for data recording and options for the growth which can be applied for purpose of advantages in BPM practices, CRM and Marketing Management. BPM can computerize this process and ensure that it remains effective and agile enough to keep pace with technological changes [37].
- Innovation: IoT offers the opportunities for businesses to deliver exciting new benefits for their customers, whether it is new product development or upgrading existing products or services [37].
- New ways of buying: The IoT facilitates the users to have a direct interface for buying process through their machines. With the applications of technology, many tasks became more easy, faster and manageable, for example the delivery of products became more effective. Also IoT aids in BPM for its working and help firms to understand the customer's demand and expectations [37].
- Customer service: IoT works in solving the problems related to products' utilization over the web. IoT helps BPM that makes sure that the customer service processes are effective, efficient and flexible aligning to the consumer's expectations and provide good SCM [37].
- Centralized BPM: Incorporating the applications of BPM with hardware that explains the analysis of data remotely and monitored centrally for results [37].

IoT can enhance PLM software because [38] of the many reasons such as it uses smart devices and sensors that are both interconnected and connected to the Internet. They can collect important information and relay it to interested parties. These gadgets can be embedded in any product, from vehicles to clothing. It can inform product developers and manufacturers in real-time how a certain product is performing out in the world. This can be valuable to engineers and designers thinking about the next iteration of the product. Companies can spot problems with the products on the field and fix them before they escalate, and the product breaks down. Such a preventive measure will help customers to avoid unnecessary downtimes and increase their bottom line. Companies can use both IoT and AI to predict future problems with a product and address them before they create inconvenience for the customer. However, companies will become more customer-centric and the manufacturers will have mountains of data on how customers are using their products, so that they can build new products in-line with the customer's behaviors and usage patterns [39].

The modern smartphone technology facilitates discussions on the potential impact of the IoT on industries, markets, companies, products, services and consumers due to the advancement of IoT. The existing economic literature investigating IoT and SP primarily aims on research questions in new business models [39], supply chain management, the fields of management, transportation, market competition, smart home and ambient assisted living, the organizational structure of companies, consumers' attitudes toward autonomously acting products, production planning and control, privacy and secrecy and wearable devices [39].

From the perspective of M.Mgnt, IoT products are of particular interest for two functionalities namely, "product analytics" (PA) and "remote access" (RA).

- PA depends on the data collected by the machines directly from the customers and provides assistance to the firms to analyze customer's pattern of usage for real products [40].
- RA offers alternatives for distant connectivity to operate for SP, varying the standards and modifying the characteristics of products, activating and deactivating product functions, and finally scheming data flowing inbound to the IoT product [40].

PA and RA can release chances and new opportunities for marketing management.

3 Research Methodology/Discussion

In the first part of the chapter we discussed the general role of IoT in ITN and presented the major success of IoT in ITN, also we discussed some of the limitations created due to IoT applications in the working of ITN. The second part of the study shows the application of IoT and its realization in MMgnt under three areas; CRM, BPM and PLC, where many research questions are answered showing the relationship between IoT and MMgnt concepts and how well they have been implemented by various firms in real scenario. We developed some proposals and recommendations for MMgnt concepts with an application of IoT. This is a qualitative analysis and various firms are studied for the application of IoT and its classification in above mentioned MMgnt sectors. Real examples will justify the usefulness of IoT and comparative analysis will prove the success of IoT.

In the final part of the discussion we presented the impact and advantages of Green manufacturing in ITN and BPM.

For the purpose of the study five proposals are given to show the impact of IoT in different fields of MMgnt. Following are the framework of proposals on which the research chapter will confer its analysis and results.

First Proposal: Customers can easily be communicated for product specification and awareness can be created through IoT, in technical terms said as creating touch lines via IoT, additionally data collection can be transformed to data handling leading to resolving CRM issues along with developing multiple interactive platforms and media between customers and firms.

Second Proposal: IoT helps developing customized pricing strategies focusing on individual customers that can result in greater return on investment and in developing smart products for building CRM.

Third Proposal: IoT products will create new kinds of customer switching costs (CSC). This is an important area of research in CRM where the application of CSC can contribute in identifying the features of CRM like building customer retention, competition, enterprise profitability and others. CSC is explained as "onetime costs that customers relate or calculate before moving from one firm's products to another firm's products. IoT help CRM to identify CSC and illustrate key performance indicators that a firm can apply to measure their results. This research proposes to apply IoT for CSC for measuring performance of firms for building their CRM.

Fourth Proposal: This suggestion focuses on PLC and we suggest to apply IoT products because it may be used without considering the stages of products. Generation substitution models (GSM) have a tendency to accentuate duplicate purchases, in contention that hi-tech period compensate the durability of products. Especially machine-based and innovation-driven products, such as personal computers and smartphones, are substitutes because they are technologically obsolete but in working conditions. The current PLC models aims and explains the features at introductory part of a product but after the application of IoT, firms can know how to enhance buying behavior, inculcate customers to buy again and slowly adopt the benefits of products.

Fifth Proposal: The IoT facilitates the application of web-based BPM with physical products. TPs are frequently promoted with a transaction-dependent revenue model. IoT products assisted joint value creation with third parties. As part of the business model concept the network model characterizes a management concepts for examining and controlling the value distribution in a joint value creation setting.

IoT plays a major role in ITN where asset-intensive industries, such as manufacturing, where the combination of OT and IT aids in improving new product developments, improving the existing product and changing the mind-set of old methods of performing business practices. IoT gives rise to opening of new channels, developing new customers, identifying new products' usages, in simple terms a complete change over in industry's way of processing. IoT gave rise to ITN, where firms are able to enhance their value BPM by using various IT-based business models and developing the products as per the customers' requirement to meet customer satisfaction and good return on investment. Application of IoT-based explanations at any level of the firm, from the operation level to strategic or production level, the objective is to cut OC and use better OT and IT so that the firm can reach to its breakeven point in less time and can increase profit. This is the most important reason for ITN to implement IoT in most of the industrial processes [41].

In the current scenario many companies use IoT solutions for receiving benefits from their investments and decreasing OC and other related expenditures. This is evident that IoT provides real-time vision and advantages to the BPM for any companies from collecting the data to analyzing and using it for decision making processes for different levels of management. Apart from these monetary gains, IoT also aids ITN in effective SCM, building relationships between customers and firms, discovering new channels for the flow of products, encouraging firms to apply personalization approaches. IoT advances internal performances at industrial levels by empowering the human resource and improving their performance, enhancing

Table 1 Threats of IoT for ITN [42]

Data security and privacy concerns are seen as the most challenging issues in IoT initiatives and strategies
Cultural change in the plants caused by implementing IoT solutions and smart machines
Data security and privacy concerns
Cost of purchase of IoT solutions
Lack of data scientists and analytics
Slow Internet connections
Cost of implementation and management of IoT solutions
Employees lack of technical development
Employee's internal organization challenges
Building the business case for IoT investment

skills and encouraging better results. As mentioned before, IoT is used for ITN for the purpose of resolving users-based problems such as problems related to services, usages, quality, durability, availability. Considering the importance of IoT in ITN, there could be two major roles of IoT; firstly focusing on over all industrial improvement and specifically targeting at the internal level of related firms and secondly aiming to facilitate for external factors such as creating needs, achieving CRM, contributing in economy of the nation and development of new as well as improved business models. When IoT has several advantages, there are also noticeable threats that industry in general has to face. Some of the threats of application of IoT in ITN are given in Table 1 [42]:

The academic literature on IoT-related topics can be traced back to early publications on ubiquitous computing that correspond to the idea of information technologies penetrating "the fabric of everyday life until they are indistinguishable from it". IoT development has affected many areas and some of these areas are given in Table 2 illustrated [42].

There are many examples and one example that can be witnessed extensively is the application of smart phones. This device is the real example of growing application and usefulness of IoT in the communication and Information Technology. Another example is automation of traffic management with the help of IoT applications. Homes are also not left behind with the successful implementation of IoT, smart homes are protected, secured and prevented from accidents with the help of IoT. It has a potential impact on industries, markets, companies, products, services and consumers, and if we refer to M.Mgnt; IoT has introduced more and better functionalities in the products which are not only beneficial to the customers but also to the company's owners for data analysis. Therefore, IoT in M. Mgnt for product development is also called as product analytics (PAs) that allows remote access and virtual use. These examples evidently show the positive impact of IoT in M.Mgnt under three areas namely; CRM, PLC and BPM. We discuss five research propositions in the fields of CRM, PLC and BPM [43].

Table 2 Development of IoT different technological areas [42]

Internet of things
Sensor technologies
Wireless communication, as well as supply
Layered architectures of digital technology
Energy consumption
Harvesting
Strategic management
Transportation
Supply chain management (scm)
Market competition and new business models
Customer relationship management (crm)
Privacy and security applications
Wearable devices

CRM has developed into market-based learning. Comparing learning connections have been recognized as the key achievement factor in CRM that improves an organization's capacity to increase selling potentially, diminish costs, give informal promotions and eventually increment exchanging expenses. This field of exploration initially rose up out of the relationship of strategic management and high profile executives, but in the current situation it manages the mix of client related connections and the utilization of frameworks that gather and dissect information across the organization. CRM frameworks hence target connecting and making both organization worth and client esteem along with the value chain. An achievement in the improvement of CRM was the worldview change from item direction to client direction. Early work on CRM hence expounded satisfying client needs rather than "just" selling items. Advances in the CRM field incorporated structure connections in such a manner that the advancement concerned structure client connections, vital organizations, coalitions and organizations, new standards from exchanges to connections, and administration connections [44]. Stream of work on CRM underscored the pertinence of business sectors and managed market direction and market concentration.

A decade ago many prestigious journals in marketing research witnessed a drastic growth in CRM by the application of IoT that explained the concept of dual creation of value. This concept under CRM elaborates the co-creation of value rests on creating and sharing economic rents for the firms and the users. The two ways communications have increased between all users in the process of marketing are with the help of the Internet and precisely with the help of social media and emails. Some of the good examples of social media platforms are Facebook, Twitter, online discussion forums, instagram, youtube, whatsapp, etc.

These platforms have immensely affected the sale and product's awareness for all the firms which are using E-commerce or IoT-based CRM solutions. One of the current topical needs in CRM is to understand the client's experience and his ventures. For example, while considering the services and information, the firms are required to know and comprehend the customer involvement in the process. To meet this need, IoT has played a major role and even today the importance of IoT is aggravating because it has aided in increasing return on investment (ROI) or to the limit of reaching breakeven in less time. Apart from benefits in sales, IoT has also affected the overall psychological behavior and understanding of the customers. IoT helped in evaluating the positive and negative customer influences and reactions for the products and particularly SP. These events of association are generally alluded to as client touch points (TPs) [45].

Research Proposition 1: When multiple channels are involved we spell them as touch lines (TL), IoTP helps customers to use this TL for various TPs therefore customers may interact with a firm through a multitude of TP in multiple TL. We have defined the purchase behavior throughout the life of a customer as customer's journey (CJ) where the set of TPs are experienced by the customers for a firm and its different products. The emergence of the Internet and the (www) with all its applications increased a company's set of options to interact with customers that indicate an explosion in potential customer TP regarding the availability of web-based and mobile applications.

Current developments suggest that the IoT will push this development process even further and turn customer TP into customer TL. IoTP regularly collects product data, its usage, and dissemination to the company in addition to enhancing the interaction with customers at different TP. PAs offer the option for the firms to check the CJ in real situations and provide connection, management and control remotely. IoTP works in coordination with PAs for one to one customer's needs and preferences and Fig. 7 shows the changes that IoT has introduced in the communication. This can be considered as an example for BPM also.

The CJ starts with a TP at the purchase stage for all products. This theoretical situation can be illustrated to explain the scenario for many products that may fail in the post-purchase stage due to the serious rejection and dissatisfaction for the benefits of products resulting in developing the negative behavior that will count for negative retrospective behavior. This example if its substituted during TP when Internet or

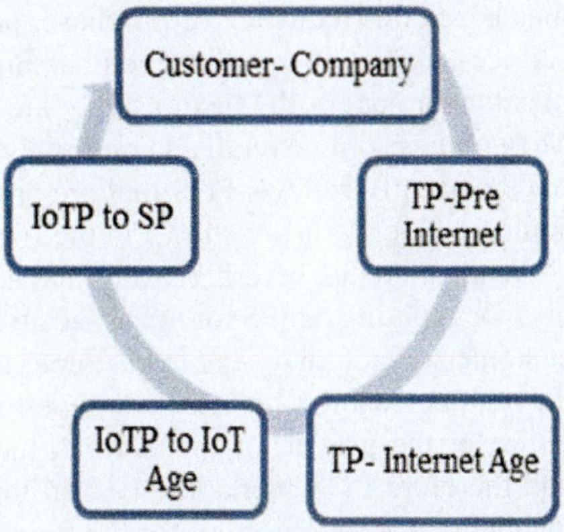

Fig. 7 Changes in medium of communication [46]

IoT did not have any importance of application, the number of customer TPs will be negligible or even least, simply because the firms are unable to receive authentic information and reasons for the failure and cannot even control negative word of mouth and therefore the product may decline and customer may not use in future [47]. This will result not in product failure but also affect the firms' image for future products. Same examples during the web era where IoT has great existence and presence TP can be increased easily because firms are able to do research and collect data on reasons and information on customers' preferences, disliking or complaints and eventually work on the improvement on the issues to achieve customer's satisfaction. A firm may have to bear a temporary loss or interruption in product's development but in the long run the product will not decline and the firm will be able to get ROI and develop good CRM. This can happen with help of IoTP that can give the firm true and authentic information on product usage. This resulted in a regular development of virtual customer's TL. Besides, PAs also contribute positively in this by allowing the firms to find or even forecast the reasons for product's failure, sometimes this process can be successful before the knowledge of a customer and therefore firms can solve the issues by the help of IoT remotely or virtually.

There are many firms who have taken the advantages of IoT but automobile firms are the pioneers in this, they used customer's TL and repaired the issues with help of IoT. Smartphones are also good examples of using IoT for achieving CRM. Nokia and Samsung are regularly exchanging data with their customers and solving detected software problems remotely [48]. In the current scenario all smartphone firms are following the same customer's TLs and offering services remotely for the solutions and updates too.

Research Proposition 2: IoTP ease the application of price discrimination strategies (PDS). The implementation of one-to-one pricing strategies (o2oPS) allows companies to maximize prices compared the price single customer is willing to pay. PDS are commonly used and found their way into online shops and online market places. During 2000 many retail firms condemned the idea of PDS and suggested the net neutrality and equality should be the aim of online retail marketing and all customers should pay same price for the same unit of products but the vision changed with more application on IoT, and the same firms introduced dynamic pricing strategies based on frequency of purchase, being the active user, using a particular web services, affiliation to online membership, online payment methods, etc. The companies that are using PDS extensively are Noon.sa, Amazon.com, ebay, Alibaba, etc. Most of these online retail and online auction firms criticize the concept of PDS [48]. IoTP currently follows PDS that are based on all the reasons mentioned above and some of the criteria are shown in the Fig. 8.

While there are several reasons and advantages for PDS, there are many concerns also for applying o2oPS for IoTP, because PDS confers the idea of inequality amongst customers which may lead to customer turnover for the firms' products and services due to this reason. This may continue for CJ and customers may choose to opt-out. However the growth of the Internet has enhanced the concept of personalization and therefore PDS works the best in this situation and customers also do not feel inequality and unimportant for the firms.

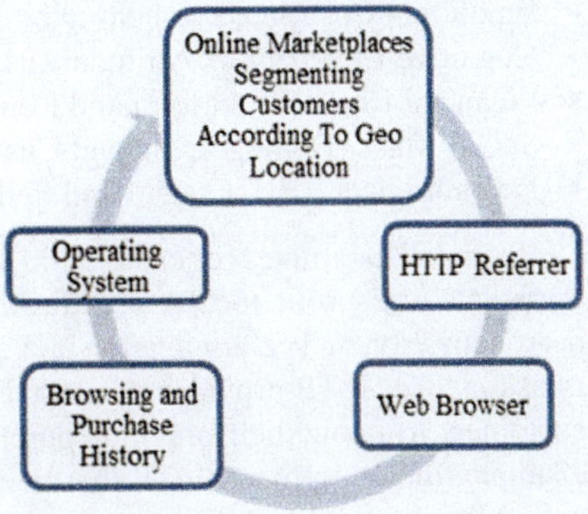

Fig. 8 Criteria for IoTP for PDS [48]

Customers are usually choosing the differentiated products and IoT has made it easier by providing remote access but at the same time customers find difficulty in making price comparisons for other customers because of personalization as well as customization.

Firms now have a complete edge on PDS from customers and customers also do not deny accepting the prices asked for. Most common examples are in online purchase of software and other programs, where software firms ask for PDS based on the individual requirements and version they are willing to buy. PAs help firms to collect data for the functionalities of their products and customers too, which can aid in deciding on a range of dynamic pricing strategies.

Research Proposition 3: Generally cost is associated as expenditure made by the firms in the process of production, managing supply chain, CRM and many more M.Mgnt functions but switching costs (SCo) are associated and understood in terms of customers' expenditure as considered as one time purchase [49]. Customers make SCo in the process of searching for the best available deal for them and value of money therefore they compare and decide on changing or switching from one firm to another. It has been seen that most of the time customers do SCo for durable or high value products. IoTP has facilitated customers in the effective decision making process for SCo also other applications such as machine learning, data analytics can be used for SCo.

IoTP aids customers in CSo and firms too for measuring customer retention, competition, enterprise profitability, satisfaction and other important aspects of M.Mgnt. There are various reasons for customers to SCo and IoT offers comprehension for this analysis. Customers can have following knowledge of various types of costs and risks through IoT applications that contribute to SCo and IoT helps in comparing the following costs and risks from one firm to another [49].

- Search Costs (SC)
- Transaction Costs (TC)
- Learning Costs (LC)
- Loyalty Discounts/offers (LDo)
- Customer style and culture (CSC)

- Emotional Cost/Sense of belongingness (ECB)
- Cognitive Effort/Skills/Attitude Risks
- Financial Risks (Short term and Long term)
- Social Risks (Demographic and Custom)
- Psychological Risk (Esteem and Self Awareness)

When IoT facilitates for customers, it is helpful for firms also as IoTP can achieve mass sale and profit too by creating awareness through digital technologies for personalization and customization and customers will acquire positive behavior and trust for SCo. IoTP provides data for R and D for firms and CRM and also helps customers to know their previous purchases, complaints and satisfactions too. For example, there are many IoTP that help customers to know their success in using particular services like services pertaining to health, security, education or even recreation. Online fitness and health clubs provide customers to know their health benefits, track their progress, requirements needed to reach a particular milestone, etc. In the education sector, for example, online learning helps users to learn remotely and can be easily accessed and eventually receive a credential for their education. Such sectors are also a part of IoTP [49]. Every product, TP or SP has a life cycle commonly known as the Product Life Cycle (PLC). PLC has stages and phases from introduction to decline but product may have different duration for each phase depending on various factors but mostly PLC has bell shape showing the movement of product from one stage to another. PLC has stages such as introduction, growth, maturity and decline [49]. Apart from bell shape there are other shapes also associated with PLC and TP or SP may have either of these shapes for the phases too. Figure 9 presents more phases as examples for TP and SP.

If we check the historical perspectives, PLC mostly concentrated on verification of functionalities of MMgnt and methods of product's introduction; these two criteria help in evaluating the shapes of PLC also. Some marketing researchers have elaborated other criteria, factors also for the shape and length of the PLC. Table 3 shows the list of these factors and criteria that affect the length of PLC.

Fig. 9 Shapes of PLC in IoT Environment [49]

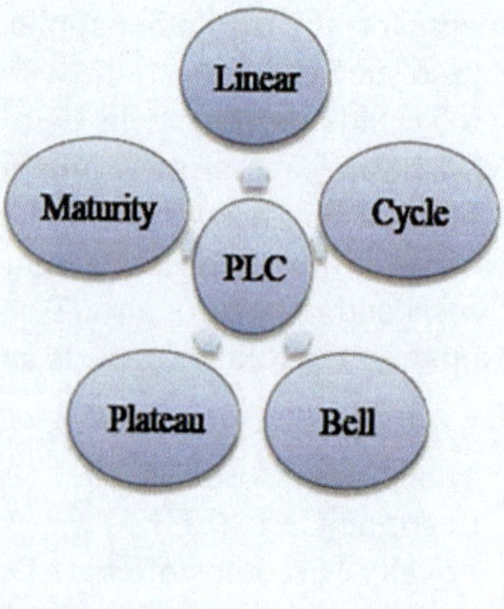

Table 3 Factors affecting the shape and length of PLC [49]

Sales of products
Changes in market and customers' needs
Changes in micro and macro economics
Level of competition for the product category
Technological changes or advancements
Existing products and threats from new entrants

In this chapter, we have tried to focus on changes introduced by IoT for the phases, shapes and length of the PLC, also we discussed how TP products had shorter PLC in comparison to SP.

Research Proposition 4: As discussed above, PLC has various stages but with the implication of IoT it may not have regular phases or lack proper generation of products. This is not because the product becomes obsolete or not usable for customers or any other complications, rather it is due to the role of technological advancement that makes good products ineffective. IoTP or SP are technology driven products and with the introduction of any new version of updated features can make existing products old although performance wise the products have no issues or complaints from customers. This is the main reason for IoTP for undefined PLC or generations [48, 49]. The main purpose of IoTP generated for CRM is increasingly based on software that is less prone to technological aging and because PAs and remote access allow the company to continuously monitor and update the software of IoTP. The more software accounts for a product's basic utility, the less relevant repurchases based on hardware upgrades become. For example some online assistants such as siri or alexa have overcome the issues of product's generations and have automatic updates in their applications and hardware too. They are excellent examples of IoTP having greater PLC. We can conclude that IoTP focuses on utilization and has wider options of scalability for their hardware and software, PLC will have longer generation and more linear shape. IoTP will also have repurchase behavior and customers will in general feel value for their money and also be willing to adopt new IoTP.

Research Proposition 5: The IoT assists the application of Internet-based business processes (IBBM) with the TP and is commonly marketed with a transaction-dependent revenue mode (TDRM).

In previous time or during pre-Internet age these BPM were facilitating the firms and customers positively because the purpose of BPM was more restricted to movement of products and maintain SPM and logistics but with the introduction of Internet and IoT, new BPM came into significance and start to execute many functions successfully in a virtual environment. Life in the cloud is a good example of IoT for BPM. Also, examples of Alphabet (Google), Amazon, Instagram, snapchat, Linkedin, WhatsApp are other examples of IoT in BPM. IoT has given rise to many BPM; some of them are listed in Table 4.

IBBM is the most cost effective BPM because of least OPC, having hardware and software scalability and of performing analysis remotely for data for M.Mgnt. There

Table 4 Types of Internet-based model BPM [49]

Internet-based business models (IBBM)
Software-as-a-service (Saas)
Product-as-a-service (Paas)
In product purchases (IPP)
The network business model (NBM)

are other BPM for IoTP such as IBBM, SaaS, PaaS, IPP and NBM. In the Internet era, companies can create value for customers through information drawn from other customers by the application of the above mentioned IoTP for BPM.

These models can be used from service industries to durable goods such as for vehicles such as use of location tracking technologies for security and other purposes [35]. Often we mention such SP as smart cars, Chevrolet and General Motors have been using IoTP by applying IBBM. Another good example is if from online purchase for retail products, where NBM works successfully for BPM using recommendation methods for defining purchase behavior and decisions. Currently all E-Commerce in retail use this method for effective BPM as well to achieve CRM. Also IBBM facilitates customers to make SCo comparison and decide for the best option and best value for their money. For firms these IoTP help in knowing customer demand, their profile and trend of the economic cycle. Also, with IoTP, firms can know the various products' usage, the right customer, the right method of accessing data and making the availability of the products or services to the end users. The feasibility does not restrict to the single firm. IBBM helps in providing learning and decision making strategies for different related firms and know how customers relate to joining benefits from products [48, 49].

GM is one of the important features of Green supply chain. To achieve eco-friendly and cost effective green supply chain, successful applications of other factors are also relevant along with GM. Figure 10 shows the list of green factors of Green supply chain. It is important to mention that all factors are interdependent and GM is the key

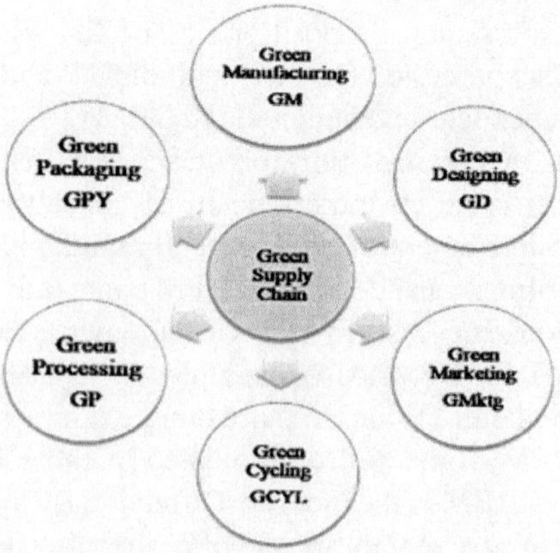

Fig. 10 List of green factors for green supply chain [50]

factor. For the purpose the study we discussed only Green Designing (GD), Green Processing (GP) and Green Packaging (GPY).

GD focuses on producing eco-friendly outputs that minimize waste and maximize materials utilization. As a part of ITN, GD is used by many firms these days for the purpose of minimizing waste. The process needs to reduce the toxics inputs used in the BPM in such a way that productivity and cost effective both can be achieved. Firms which have beliefs in environment preservation use GD in all the phases of PLC and the purpose is to maximize materials utilization, product recycling, reuse, ease of re-production and ensuing energy recycling. Firms using GD in GM is focused on providing an environmentally aware that measures the product green quality, best use of product design, coordinate different product development phases and satisfy the green needs of consumers, companies, the environment and society. Eventually GD in the GM aids in developing good product and achieves company's competitive advantage [50].

When the firms more focused in using GM in their BPM, GP helps the firms to achieve their image of eco-friendly because GP tries to reduce all manufacturing processes the environmental burden in such areas as input resources, chemical substances used and energy consumption. GP focuses on the applications of environmental technologies that minimize the environmental hazards through sustainable product and process design to conserve energy and reduce reliance on non-replaceable raw materials therefore if firms use GP as one of the factor for GM, firm will be able to use GM in BPM control pollution at all the phases of PLC and reduce the waste through recycling. GM and technologies are applied in ITN and M.Mgnt at all the levels such as in manufacturing, operational processes, BPM, production technologies and management-oriented factors like CRM [51]. GPY has a major role in ITN and in Green marketing because by adopting the concept of GPY, firms can minimize the environmental pollution and encourage using the material which is GCYL. Many firms conduct their marketing campaign on use of GPY to achieve competitive advantages and also to build CRM [51].

4 Results

Innovation has become an imperative for the entire world's industrial companies. Dealing with cost pressure is the most crucial issue for most of the companies. IoT has provided solutions to such critical issues and resolved OC and IT related problems at internal as well as external levels. IoT has provided major advantages such as increasing productivity, reducing OC, improving operational efficiency by introducing new and advanced business models, effective data analysis and achieving CRM.

The majority of companies is already using IoT applications in different phases of production, investment decisions and analyzing customers' data and reached to the beneficial solutions. Although there are many limitations and threats of IoT applications pertaining to privacy and security but industries for its transformation will continue using IoT because advantages are much more and limitations can be solved and prevented by the applications of other IS and IT tools.

The Internet has made fundamental changes in all the functions of M.Mgnt and with the IoT online environment has expanded to a greater extent. All major branches of M.Mgnt got affected and CRM, PLC and BPM also received a complete change over for customers and firms. The notion of joining the TP to IoTP, growth of new product mix for collecting, analyzing the data remotely and automatically that eventually aids CRM. PAs are done by the IoTP for product usage and its value for the customers and sharing this information with other customers, firms and their competitors. This process constitutes enhancing the PLC and solving SP related issues more efficiently and achieving CRM more easily. With IoT products, customers are always checked by many sensors and TP giving rise to the TL for. Also IoTP challenges the traditional CRM systems and introduces a change from only data collection to data analysis remotely. IoTPs have also witnessed the increase of strategic relevance for CRM. IoTP has also facilitated the application of PDS and SCo and developed new performance indicators for CRM.

PLC is adopting IoTP at a slower pace because IoTP usually does not have a product's age. The IoT facilitates the application of IBBM with SP as well as joint value creation for competitors.

The growth of the IoT era gave new direction to MMgnt where customers use more IoTP or a SP. The CJ from the view point of a single firm may therefore evolve to a CJ in an environment and operated by a network of many related firms.

IoTPs are likely to affect theories and concepts from other marketing fields in similar ways such as for marketing mix and brand management and the practical aspect of IoT can be seen in CRM and increase in strategic customer behavior. Further the impact of SP is immense and significant on given MMgnt such as CRM, PLC and BPM by connecting the SP in process of decision making for purchase behavior and product's usage. Finally, IoT has not only affected the B2C but also in the B2B or C2C categories of MMgnt.

The implementation of a GM means firms are using eco-friendly material in the BPM and all the phases of PLC, which clearly means that all goods are made from non-toxic materials and biodegradable materials. Also it is noticed that change in material has not only affected the finished products but also affected the PD and its SCM, and may increase in cost for some materials but in return advantage of GM are so high that these issues were small and easy to manage. Also firms applying GM in the process of BPM are acknowledged as responsible and show care for the nature and customers and contribute to the social benefits as a result they achieve CRM and increase in profit.

5 Conclusion

The industries will increase their IoT spending in future for ITN and currently most of the firms are already investing in IoT applications for ITN because of high productivity and low OC.

Customer Relationship Management: IoT gives smart nature to existing products and customers are facilitated in having multiple channels for completing purchase behavior, building customer relationships with the firms and developing strategies to enhance awareness and brand loyalty. IoT can develop price discrimination policy focusing of individual customer resulting in achieving profit maximization.

Product Life Cycle Management: Smart products developed by IoT products will have steep growth and longer maturity and before decline stage IoT contribute in new product development to retain its PLC.

Business Process Model Development: The IoT facilitates the application of Internet-based business models with physical products. IoT products facilitate joint value creation, developing automatic options for business environments. Firms who have implemented GM in BPM and all phases of PLC have made major contribution in preserving the environment but the process was not easy. They have to create and choose options which reduced wastes and pollution in the BPM. Sometime the inputs for GM cost higher and also firms have to change or modify the entire PD to produce desired finished products. All the efforts for applying GM are rewarding in terms of customer satisfaction, increasing productivity and environmental care.

References

1. Lee, I., Lee, K.: The internet of things (IoT): Applications, investments, and challenges for enterprises. Bus. Horiz. **58**(4), 431–440 (2015)
2. Ma, X.: Influence research on the industrial transformation and upgrading based on internet plus strategy. J. Comput. Theor. Nanosci. **14**(9), 4384–4390 (2017)
3. Piper, L.: The Industrial Transformation of Subarctic Canada. ubc Press (2010)
4. Khan, M.A., Salah, K.: IoT security: review, blockchain solutions, and open challenges. Futur. Gener. Comput. Syst. **82**, 395–411 (2018)
5. Nguyen, B., Simkin, L.: The Internet of Things (IoT) and Marketing: the State of Play, Future Trends and the Implications for Marketing (2017)
6. Decker, R., Stummer, C.: Marketing management for consumer products in the era of the internet of things. Adv. Internet Things **7**(3) (2017)
7. Buttle, F., Maklan, S.: Customer Relationship Management: Concepts and Technologies. Routledge (2019)
8. Kumar, V.: Customer relationship management. Wiley international encyclopedia of marketing (2010)
9. Simões, D., Filipe, S., Barbosa, B.: An overview on IoT and its impact on marketing. In: Smart Marketing with the Internet of Things, pp. 1–20 (2019)
10. Thoma, M., Meyer, S., Sperner, K., Meissner, S., Braun, T.: On iot-services: Survey, classification and enterprise integration. In: 2012 IEEE International Conference on Green Computing and Communications, pp. 257–260. IEEE (2012)

11. Shukla, G.P., Adil, G.K.: A conceptual four-stage maturity model of a firm's green manufacturing technology alternatives and performance measures. J. Manuf. Technol. Manag. (2021)
12. Kotabe, M.M., Helsen, K.: Global Marketing Management. John Wiley & Sons (2020)
13. Rizvi, M.: Implications of Internet of Things (IoT) for CRM (2017)
14. Yerpude, S., Singhal, T.K.: Internet of things based customer relationship management–a research perspective. Int. J. Eng. Technol. **7**(2.7), 444–450 (2018)
15. Junior, M.R.F.B., Batista, C.L., Marques, M.E., Pessoa, C.R.M.: Business models applicable to IoT. In: Handbook of Research on Business Models in Modern Competitive Scenarios, pp. 21–42. IGI Global (2019)
16. Hashem, D.T.N.: The reality of internet of things (Iot) in creating a data-driven marketing opportunity: mediating role of customer relationship management (Crm). J. Theor. Appl. Inf. Technol. **99**(2) (2021)
17. Ghazaleh, M.A., Zabadi, A.M.: Promoting a revamped CRM through internet of things and big data: an AHP-based evaluation. Int. J. Organ. Anal. (2020)
18. Janiesch, C., Koschmider, A., Mecella, M., Weber, B., Burattin, A., Di Ciccio, C., Zhang, L.: The internet-of-things meets business process management: mutual benefits and challenges (2017). arXiv preprint arXiv:1709.03628
19. Chiu, H.H., Wang, M.S.: A study of iot-aware business process modeling. Int. J. Model. Optim. **3**(3), 238 (2013)
20. Suri, K., Gaaloul, W., Cuccuru, A., Gerard, S.: Semantic framework for internet of things-aware business process development. In: 2017 IEEE 26th International Conference on Enabling Technologies: Infrastructure for Collaborative Enterprises (WETICE), pp. 214–219. IEEE (2017)
21. Janiesch, C., Koschmider, A., Mecella, M., Weber, B., Burattin, A., Di Ciccio, C., Zhang, L.: The internet of things meets business process management: a manifesto. IEEE Syst Man Cybern. Mag. **6**(4), 34–44 (2020)
22. Liu, Y., Zhang, Y., Ren, S., Yang, M., Wang, Y., Huisingh, D.: How can smart technologies contribute to sustainable product lifecycle management? J. Clean. Prod. **249**, 119423 (2020)
23. Marolt, M., Zimmermann, H.D., Žnidaršič, A., Pucihar, A.: Exploring social customer relationship management adoption in micro, small and medium-sized enterprises. J. Theor. Appl. Electron. Commer. Res. **15**(2), 38–58 (2020)
24. Naim, A., Khan, M.F.: Consumer behavior for health services: a psychological approach. SPR **1**(4), 356–367 (2021). https://doi.org/10.52152/spr/2021.155
25. Xin, Y., Ojanen, V.: The impact of digitalization on product lifecycle management: How to deal with it? In: 2017 IEEE International Conference on Industrial Engineering and Engineering Management (IEEM), pp. 1098–1102. IEEE (2017)
26. Mendling, J., Simon, C.: Business process design by view integration. In: International Conference on Business Process Management, pp. 55–64. Springer, Berlin, Heidelberg (2006)
27. Allmendinger, G., Lombreglia, R.: Four strategies for the age of smart services. Harv. Bus. Rev. **83**, 131–134 (2005)
28. Dhebar, A.: Information technology and product policy: "smart" products. Eur. Manag. J. **14**, 477–485 (1996)
29. Korling, M.: Smart products: why adding a digital side to a toothbrush could make a lot of sense. Ericsson Bus. Rev. **18**, 26–31 (2012)
30. Kowatsch, T., Maass, W., Filler, A., Janzen, S.: Knowledge-based bundling of smart products on a mobile recommendation agent. In: Proceedings of the 7th International Conference on Mobile Business (ICMB 08), Barcelona, 7–8 July 2008, pp. 181–190 (2008)
31. Mayer, P.: Economic Aspects of Smart Products. White Chapter, Institute of Technology Management, University of St. Gallen, St. Gallen (2010)
32. Resatsch, F.: Ubiquitous Computing: Developing and Evaluating Near Field Communication Applications. Springer, Wiesbaden (2010)
33. Meyer, G.G., Buijs, P., Szirbik, N.B., Wortmann, J.C.: Intelligent products for enhancing the utilization of tracking technology in transportation. Int. J. Oper. Prod. Manag. **34**, 422–446 (2014)

34. Zhou, L., Chong, A.Y.L., Ngai, E.W.T.: Supply chain management in the era of the internet of things. Int. J. Prod. Econ. **195**, 1–3 (2015)
35. Glova, J., Sabol, T., Vajda, V.: Business models for the internet of things environment. Procedia Econ. Financ. **15**, 1122–1129 (2014)
36. Rijsdijk, S.A., Hultink, E.J.: How today's consumers perceive tomorrow's smart products. J. Prod. Innov. Manag. **26**, 24–42 (2009)
37. Porter, M.E., Heppelmann, J.E.: How smart, connected products are transforming companies. Harv. Bus. Rev. **93**, 1–37 (2015)
38. Meyer, G.G., Wortmann, J.C., Szirbik, N.B.: Production monitoring and control with intelligent products. Int. J. Prod. Res. **49**, 1303–1317 (2011)
39. Weinberg, B.D., Milne, G.R., Andonova, Y.G., Hajjat, F.M.: Internet of things: convenience vs. privacy and secrecy. Bus. Horiz. **58**, 615–624 (2015)
40. Robson, K., Pitt, L.F., Kietzmann, J.: APC forum: extending business values through wearables. MIS Q. Exec. **15**, 167–177 (2016)
41. Rocker, C.: Intelligent environments as a promising solution for addressing current demographic changes. Int. J. Innov. Manag. Technol. **4**, 76–79 (2013). https://doi.org/10.1509/jmkg.2005.69.4.155
42. Riva, G., Vatalaro, F., Davide, F., Alcañiz, M.: Ambient Intelligence. The Evolution of Technology, Comm. and Cognition Towards the Future of Human–Computer Interaction. IOS Press, pp. 1–320 (2005)
43. Heinze, T.S., Amme, W., Moser, S.: Static analysis and process model transformation for an advanced business process to Petri net mapping. Softw. Pract. Exp. **48**(1), 161–195 (2018)
44. Naim, A., Alahmari, F., Rahim, A.: Role of artificial intelligence in market development and vehicular communication. Smart Antennas Recent Trends Des. Appl. **2**, 28 (2021)
45. Weske, M.: Business Process Management Architectures, pp. 305–343. Springer, Berlin Heidelberg (2007)
46. Chatterjee, S., Chaudhuri, R., Vrontis, D., Thrassou, A., Ghosh, S.K., Chaudhuri, S.: Social customer relationship management factors and business benefits. Int. J. Organ. Anal. (2020)
47. Naim, A., Khan, M.F., Hussain, M.R., Khan, N.: "Virtual Doctor" management technique in the diagnosis of ENT diseases. JOE **15**(9), 88 (2019)
48. Zhironkin, S., Gasanov, M., Barysheva, G., Gasanov, E., Zhironkina, O., Kayachev, G.: Sustainable development vs. Post-industrial transformation: possibilities for Russia. In: E3S Web of Conferences, vol. 21, p. 04002. EDP Sciences (2017)
49. Porter, M.E., Heppelmann, J.E.: How smart, connected products are transforming competition. Harv. Bus. Rev. **92**, 64–88 (2014)
50. Acquah, I.S.K., Essel, D., Baah, C., Agyabeng-Mensah, Y., Afum, E.: Investigating the efficacy of isomorphic pressures on the adoption of green manufacturing practices and its influence on organizational legitimacy and financial performance. J. Manuf. Technol. Manag. (2021)
51. Bustinza, O.F., Vendrell-Herrero, F., Sánchez-Montesinos, F.J., Campos-Granados, J.A.: Should manufacturers support the entire product lifecycle with services? Sustainability **13**(5), 2493 (2021)